POWER HYDRAULICS

POWER HYDRAULICS

MICHAEL J. PINCHES
JOHN G. ASHBY

Automation Advisory Service,
Faculty of Technology,
Sheffield City Polytechnic, UK

PRENTICE HALL
NEW YORK   LONDON   TORONTO   SYDNEY   TOKYO

First published 1988 by
Prentice Hall International (UK) Ltd,
66 Wood Lane End, Hemel Hempstead,
Hertfordshire, HP2 4RG
A division of
Simon & Schuster International Group

© 1989 Prentice Hall International (UK) Ltd

All rights reserved. No part of this publication may be
reproduced, stored in a retrieval system, or transmitted, in
any form or by any means, electronic, mechanical,
photocopying, recording or otherwise, without the prior
permission, in writing, from the publisher.
For permission within the United States of America contact
Prentice Hall Inc., Englewood Cliffs, NJ 07632.

Printed and bound in Great Britain at the
University Press, Cambridge.

Library of Congress Cataloging-in-Publication Data

Pinches, Michael, 1931–
  Power hydraulics/Michael Pinches and John G.
Ashby.
   p. cm.
  Includes index.
  ISBN 0-13-687443-6 : $57.95 (est.). $25.95 (pbk. : est.)
  1. Fluid power technology.   I. Ashby. John G., 1932–
II. Title.
TJ843.P46 1988
621.2–dc19   87–29268

1 2 3 4 5 92 91 90 89 88

ISBN 0-13-687443-6

TO THE MEMORY OF BRENDA

CONTENTS

PREFACE   xi

1   INTRODUCTION   1

   1.1   Hydraulic principles   1
      1.1.1   Properties of fluids   1
      1.1.2   Units   1
      1.1.3   Pressure of a liquid   1
      1.1.4   Flow of a fluid   5
      1.1.5   Work done   8
   1.2   Hydraulic symbols   9

2   PUMPS   15

   2.1   Types of pump   15
      2.1.1   Rotary pumps   18
      2.1.2   Reciprocating pumps   22
      2.1.3   Variable-displacement pump-control systems   25
      2.1.4   Pump selection   30
   2.2   Pumping circuits   34
      2.2.1   Single fixed-displacement pump   35
      2.2.2   Single fixed-displacement pump with accumulator   36
      2.2.3   Multi-pump   38
      2.2.4   Variable-delivery pump   41
   2.3   Pump drives   41
   2.4   Pump circuit design study   43

3   HYDRAULIC VALVES   57

   3.1   Pressure-control valves   57
      3.1.1   Relief valves   58
      3.1.2   Counterbalance valves   64
      3.1.3   Pressure sequence valves   67
      3.1.4   Pressure-reducing valves   69
   3.2   Flow-control valves   70
      3.2.1   Speed control of a cylinder   75
      3.2.2   Three-port or bypass-type flow-control valve   83
      3.2.3   Priority flow control   85
      3.2.4   Bridge networks   86
      3.2.5   Multi-speed systems using flow-control valves   88
      3.2.6   Flow dividers   89

	3.3 Directional control valves	93
	3.3.1 Check valves	93
	3.3.2 Poppet valves	98
	3.3.3 Sliding spool-type directional control valves	100
	3.3.4 Two-stage directional control valves	108
	3.3.5 Valve sizes and nomenclature	110
	3.4 Cartridge valves	112
	3.4.1 Poppet-type cartridge valves	113
	3.4.2 Spool-type cartridge valves	121
	3.5 Mobile hydraulic valves	123
	3.5.1 Valve arrangements	124
4	**ACTUATORS**	**127**
	4.1 Hydraulic cylinders	127
	4.1.1 Displacement cylinders	128
	4.1.2 Single-acting cylinders	133
	4.1.3 Double-acting cylinders	133
	4.1.4 Acceleration and deceleration of cylinder loads	140
	4.1.5 Cylinder mountings and strength calculations	146
	4.2 Semi-rotary actuators	152
	4.2.1 Vane-type actuators	152
	4.2.2 Piston-type actuators	152
	4.2.3 Helical screw actuator	154
	4.2.4 Control of semi-rotary actuators	155
	4.3 Hydraulic motors	155
	4.3.1 Generated form types	155
	4.3.2 Piston-type motors	159
	4.4 Hydraulic motor circuits	163
	4.4.1 Open circuit transmissions	163
	4.4.2 Closed-loop transmissions	168
	4.4.3 Multi-motor circuits	170
	4.5 Motor circuit design examples	172
5	**FLUIDS FOR HYDRAULIC SYSTEMS**	**180**
	5.1 Hydraulic fluids	180
	5.1.1 An outline of the development of hydraulic fluids	180
	5.1.2 Properties of hydraulic fluids	180
	5.1.3 Hydraulic fluids survey	185
	5.1.4 Future developments	187
	5.2 Fluid contamination control	189
	5.2.1 Energy contamination	189
	5.2.2 Gaseous contamination	190
	5.2.3 Liquid contamination	192
	5.2.4 Microbiological contamination	194
	5.2.5 Particulate contamination (dirt)	194
	5.3 Filter construction and filtration technology	200
	5.3.1 Filter construction	200
	5.3.2 Filtration technology	202
	5.3.3 Filter location	207
	5.3.4 Filter sizing	212

		5.4 Leakage control	214
		5.4.1 Hydraulic conduits	214
		5.4.2 High-pressure tube connections	217
		5.4.3 Hydraulic hoses	225
		5.4.4 Supports and clamps	230
		5.4.5 Circuit accessories	230
		5.4.6 Pipework elimination and simplification	232
		5.4.7 Seal protection	233
		5.4.8 Rules for leakage control	234

6 HYDRAULIC SYSTEM DESIGN — 235

    6.1 Design criteria — 235
        6.1.1 Design information required — 236
    6.2 Summary of basic formulae and rules — 237
        6.2.1 Fluid flow — 237
        6.2.2 Pressure losses — 240
        6.2.3 Cylinder formulae — 241
        6.2.4 Pump formulae — 242
        6.2.5 Hydraulic motor formulae — 244
    6.3 Power-pack design — 245
        6.3.1 Layout — 245
        6.3.2 Reservoirs — 247
        6.3.3 Centralized hydraulic systems — 253
    6.4 Hydraulic accumulators — 254
        6.4.1 Types of accumulators — 254
        6.4.2 Accumulator applications — 259
        6.4.3 Operation and safety precautions — 265
    6.5 Hydraulic intensifiers — 267
    6.6 Design study—a simple hydraulic press — 268
    6.7 Design study—conveyor feed system — 287

7 HYDRAULIC SYSTEM MAINTENANCE — 296

    7.1 Equipment and practices which benefit maintenance — 296
        7.1.1 Good housekeeping practice — 296
        7.1.2 Fluid storage and handling — 300
        7.1.3 Installations and commissioning of hydraulic systems — 301
        7.1.4 Routine maintenance — 303
    7.2 Trouble shooting in hydraulic systems — 304
        7.2.1 Test equipment — 304
        7.2.2 General rules for hydraulic maintenance engineers — 305
        7.2.3 The concept of logical fault finding — 305
    7.3 Solutions to fault-finding exercises — 323

8 CONTROL SYSTEMS — 328

    8.1 Servo control — 329
    8.2 Valve servo systems — 330
        8.2.1 Valve lap — 330
        8.2.2 Mechanical feedback — 331
        8.2.3 System response — 333

	8.2.4 Electro-hydraulic servo valves	338
	8.2.5 System response and stability	340
8.3	Pump servo systems	340
	8.3.1 Effect of leakage	341
	8.3.2 Effect of compressibility	342
	8.3.3 Natural frequency	344
	8.3.4 Hydraulic stiffness	344
	8.3.5 Damping ratio	345
8.4	Proportional valves	350
	8.4.1 Force control	350
	8.4.2 Force position control	351
	8.4.3 Spool positional control	352
	8.4.4 Proportional pressure control	354
	8.4.5 Two-stage proportional valves	356
	8.4.6 Proportional flow control	358
	8.4.7 Electrical control of proportional valves	360
8.5	Proportional versus servo valves	360
	8.5.1 Response speed and dynamic characteristics	360
	8.5.2 Hysteresis effect	361
	8.5.3 Null position	361
8.6	Some applications of proportional control valves	362
	8.6.1 Control of actuators	362
	8.6.2 Pump control systems	363

APPENDIX EXERCISES AND SOLUTIONS   367

A.1	Circuitry questions	367
A.2	Hydraulic calculations	376
A.3	Design problems	383
A.4	Solutions to Section A.2	388

FURTHER READING   391

INDEX   393

PREFACE

The authors have considerable experience as engineers in industry and education. For over twenty years they have operated the Automation Advisory Service at Sheffield City Polytechnic, providing for industry a consultancy and training service specializing in fluid power.

Among existing books which are applicable to the study of power hydraulics there are at one extremity many highly respectable tomes on the theory of fluid mechanics, hydraulic principles and control engineering. At the other end of the scale are several excellent books produced by hydraulic equipment manufacturers describing the construction and operation of hydraulic components, including simple circuitry but naturally concentrating on their own products.

This book is an attempt to fill the void between these extremes. It covers simple hydraulic theory, types and principles of operation of many components but tries to concentrate on their application. It will be of value not only to lecturers and students but to engineers involved in the design, purchase, operation and maintenance of plant and machines which are hydraulically powered or contain a proportion of hydraulic equipment.

It demonstrates from a practical viewpoint the simple calculations, circuitry and component selection involved in system design.

Users of hydraulic equipment will find it beneficial in understanding the working of plant and machinery under their control. Equipment purchasers will find that they are better able to prepare specifications and make valid judgements on the suitability of systems and components offered. A chapter on maintenance gives guidance on troubleshooting, practices and procedures to minimize breakdowns.

The presentation is both descriptive and quantitative with numerous diagrams. It is intended to form a useful reference book and teaching aid for lecturers and students. Whilst not necessarily following the format of the syllabuses of the BTec and CGI modules in hydraulics, it more than adequately covers the subject matter of these technician courses as well as that of most degree courses. Worked examples are included in the text together with exercises for the students to undertake themselves.

Acknowledgements

The authors particularly wish to thank Lynne Thornhill for her considerable efforts in typing and preparing the manuscript, and Crystal Ashby for her support and encouragement and for putting up with the consequential disruption to family life, whilst the work was being undertaken. The various manufacturers who have provided or permitted diagrams and tables to be copied from their catalogs and technical manuals are kindly acknowledged.

CHAPTER ONE
INTRODUCTION

The transmission and control of power by means of fluid under pressure is becoming increasingly used in all branches of industry. Pneumatics deal with the use of compressed air as the fluid whilst hydraulic power covers the use of oils and other liquids.

Pneumatics are generally used when relatively low forces up to 10 kN (1 ton) and fast cycling speeds are wanted. Where high forces, precision speed control and high power weight ratios are needed, hydraulic systems are used. 'Fluid power' is a term covering both pneumatic and hydraulic power.

Applications of hydraulic power range from car jacks to hospital beds, from presses exerting forces of thousands of tonnes to robots with accuracies measured in micrometers.

1.1 HYDRAULIC PRINCIPLES

1.1.1 Properties of fluids

A fluid covers both liquids and gases and is a substance in which the molecules can move about with freedom. A gas is a fluid which will expand to completely fill the available space; its density varies considerably with its temperature and pressure. A liquid is a fluid which will flow under gravity to take up the shape of the containing vessel in such a way that it reduces its potential energy to a minimum. The density of a liquid changes only very slightly with changes in temperature and pressure.

1.1.2 Units

There are many systems of units currently in use, the three most popular being: (1) the Metric System based on meter, kilogram and second; (2) the Imperial System using the foot, pound, second; and (3) the SI System using meter, newton, second as a base. Table 1.1 gives a comparison of some of the more common quantities in the three systems of unit.

1.1.3 Pressure of a liquid

Pressure is the force per unit area, i.e.

   Pressure = Force/Area

Table 1.1 Currently used systems of units.

Quantity	Symbol	System		
		SI System	Imperial System	Metric System
Length	l	meter 1 m = 39.37 in 1 m = 3.281 ft micron 1 μm = 10^{-6} m	inch 1 in = 0.0254 m foot 1 ft = 0.3048 m	centimeter 1 cm = 10^{-2} m millimeter 1 mm = 10^{-3} m
		m μm		cm mm
Area	A	square meter 1 m^2 = 1550 in^2	square inch 1 in^2 = 0.645 × 10^{-3} m^2 1 in^2 = 6.45 cm^2	square centimeter 1 cm^2 = 10^{-4} m^2
		m^2	in^2	cm^2
Volume	V	cubic meter 1 m^3 = 220 gal 1 m^3 = 10^3 liters	cubic inch 1 in^3 = 16.39 × 10^{-6} m^3 Gallon 1 gal = 277.4 in^3 = 0.00454 m^3 cubic foot 1 ft^3 = 6.24 gal	Cubic centimeter 1 cm^3 = 10^{-6} m^3 Liter 1 l = 10^{-3} m^3
		m^3	in^3 gal ft^3	cm^3 l
Time	t	second	minute	minute
		s	min	min
Volumetric flow rate	q	cubic meters per second 1 m^3/s = 13.2 × 10^3 gal/min	cubic inches per minute Gallons per minute	liters per minute
		m^3/s	in^3/min gal/min	l/min

Quantity	Symbol	SI		Imperial		Other metric	
Velocity	v	meters per second	m/s	feet per second	ft/s	meters per minute	m/min
Acceleration	a	meters per second squared	m/s^2	feet per second squared	ft/s^2	meters per second squared	m/s^2
Mass	M	kilogram $1\text{ kg} = 2.2\text{ lb}$	kg	pound mass $1\text{ lb} = 0.4536\text{ kg}$	lb		kg·s^2/m $\text{kg·s}^2/\text{m} = 9.807\text{ kg}$
Force or weight	F.P	newton	N	pound force $1\text{ lbf} = 4.45\text{ N}$	lbf	kp kgf	$1\text{ kp} = 1\text{ kgf} = 9.81\text{ N}$
Torque	M_T	newton meter	Nm	foot pound force $1\text{ ft lbf} = 1.356\text{ Nm}$	ft lbf	kpm kg fm	$1\text{ kpm} = 1\text{ kgfm} = 9.81\text{ Nm}$
Pressure	P	newton per square meter $1\text{ bar} = 10^5\text{ N/m}^2$ $1\text{ Pa (Pascal)} = 1\text{ N/m}^2$	N/m^2	pound force per square inch $1\text{ lbf/in}^2 = 6897\text{ N/m}^2$	lbf/in^2	kgf/cm^2 $1\text{ kgf/cm}^2 = 9.81 \times 10^4\text{ N/m}^2$	kp/cm^2
Work	A, W	joule $1\text{ J} = 1\text{ Nm}$	J	foot pound force $1\text{ ft lbf} = 1.356\text{ J}$	ft lbf	kilogram force meter $1\text{ kgfm} = 9.81\text{ J}$	kgfm kpm
Power	P, N	watts $1\text{ W} = 1\text{ Nm/s}$	W	foot pound force per second $1\text{ ft lbf/s} = 1.356\text{ W}$ horse power $1\text{ hp} = 745.7\text{ W}$	ft lbf/s hp	metric horse power $1\text{ PS} = 1\text{ ch}$ $= 75\text{ kpm/s}$ $= 735.5\text{ W}$	PS ch

Pascal's laws relating to the pressure in a fluid are as follows:

1. The pressure will be the same throughout a fluid which is at rest provided the effect of the weight of the fluid is neglected.
2. This static pressure acts equally in all directions at the same time.
3. This pressure always acts at right angles to any surface in contact with the fluid.

In the simple hydraulic system shown in Figure 1.1, a large load W is balanced by a small force F at the pumping ram. Consider the pressure due to load W:

$$\text{Pressure} = \text{Load}/\text{Area} = W/A$$

The pressure due to the force F:

$$\text{Pressure} = \text{Force}/\text{Area} = F/a$$

For the system to be in equilibrium the pressure must be the same in the large and small rams, i.e.

$$W/A = F/a$$

or

$$W/F = A/a$$

So for balance the ratio of the loads and areas are equal. This can be compared to a lever system. Taking moments about the pivot:

$$Wa = FA$$

or

$$W/F = A/a$$

To lift the load W by the hydraulic system there must be a flow of liquid from the smaller to the larger ram. To obtain this flow there has to be a pressure difference across the pipe connecting the two rams, so to raise W the force F must increase by a small amount ΔF.

To raise the load W by a distance L, fluid must be displaced from the small ram to the large ram.

$$\text{Volume displaced } V = A \times L = a \times l$$

Work done is equal to the force times the distance moved through; in this case the weight times the height lifted.

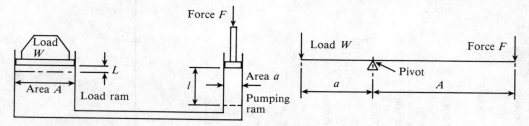

Figure 1.1 Hydraulic force transmission.

Section 1.1 Hydraulic principles

Work done at load = $W \times L$

but

Pressure $P = W/A$

therefore,

$$W = P \times A$$

so work done at load is

$P \times A \times L$

which is pressure times volume.

Pressure head of a liquid

A column of liquid will cause a pressure at its base owing to its own weight, the pressure increasing the higher the column of liquid. Consider the pressure at the bottom of a column of fluid of a cross-section area A and height h. Let the weight per unit volume be w.

Weight of column = Volume × Weight per unit volume
$= Ah \times w$

Pressure $= \text{Weight}/\text{Area} = Ahw/A = wh$

EXAMPLE 1.1

The inlet to a hydraulic pump is 0.6 m below the top surface of the oil reservoir. If the specific gravity of the oil used is 0.86 determine the static pressure at the pump inlet.

Pressure $= wh$
Density of water is 1 g/cm^3 or 1000 kg/m^3
Therefore the density of oil is 0.86×1 g/cm^3 or 860 kg/m^3
Pressure at pump inlet $= 860 \times 0.6$ kg/m^2
$\qquad = 516$ kg/m^2
$\qquad = 0.0516$ kg/cm^2
$\qquad = 0.0516 \times 0.981$ bar
$\qquad = 0.0506$ bar

Note 1 kg/cm^2 = 0.981 bar.

1.1.4 Flow of a fluid

In any system friction opposes motion. In order to cause an object to move a force has to be applied to overcome friction. A similar state exists in fluid flow. In a pipe or passage containing fluid, there must be a pressure difference between the ends of the pipe for flow to occur and the direction of flow is from the higher to the lower pressure.

The greater the pressure difference, the higher the flow rate. Whenever there is a pressure drop in a pipe there will be flow and conversely any flow will be accompanied by a pressure drop. At low velocities the flow in the pipe will be streamlined, all the molecules of fluid moving in the same direction. When the flow velocity exceeds a certain value, the flow pattern changes to turbulent in which the molecules of fluid no longer all move in the same direction.

For a streamline flow the pressure drop or frictional resistance of a pipe is:

(a) Proportional to the length and diameter of the pipe.
(b) Proportional to the quantity of fluid flowing.
(c) Independent of system pressure.
(d) Independent of surface roughness of the pipe.
(e) Dependent upon the viscosity of the fluid, which is a function of temperature.

Under turbulent flow conditions the pressure drop in the pipe is:

(a) Proportional to the length and diameter of the pipe.
(b) Proportional to the square of the quantity of fluid flowing.
(c) Independent of system pressure.
(d) Dependent upon surface roughness of the pipe.
(e) Independent of fluid viscosity.

In order to achieve maximum efficiency in a hydraulic system the pipe sizes are selected to give streamline flow. A full account of fluid flow in pipes can be found in any standard hydraulics text book.

A typical set of curves relating pressure drop to flow rate for a series of pipe diameters is shown in Figure 1.2. Full curves relating pressure drop to flow rate for different viscosity and specific gravity are given in detail in reference books (see further reading at the end of this book).

As a rough guide, the flow velocity in a pump suction line should be between 0.6 and 1.2 m/s (2 and 4 ft/s), and in the pressure and return lines lie between 2.1 and 4.6 m/s (7 and 15 ft/s). The flow velocity in valves and through orifices can be greatly in excess of these values.

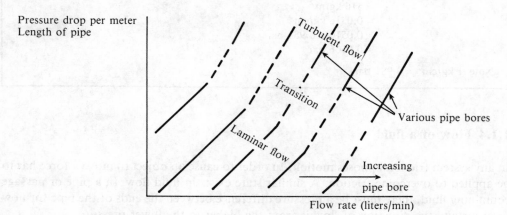

Figure 1.2 Relationship between pressure drop and flow rate in various pipe bores.

EXAMPLE 1.2

Calculate the pipe bores required for the suction and pressure lines of a pump delivering 40 l/min using a maximum flow velocity in the suction line of 1.2 m/s and a maximum flow velocity in the pressure line of 3.5 m/s.

Consider the suction line

$$\text{Flow} = \text{Average velocity} \times \text{Flow area}$$

$$\text{Area of pipe} = \frac{\text{Flow through pipe}}{\text{Velocity of flow}}$$

$$\begin{aligned}\text{Flow} &= 40 \text{ l/min} \\ &= 40/60 \text{ l/s} \\ &= 40/60 \times 10^{-3} \text{ m}^3/\text{s}\end{aligned}$$

$$\text{Area of pipe} = \frac{40 \times 10^{-3}}{60 \times 1.2} \text{ m}^2$$

$$= 0.555 \times 10^{-3} \text{ m}^2$$

Let the bore of pipe be of diameter D

$$\text{Area of pipe} = \pi D^2/4 = 0.555 \times 10^{-3} \text{ m}^2$$

therefore,

$$\begin{aligned}D &= (4/\pi \times 0.555 \times 10^{-3})^{1/2} \\ &= 0.0266 \text{ m}\end{aligned}$$

Minimum bore of suction pipe = 0.0266 m = 26.6 mm

Note In all calculations great care must be taken to ensure that units are correct.

Alternatively, if a flow velocity of 1 m/s is used then suction pipe bore can be shown to be of diameter 29 mm.

The required diameter of the pressure line can be calculated in a similar manner taking the flow velocity as 3.5 m/s. Here, minimum bore of pressure pipe = 15.6 mm

It is unlikely that a pipe having the exact bore will be available, in which case select a standard pipe having a larger bore. Alternatively a smaller bore pipe may be chosen but it will be necessary to recheck the calculation to ensure that the flow velocity falls within the recommended range, i.e. a standard pipe with an outside diameter of 20 mm and a wall thickness of 2.5 mm is available. This gives an internal diameter of 15 mm.

$$\text{Flow velocity} = \text{Flow through pipe}/\text{Area of pipe bore}$$

$$\text{Thus area of pipe bore} = \frac{\pi}{4} 15^2 \text{ mm}^2 = 177 \text{ mm}^2$$

$$= 177 \times 10^{-6} \text{ m}^2$$

$$\text{Flow velocity} = \frac{40 \times 10^{-3}}{60 \times 177 \times 10^{-6}} \left(\frac{\text{m}^3}{\text{s m}^2}\right)$$

$$= 3.77 \text{ m/s}$$

which is satisfactory.

It is also important to ensure that the wall thickness of the pipe is sufficient to withstand the working pressure of the fluid.

1.1.5 Work done

The work done by a force is defined as:

Work done = Force × Distance moved through

Let the area of the piston of a hydraulic cylinder be A, the effective pressure acting on the piston is P and the stroke of the piston is L. Then

Force on the piston = Pressure × Area = $P \times A$

So the work done is $P \times A \times L$.

$A \times L$ is the volume V of the fluid flowing into the cylinder to drive the piston forwards. So

Work done = $P \times V$

= Pressure × Volume

If the pressure is measured in pascals (N/m^2) and the volume in cubic meters (m^3) then

Work done = $P \times V \left(\dfrac{N}{m^2} \times m^3 \right)$

= $P \times V$ (Nm)

Thus work done is in newton meters (Nm).

Power is the rate of doing work, i.e. work done per unit time or $P \times V$ per unit time. The volume V per unit time is the flow rate Q.

Thus

Hydraulic power = Pressure × Flow rate

If the pressure is in pascals (N/m^2) and the flow rate in cubic meters per second (m^3/s), then

Hydraulic power = $P \times Q \left(\dfrac{N}{m^2} \times \dfrac{m^3}{s} \right)$

= $P \times Q$ (Nm/s)

= $P \times Q$ (watts)

Note 1 Nm/s = 1 watt.

It is usual to express flow rates in liters/minute and pressures in bars. To calculate the hydraulic power using these units a conversion has to be made. Thus

Q (l/min) = $Q/60$ (l/s)

= $\dfrac{Q}{60 \times 10^3}$ (m^3/s)

Section 1.2 Hydraulic symbols

and

$$P \text{ (bar)} = P \times 10^5 \text{ (N/m}^2\text{)}$$

Hydraulic power is:

$$Q \text{ (l/min)} \times P \text{ (bar)} \times \frac{1 \times 10^5}{60 \times 10^3} \left(\frac{m^3}{s} \times \frac{N}{m^2} \right)$$

$$= Q \times P \times \frac{10^3}{600} \left(\frac{Nm}{s} \right)$$

$$= Q \times P \times \frac{10^3}{600} \text{ (watts)}$$

$$= \frac{Q \times P}{600} \text{ (kW)}$$

Thus hydraulic power (kW) is

$$\frac{\text{Flow (l/min)} \times \text{Pressure (bar)}}{600}$$

EXAMPLE 1.3

A hydraulic pump delivers 12 liters of fluid per minute against a pressure of 200 bar.
1. Calculate the hydraulic power.
2. If the overall pump efficiency is 60%, what size electric motor would be needed to drive the pump?

$$\text{Hydraulic power (kW)} = \frac{12 \text{ (l/min)} \times 200 \text{ (bar)}}{600}$$

$$= 4 \text{ kW}$$

$$\text{Pump overall efficiency} = \frac{\text{Power output of pump}}{\text{Power input}}$$

$$\text{Motor power (power input)} = \frac{\text{Output power}}{\text{Overall efficiency}}$$

$$\text{Electric motor power} = 4/0.6$$
$$= 6.67 \text{ kW}$$

A summary of the formula used in hydraulic power is given in section 6.2 of Chapter 6.

1.2 HYDRAULIC SYMBOLS

Hydraulic components are represented by symbols to facilitate the drawing of hydraulic circuits. The symbols used in this book are based on British Standard BS 2917 1977 (ISO

1219: 1976). The symbol is intended to show the type or function of the component connections and flow paths. Basic symbols may be combined together to make a composite symbol. They do not give any indication of size and are not orientated in any particular direction. Where a control element is shown on a component it does not represent the actual physical location.

An arrow ╱ across an element indicates adjustment or variability.

It is usual for components to be shown in the unoperated position.

A full line represents a flow line; it does not give any indication of the pressure in the line. The line may be a suction, pressure or return line:

——————————— Main flow line

A drain line which takes leakage fluid from a component back to the reservoir is shown as a series of short dashes:

- - - - - - - - - - - - - - - - - - Drain line

A pilot line which is used to transmit a pressure signal from one point to another with minimal flow is shown as a series of long dashes:

— — — — — — — — Pilot line

On many hydraulic circuit diagrams including those in this book, the pilot and drain lines are drawn in the same way, because it is easy to differentiate between them by their application; a drain line nearly always returns to the reservoir.

A non-return valve or check valve consists of a ball or poppet held closed by a spring. It is represented as:

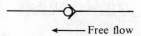
← Free flow

If the pressure at which the check valve opens is critical to the function of the circuit the spring which holds the poppet on its seat is shown:

The check valve can be remotely piloted:

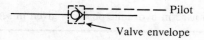
Pilot
Valve envelope

In this case the pilot is used to lift the poppet off its seat to allow free flow in the normally blocked direction. The valve is known as a pilot-operated check valve.

Similarly, the pilot may be used to prevent the valve opening:

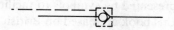

Directional control valves are shown as a number of rectangles; if there are two rectangles the valve has two states or positions it can adopt:

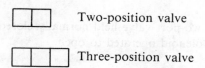

The *pipework connections are shown to one rectangle only*; this is usually the unoperated state of the valve. A two-port valve has two connections and can be either open or closed. The two states are:

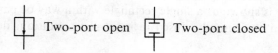

Combining these together and showing a spring normally holding the valve open

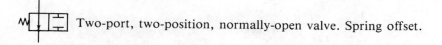

The most common directional control valve is the four-port valve

The ports are identified by letters: P is the supply or pressure; T is the return or tank; and A and B are the service or output ports. In the left-hand rectangle, P is connected to A and B to T; this is sometimes referred to as 'tramline connection'. In the right-hand rectangle, P is connected to B and A to T; this reverses the connection and is sometimes called 'crossover'. To visualize the operation of a directional control valve imagine the pipework to the valve remaining fixed and the rectangles moving across.

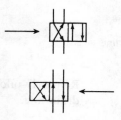

Directional control valves can be manually, mechanically, electrically, pneumatically or hydraulically operated. The method of operation is shown at the end of the rectangle to which it operates the valve, although this may not be its physical position.

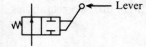
Two-port valve held normally open by a spring. Lever operated to close.

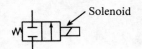
Two-port valve held normally closed by a spring. Solenoid operated to open.

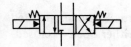
Four-port, three-position, spring-centered, solenoid-controlled, hydraulic pilot operated; with pressure port blocked and A, B and T interconnected in the center condition.

A pressure control valve can take up any position between fully open and fully closed. A pressure-control valve is shown as a single rectangle with a way or passage through it. It may be normally open or normally closed in its at-rest condition, dependent upon the function of the valve.

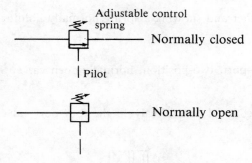

A spring which is adjustable keeps the valve biased to its normal position. A pilot signal acts against the spring to change the state of the valve when the pilot pressure exceeds the equivalent setting of the control spring. The pilot may be taken internally from the valve block, or from a remote source.

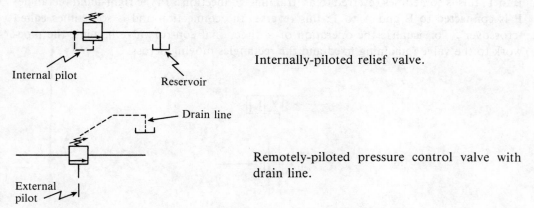

A flow-control valve is shown as a restriction on the flow line,

Section 1.2 Hydraulic symbols

If the flow control is adjustable this is indicated by an inclined arrow

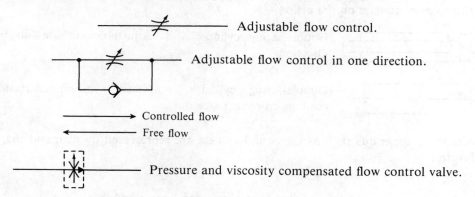

Adjustable flow control.

Adjustable flow control in one direction.

Controlled flow
Free flow

Pressure and viscosity compensated flow control valve.

Any symbol based on a circle represents a rotational unit such as a hydraulic pump or motor. A 'blocked in' triangle shows the direction of fluid flow, out of the unit for a pump and into the unit for a motor:

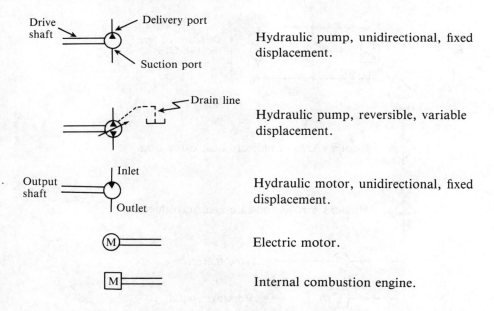

Hydraulic pump, unidirectional, fixed displacement.

Hydraulic pump, reversible, variable displacement.

Hydraulic motor, unidirectional, fixed displacement.

Electric motor.

Internal combustion engine.

A hydraulic cylinder is shown diagrammatically with the cylinder body, piston and piston rod

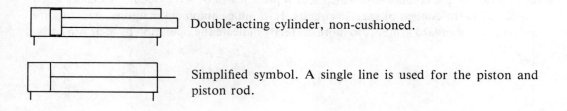

Double-acting cylinder, non-cushioned.

Simplified symbol. A single line is used for the piston and piston rod.

Cushions are fitted to a cylinder to slow the piston down at the extremes of its stroke, and are shown as rectangles on the piston.

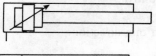 Double-acting cylinder with adjustable cushions in both directions.

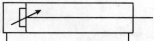 Double-acting cylinder (simplified symbol). Adjustable cushion on retract side only.

Conditioning apparatus such as filters and coolers are represented by diamond-shaped enclosures:

 Filter or strainer (non-bypassing)

A filter with a bypass valve and electrical indicator is represented by

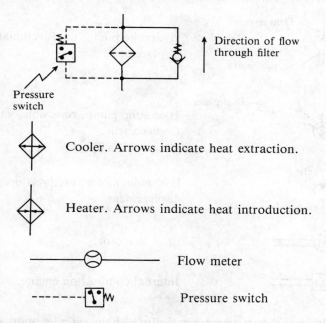

These are just a few of the symbols in general use. Many variations and combinations will be found among the illustrations in this book. Once the basic principles are understood the function of the components which they represent will be self evident, especially when considered in the context of specific circuits. Hydraulic equipment manufacturers modify and combine standard symbols to more correctly indicate the operation of their products.

CHAPTER TWO
PUMPS

2.1 TYPES OF PUMP

In any hydraulic system the pump creates a flow of fluid. The pump does not create pressure but has to overcome the resistance to flow in the circuit. There are two basic groups of pumps – non-positive displacement and positive displacement.

Non-positive displacement pumps

A typical non-positive displacement unit is a centrifugal pump where the delivery from the pump reduces as the pressure against which it has to operate increases. If the discharge side of a centrifugal pump is completely blocked the pump 'stalls off', and its delivery falls to zero. A centrifugal pump is shown diagrammatically in Figure 2.1 together with its operating characteristics. The impeller rotates and causes fluid to be sucked through the inlet port and flow to the outlet port by the action of centrifugal force.

The use of non-positive pumps in power hydraulic circuits is limited to providing a boosted supply to the main positive displacement pumps, for fluid transfer systems or for cooling and conditioning systems.

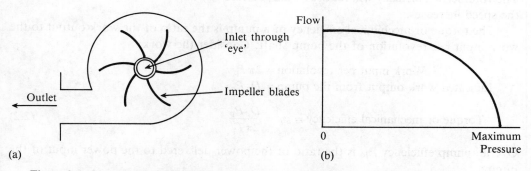

Figure 2.1 Centrifugal pump. (a) Diagrammatic section. (b) Flow/pressure characteristics.

Positive displacement pumps

A simple fixed positive displacement pump is shown in Figure 2.2.

Let the pump stroke be L and pump speed n_p revs/minute. Then displacement per revolution $D_p = \pi d^2/4 \times L$.

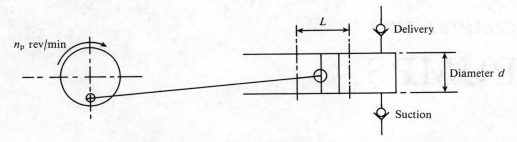

Figure 2.2 Fixed positive displacement pump.

Let Q_p be actual pump delivery per minute, T_p be the average input torque at the pump shaft, and P_p be the pressure rise across the pump. Then

Theoretical pump delivery = Displacement per revolution
× number of revolutions per minute
$= D_p \times n_p$

The actual pump delivery will be less than the theoretical delivery owing to internal leakage and slippage:

$$\frac{\text{Actual pump delivery}}{\text{Theoretical pump delivery}} = \text{Pump volumetric efficiency}$$

$$= {}_p\eta_v$$

where

$${}_p\eta_v = \frac{Q_p}{D_p \times n_p} \tag{2.1}$$

The volumetric efficiency will reduce as the pressure rise across the pump increases and as the speed increases.

The torque or mechanical efficiency of a pump is the ratio of the work output to the work input per revolution of the pump shaft. Consider the work:

Work input per revolution $= 2\pi T_p$
Indicated work output from the pump $= D_p P_p$

$$\text{Torque or mechanical efficiency} = {}_p\eta_t = \frac{D_p P_p}{2\pi T_p} \tag{2.2}$$

Overall pump efficiency ${}_p\eta_o$ is the ratio of the power delivered to the power input of the pump:

$${}_p\eta_o = \frac{Q_p \times P_p}{2\pi T_p \times n_p} \tag{2.3}$$

From equation (2.3) multiplying top and bottom by D_p

$${}_p\eta_o = \frac{Q_p}{n_p \times D_p} \times \frac{D_p \times P_p}{2\pi T_p} \tag{2.4}$$

Section 2.1 Types of pump

From equations (2.1), (2.2) and (2.4)

$$_p\eta_o = (_p\eta_v) \times (_p\eta_t)$$

EXAMPLE 2.1

A pump having a displacement of 14 cm³/rev is driven at 1440 rev/min and operates against a maximum pressure of 150 bar. The volumetric efficiency is 0.90 and the overall efficiency is 0.80. Calculate:

(i) The pump delivery in liters per minute.
(ii) The input power required at the pump shaft in kilowatts.
(iii) The drive torque at the pump shaft.

Pump delivery is:

$$Q_p = \text{Volumetric efficiency} \times \text{displacement per revolution} \times \text{pump speed}$$

$$= 0.9 \times 14 \times 10^{-3} \times 1440 \quad \left(\text{cm}^3 \times \frac{\text{liter}}{\text{cm}^3} \times \frac{\text{rev}}{\text{min}}\right)$$

$$= 18.14 \text{ l/min}$$

$$\text{Input Power} = \frac{\text{Hydraulic power}}{\text{Overall efficiency}}$$

If the flow Q is in liters per minute and the pressure P in bar then,

$$\text{Hydraulic power} = \frac{Q \times P}{600} \text{ (kW)}$$

$$\text{Input power} = \frac{1}{0.8} \times \frac{18.14 \times 150}{600}$$

$$= 5.67 \text{ kW}$$

$$\text{Torque efficiency} = \frac{\text{Overall efficiency}}{\text{Volumetric efficiency}}$$

$$= \frac{0.8}{0.9}$$

$$= 0.89$$

$$_p\eta_t = \frac{D_p \times P_p}{2\pi T_p}$$

Torque at pump shaft, $T_p = \dfrac{D_p \times P_p}{2\pi \times _p\eta_t}$

$$= \frac{14 \times 10^{-6} \times 150 \times 10^5}{2 \times 0.89} \quad (\text{m}^3 \times \text{N/m}^2)$$

$$= 37.6 \text{ Nm}$$

A positive displacement pump is one which theoretically delivers a fixed volume of fluid per revolution provided there are no displacement controls incorporated in the pump. If the discharge side of a positive displacement pump is blocked, the delivery pressure will rise very rapidly to a value at which the pump suffers mechanical breakdown.

EXAMPLE 2.2

A positive displacement pump with a delivery of 1 l/min is fed into a pipe with a total volume of 1 liter. If the end of the pipe is suddenly blocked, calculate the rise in pressure after 1 second.

(The bulk modulus of the fluid being pumped may be taken as 2000 MPa (20 000 bar); neglect any change in volume of the pipe.)

Note Pascal (Pa) is another name for the unit of pressure N/m^2. 1 MPa (Mega Pascal) = 1,000,000 N/m^2 = 10 bar.

Bulk modulus is:

$$B = \frac{\text{Volumetric stress}}{\text{Volumetric strain}}$$

$$B = \Delta P / (\Delta V / V)$$

where ΔP is the change in pressure, ΔV is the change in volume, and V is the original volume.

ΔV = Pump flow in one second
   = 1/60 liters
$\Delta P = B \Delta V / V$

$$= 2000 \times \frac{1/60}{1} (\text{MPa})$$

= 33.3 MPa
= 333 bar

This rapid rise in pressure illustrates the necessity of having some form of control to limit the rise in pressure in a system should a pump be deadheaded. The control may be built into the pump or may be an external pressure-limiting device such as a relief valve.

Positive displacement pumps are invariably used as the main pumps in hydraulic circuits. There are two major categories – rotary and reciprocating – and within these groups a wide variety of designs.

2.1.1 Rotary pumps

Generated form types

These have as the pumping elements two-dimensional generated forms which are cut to various widths to give different swept displacements.

PRECISION GEAR PUMP WITH EXTERNAL GEARS
This consists of a pair, or number of pairs, of external form intermeshing gears running in a casing with very small clearances between the tips of the teeth and the internal diameter of the casing. End plates which are either pressure or spring-loaded, seal the gear faces.

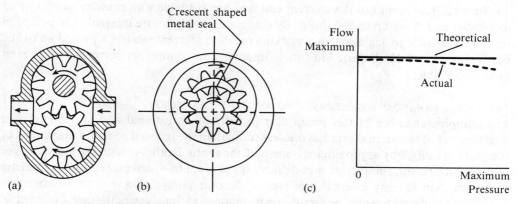

Figure 2.3 (a) External gear pump. (b) Internal gear pump. (c) Flow/pressure characteristics.

Straight gears can be used up to pressure of about 210 bar, but they tend to be noisy. The noise level can be reduced by using helical gears but owing to sealing difficulties the volumetric efficiency reduces and consequently the units have a lower maximum operating pressure. Diagrammatic sections of external and internal gear pumps are shown in Figure 2.3, together with characteristic flow/pressure curves. For a particular gear size, various displacements are obtained by using different widths of body and working elements. In each range a number of components such as end caps, bearings, seals, etc. will be common. Currently available external type gear pumps have displacements ranging from 0.2 to 400 cm^3 per revolution and speed ranges from 500 to 6000 rev/min − special designs can be much higher. Overall efficiency varies considerably depending upon the manufacturing tolerances and design detail but some models can exceed 90% and maximum operating pressure may be up to 300 bar.

The theoretical displacement (swept volume per revolution) of a gear pump with external type gears is given by:

$$D_p = \frac{\pi}{4}(d_a^2 - d_d^2)w$$

where d_a is the addendum circle diameter, d_d is the dedendum circle diameter and w is the face width of the gears.

The actual displacement will be less owing to the leakage across the gears and the fluid volume trapped at the root of the teeth.

INTERNAL GEAR PUMP WITH CRESCENT SEAL
In this pump an internal gear is driven by an external gear. The centers of the two gears are offset in one direction, and a fixed crescent shape located between the inlet and outlet of the pump acts as a sealing element. As the teeth come out of mesh, a suction zone is formed. Because this is spread over several teeth the flow velocities and consequential noise is considerably lower than with a conventional external gear pump. The discharge from the pump occurring as the teeth go into mesh is also spread over several teeth. The lengths of the suction and pressure zones of an internal gear pump are approximately three times greater than those of an external gear pump.

Internal gear pumps of the crescent seal type are available with pressure ratings of up to 100 bar and flows up to 200 l/min. By combining two or more internal gear pumps in series it is possible to achieve higher working pressure (current models are rated up to 300 bar with flows of 125 l/min). Multistage internal gear pumps are some of the quietest pumps available.

INTERNAL GEAR PUMP – GENERATED ROTOR TYPE (GEROTOR)
The pumping chamber of this pump also consists of an external and an internal gear intermeshing. The external gear has one less tooth than the internal gear and the centers of the gears are offset by approximately one-half the tooth depth. As with the crescent-type internal gear pump, the suction and delivery takes place over several teeth giving quieter operation than for any external gear pump. Gerotor pumps also tend to be extremely dirt-tolerant. As with other generated-form pumps, various capacities are obtained in each size range by using different widths of working element. These may be fitted into an eccentric ring which simplifies the machining of the pump housing. The eccentric ring is normally pinned in the housing to prevent rotation but a variation on the design shown in Figure 2.4 permits the ring to rotate through 180° reversing the eccentricity. Consequently, direction of flow is the same for both clockwise and counterclockwise rotation.

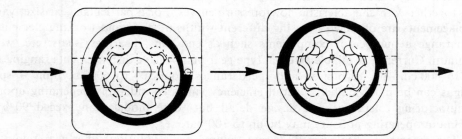

Nichols Portland Division of Parker Hannifin Corporation

Figure 2.4 Gerotor pump

Vane pumps

SIMPLE VANE PUMP
A fixed displacement simple vane pump is shown diagrammatically in Figure 2.5(a). It has a slotted rotor carrying vanes which can slide radially in and out. The rotor assembly is mounted eccentrically in a ring housing and the end faces are sealed by plates. As the rotor rotates the volume trapped between adjacent blades and the ring varies, which results in a pumping action. The vanes are thrown out by centrifugal force onto the ring and this force is increased by pressure oil applied to the underside of the vanes improving the sealing characteristics. Vane pumps have a minimum operating speed of about 600 rev/min in order to produce sufficient centrifugal force to cause the vanes to seal effectively on the ring.

The approximate displacement per revolution of a vane pump neglecting the thickness of the vanes is:

$2\pi Dew$

where D is the internal diameter of the housing, e is the eccentricity of the rotor relative to the housing, and w is the width of the vane.

BALANCED VANE PUMP

If the vane pump ring is made oval in shape, the vanes reciprocate twice during one revolution of the rotor thus giving two pumping actions per revolution. This double action has the advantage of balancing out the pressure forces on the rotor and such a pump, which is known as a 'balanced vane pump', is shown in Figure 2.5(b).

A further refinement to the balanced vane pump is the 'intra-vane' principle. Pressure oil is fed to the underside of the vane in such a manner that maximum force occurs on the vane when it is having to seal the highest pressure; at other times a lower force is applied. This reduces wear and extends the pump life.

Balanced vane pumps are often repairable *in situ* by direct replacement of the pumping cartridge. The cartridge comprises rotor, vanes and eccentric ring, which are the parts normally subject to wear.

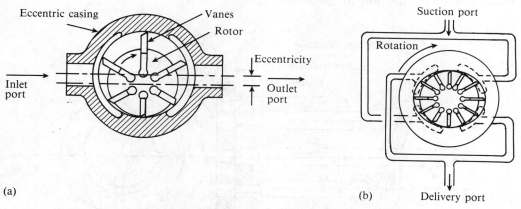

Figure 2.5 (a) Simple vane pump. (b) Balanced vane pump.

VARIABLE-DISPLACEMENT VANE PUMP

If in an unbalanced vane pump the ring is moved relative to the rotor, the eccentricity is altered but there is a natural tendency for pressure within the pumping cavities to cause the rotor and ring to centralize. The usual method of control, known as 'pressure compensation', is to oppose this movement by means of a spring. As pressure increases the spring is compressed; eccentricity and hence the flow are reduced accordingly. Maximum delivery and, in some cases, minimum delivery, can be controlled by adjustable screwed stops which limit the eccentricity. Variable vane pumps of this type are normally limited to a maximum working pressure of 70 bar with deliveries up to 350 l/min.

In some pressure-compensated variable vane pumps the spring is replaced by a piston and pressure-control valve. When system pressure reaches the setting of the control valve, it is applied to the piston centralizing the ring and rotor, reducing pump displacement to zero. A variety of control devices are available both pressure- and flow-sensitive, and internally or externally piloted, including proportional and step response. Such energy-saving controls are discussed later in this chapter and in Chapter 8.

2.1.2 Reciprocating pumps

Piston pumps are mainly used in systems where the operating pressure is 140 bar and above. The prime characteristic of piston pumps is their high efficiency at high pressures. This is of great importance when a constant flow is required, independent of pressure variations.

There are three major types of multi-piston hydraulic pump – axial, radial and plunger in-line.

Axial piston pumps

These consist of a number of pistons which are caused to reciprocate by the relative rotation of an inclined plate or by angling the piston block with regard to the drive shaft. By the use of suitable valves or a valve plate assembly, the pistons can be made to pump fluid from the suction port to the delivery port.

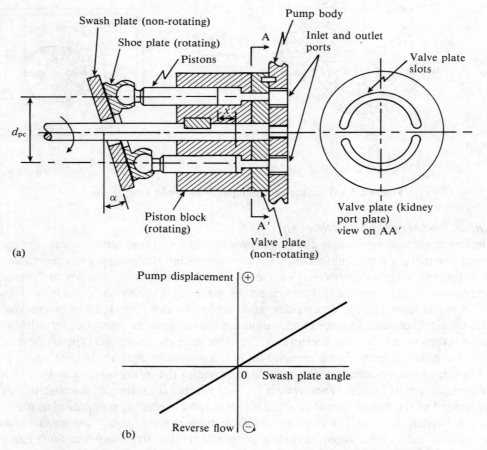

Figure 2.6 Principle of axial piston pump. (a) Section. (b) Characteristics.

Figure 2.6 illustrates the working elements of an axial piston pump with a slotted valve plate to port the fluid. The inclined cam-plate which causes the pistons to reciprocate is non-rotating and is known as a swash plate. In some configurations the cylinder block is stationary and the cam-plate rotates with the drive shaft; in such cases it is often described as a wobble plate.

If the inclination of the swash-plate is altered, the distance moved by the piston changes and so does the quantity of fluid pumped. This is the principle of a variable-delivery axial piston pump. Taking the swash-plate past zero swash angle (i.e. overcenter) reverses the direction of flow even though the direction of rotation of the drive shaft remains unchanged.

Axial piston pumps which have seated valves are capable of operating at higher pressures than those using port plates but are not reversible. Currently available models are suitable for pressures up to 700 bar and others for flow rates of 640 l/min. Models using port plates have flows of up to 1400 l/min or pressures up to 350 bar. Figure 2.7 shows the principle of a fixed-displacement bent axis pump where reciprocation of the pistons results from the cylinder block being positioned at an angle to the drive shaft. The drive to the pistons and cylinder block is through a bevel gear but other designs use connecting rods. Flow into and out of the pumping chambers is through a valve port plate. Again bent axis pumps which use port plates rather than seated valves have a lower maximum operating pressure (usually about 350 bar maximum) but tend to be quieter in operation. Bent axis pumps are available with flow rates up to 3500 l/min. The pump shown in Figure 2.7 will also function as a motor.

In certain designs it is possible to vary the angle between the axes; this alters the distance moved by the piston and consequently the amount of fluid pumped per revolution. The basic variable displacement pump can have a number of control devices which enable different pumping characteristics. These are discussed later in Section 2.1.3.

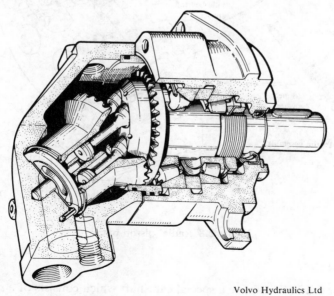

Volvo Hydraulics Ltd

Figure 2.7 Fixed displacement bent axis piston pump/motor.

Radial piston pumps

The pump shown in Figure 2.8 has pistons mounted radially in the main housing and spring-loaded onto an eccentric bearing on the drive shaft. Fluid is fed to and from the pistons through non-return valves and in this particular design the output from individual pistons can be used separately.

Another form of radial piston pump has pistons housed within a block rotating in an eccentric ring. Fluid is ported through the central shaft or pintle and variable delivery can be achieved by altering the eccentricity of the ring.

Certain models operate at pressures up to 1700 bar and flow rates range up to 1000 1/min.

Dual-pressure radial piston pumps consist of large low-pressure pistons and small high-pressure pistons reciprocated by the same eccentric. They have been developed particularly for press applications to provide a high-speed low-pressure advance followed by a slow-speed high-pressure operation. This type of circuit is considered in detail in Section 2.2.3.

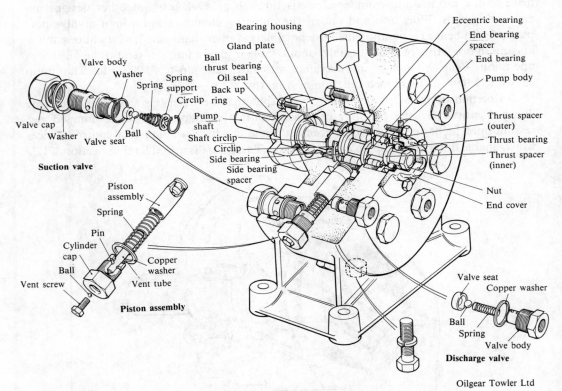

Figure 2.8 Radial piston pump.

Plunger pumps

Pistons are arranged in line above a special camshaft which consists of a shaft carrying a number of eccentric roller bearings (Figure 2.9). Fluid is fed to, and discharged from, the

Section 2.1 Types of pump

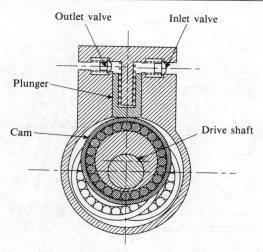

Figure 2.9 Plunger pump.

cylinders through spring-loaded poppet valves. The output of a plunger pump cannot be reversed and fixed-delivery types are the most common. They have a high volumetric efficiency and some will work against pressures in excess of 1000 bar. Flow rates are up to 600 l/min.

2.1.3 Variable-displacement pump-control systems

The delivery of a variable pump is adjusted by altering the pump displacement. This is accomplished in vane and radial piston pumps by varying the eccentricity between the rotor and stator. Axial piston pumps are altered by either bending the axis of the cylinder barrel out of line with respect to the drive shaft (bent axis pumps) or by adjusting the angle of the swash (swash plate pumps).

Only the control of swash plate-type axial piston pumps will be considered but similar methods of control can be applied to the other types of variable flow pump.

Referring to the pump shown in Figure 2.6, the piston displacement is dependent upon the swash plate angle α and the piston stroke $x = d_{pc} \tan \alpha$ where d_{pc} = pitch circle diameter of the pistons and α = tilt or swash angle.

The pump theoretical displacement per revolution = $nAd_{pc} \tan \alpha$ where n = number of pistons and A is the area of one piston.

Taking the swash plate past zero swash angle (i.e. over center) reverses the direction of flow. Because of the pressure against which the pump operates, a comparatively high force is needed to adjust the swash plate angle and a servo-system is often used.

Manual servo control

In the control shown in Figure 2.10 the pilot pressure is continually applied to port A. As the manual control lever is moved to the right it pivots on the piston taking the valve spool

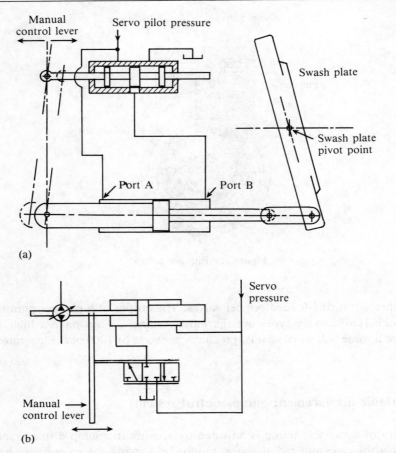

Figure 2.10 Manual servo control. (a) Diagrammatic section. (b) Symbol.

to the right. This feeds servo pilot pressure to the cylinder port B forcing the piston to the left. The piston causes the manual control lever to move to the left, resetting the spool back to the closed condition. Thus the swash plate movement is proportional to that of the control lever.

If the lever is moved to the left, the spool opens cylinder port B to the tank line. Pressure fluid constantly applied to port A of the cylinder forces the piston to the right, which carries the control lever with it, resetting the valve spool to the neutral condition.

Physical stops are incorporated into the mechanism to limit the maximum swash plate angle. This method of control can be used for reversible flow pumps or non-reversible pumps. The characteristic of a manual servo is that flow is proportional to control lever movement (Figure 2.11).

With a non-reversing variable displacement pump, it is usual to have a minimum flow to provide lubrication and cooling (i.e. the swash plate angle is never zero). Reversible flow pumps which form part of a closed loop hydrostatic transmission have a separate make up or boost pump to provide lubrication and cooling flow even when the main pump is at zero swash angle.

Section 2.1 Types of pump

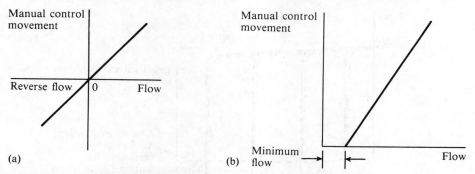

Figure 2.11 Manual servo characteristics curves. (a) Reversible flow pump. (b) Non-reversible pump.

Pressure-compensated control

The swash plate angle is automatically adjusted to set the pump delivery to maintain the set pressure. A pressure-compensated pump is shown diagrammatically and symbolically in Figure 2.12.

As the system pressure reaches the spring setting of the control valve it opens, feeding pressure fluid to the full bore end of the cylinder. Owing to the difference in full bore and annulus areas, the cylinder extends reducing the swash plate angle. If the pressure falls the valve spool is moved to the left by the control spring, connecting to tank the full bore end of the cylinder which now retracts increasing the swash plate angle. A stroke limiter is fitted to some pumps to give an adjustable maximum swash plate angle and hence an adjustable maximum pump delivery at a given rotational speed.

As there is some delay between the pump delivery pressure reaching the maximum setting and the swash angle being reduced, it is advisable to use a rapid-acting relief valve set at a pressure approximately 20% above the pump compensator setting immediately after the pump, to prevent pressure surges which can take place when the pump is 'dead-headed'. In some designs of pump the swash plate angle never becomes zero, so as

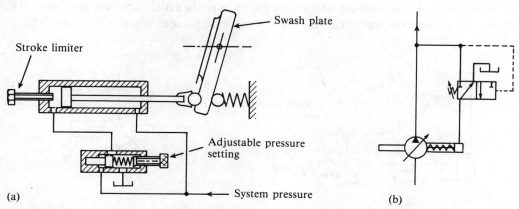

Figure 2.12 Pressure-compensated pump control. (a) Diagrammatic section. (b) Symbol.

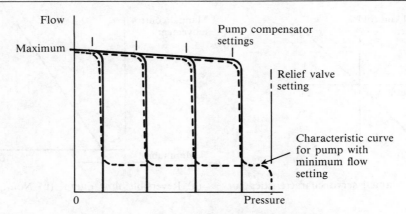

Figure 2.13 Curves for pressure-compensated pump.

to provide a flow of lubricating oil within the case. Here it is essential to have a relief valve. Figure 2.13 shows characteristic curves for a pressure-compensated pump.

The swash plate angle and hence the pump delivery is automatically adjusted in such a way that the system pressure remains constant at the value set by the compensator spring. If the pressure falls below the set value, the swash plate angle increases and consequently the pump delivery.

Pressure-compensated controls can only be used on non-reversible pumps.

Constant power control

This limits the maximum power input to the pump. As the pump delivery pressure increases, the swash plate angle reduces so that the product of pump delivery and system pressure is constant. The control is shown diagrammatically and symbolically in Figure 2.14. As the system pressure increases the cylinder extends against the spring. Provided the prime mover operates at constant speed its power output will be reasonably constant within limits, as can be seen in Figure 2.15.

With a constant power control, as the pressure rises the flow reduces up to the relief valve setting, at which pressure all the flow passes over the relief valve. At this point, the energy dissipated over the relief valve is the full pump input power. To avoid this, a

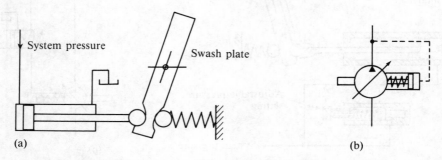

Figure 2.14 Constant power control. (a) Diagrammatic. (b) Symbol.

Section 2.1 Types of pump

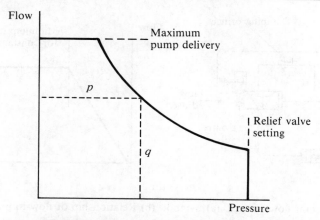

Figure 2.15 Flow pressure relationship with constant power control. At any point on the control curve, power = pq = constant.

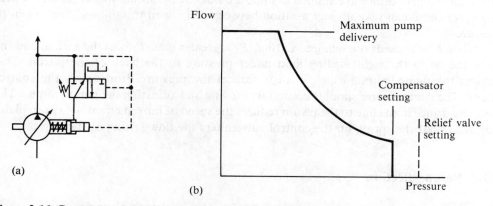

Figure 2.16 Constant power control with pressure compensation. (a) Symbol. (b) Flow/pressure characteristics.

pressure-compensation control can be used in conjunction with the constant power control, as shown symbolically in Figure 2.16 together with its characteristic curves.

The pump flow progressively reduces as system pressure increases up to the setting of the pressure compensator. When this pressure is reached, the compensator automatically reduces the pump output flow to just sufficient to maintain that pressure. No flow goes over the relief valve.

Constant-flow control

This enables a constant pump output to be sustained independent of changes in volumetric efficiency and pump drive speed. This method of control is used on internal combustion-engine pump drives to maintain a constant flow delivery as the engine speed varies. It is shown symbolically in Figure 2.17 together with its characteristic curves. A control orifice in the pump discharge line causes a pressure drop which is a function of the

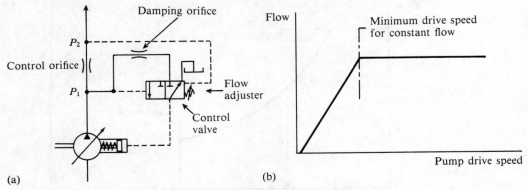

Figure 2.17 Constant flow control. (a) Symbol. (b) Relationship of flow to pump drive speed.

quantity of fluid flowing, but is independent of system pressure. The pressures on either side of the control orifice are applied to opposite sides of the spring-loaded control valve. The valve spool can take up any position between fully open to tank and fully open to pressure.

When flow exceeds the required setting, P_1 is greater than P_2 plus the spring, and the spool moves to the right feeding fluid under pressure to the back of the piston. This reduces the pump output which in turn reduces the pressure drop across the control orifice. The control valve spool is constantly moving and adjusting the pump output. The damping orifice in the line to the piston reduces the speed of movement of the swash plate. Adjustment of the spring on the control valve alters the flow setting of the pump.

2.1.4 Pump selection

The main parameters affecting the selection of a particular type of pump are:

Maximum operating pressure
Maximum delivery
Type of control
Pump drive speed
Type of fluid
Pump contamination tolerance
Pump noise
Size and weight of pump
Pump efficiency
Cost
Availability and interchangeability
Maintenance and spares.

Maximum operating pressure

This is determined by the power requirements of the circuit, the particular application, availability of components, type of fluid and to some extent, the environment and level of labor both using and maintaining the equipment.

In general, the higher the operating pressure the higher the component cost and the lower the choice of components. The main advantage of higher working pressures is the reduction in fluid flow rates for a given system power, resulting in smaller pumps, smaller bore pipes and smaller components.

The disadvantages are that at higher working pressures, the compressibility of the

Table 2.1 System maximum pressure in relation to application.

Application	Pressure	
	bar	psi
Mechanical handling	250	3700
Machine tools	200	3000
Mobile	300	4500
Press work	800	12 000

Table 2.2 Operating pressures and size ranges for hydraulic pump types.

Pump type	Operating pressure (continuous) (bar)		Flow range maximum (liters/min)	
	Maximum	Normal	From	To
Precision gear	300	170	0.25	760
Internal gear				
single-stage	210	100		
multi-stage	300	200	0.6	740
Balanced vane	175	100	2	620
Pressure compensated vane	175	100	6	360
Cam rotor	175	120	1	400
Axial piston swash plate				
port plates	350	200	0.7	600
seated valves	700	350	1	760
variable displacement	350	200	1	1450
Axial piston bent axis	350	300	7.5	3500
variable displacement	350	210	17	3500
Radial piston	1720	300	0.3	1000
variable displacement	350	175	1	580
Plunger in-line	1000	400	0.1	600

fluid used can have considerable adverse effects where precision control is required over a wide range of loads.

The general tendency is towards increased operating pressures. Typical maximum pressures for various applications are given in Table 2.1. The operating pressures of pumps depend to some extent on the fluid used — a fire-resistant fluid is generally not as good a lubricant as a mineral oil. So to give a reasonable pump life expectancy when using fire-resistant fluid, the maximum operating pressure must be reduced and it is advisable to consult the pump manufacturer.

The maximum operating pressures and range of flow rates for different types of currently available hydraulic pump are shown in Table 2.2. Figures given cover a range of sizes and makes; maximum values of delivery and pressure will not be applicable to one pump. For example some multi-stage internal gear pumps are available with flow rates of 740 l/min. Others are capable of operating at 300 bar, but 740 l/min at 300 bar is not obtainable from any one pump in this category.

Maximum delivery

The pump system selected must be capable of delivering the maximum flow rate demanded by the circuit. If the circuit demand is reasonably constant, choose a fixed displacement pump. When the demand is at a series of fixed levels, a multi-pump system can be used. For demands which vary within a relatively narrow band, use a variable displacement pump. If there is a wide variance in system demand, an accumulator circuit may best satisfy the requirements.

The different types of pumping circuits are described in detail later in Section 2.2.

Pump capacities are stated by manufacturers for a particular viscosity fluid at given operating temperatures and pressures. Any increase in temperature and hence a reduction in viscosity or an increase in operating pressure will cause more leakage across the pump and consequently reduce the pump delivery. As the pump wears the leakage will increase. When determining the pump delivery required for a particular application, the previous points must be considered. It is usual to select a pump with a capacity about 10% higher than required to make an allowance for the reduction in volumetric efficiency with wear.

Pumps are available with flows from a fraction of a liter per minute to 1000 l/minute and above.

Type of control

The various types of pump control have already been discussed; the choice of control is dependent upon the circuit requirements.

Pump drive speed

The majority of pumps are driven directly from the prime mover — electric motor or internal combustion engine — so the proposed drive speed will be known. The fluid delivery rate is proportional to speed of rotation. Each design has a minimum and maximum operating speed; the faster the pump is run, the shorter will be its life.

Type of fluid

Pumps are designed to operate within a particular range of fluid viscosity. Mineral oils of the correct viscosity will work satisfactorily with most pumps provided the oil is clean.

Operating with synthetic or water-based fluids reduces the working life of a pump which relies on the hydraulic fluid to lubricate the bearings and moving parts. When any fluid other than a mineral oil is to be used it is advisable to seek the pump manufacturer's advice.

FLUID CONTAMINATION

Any fluid contamination will cause pump damage. Precision pumps with very fine clearances are more susceptible to damage. If contaminated fluid has to be pumped, such as in a clean-up loop, particular attention must be paid to pump selection.

Non-precision gear pumps, lobe pumps and gerotor pumps are the most dirt-tolerant. Whichever type is used a strainer must be fitted in the suction line.

In the case of precision pumps, the manufacturer's recommendations on filtration must be followed otherwise the life of the pump will be drastically reduced and the makers warranty voided.

Pump noise

Noise is becoming increasingly important environmentally. Operating levels vary considerably between pumps of the same type but of different makes, and the manufacturers are working on those aspects which most affect its emission – port plate design, bearings, flow passages, pressure controls, materials and methods of mounting. Generally, the sound generated increases with speed and pressure. Certain kinds do, however, propagate lower noise levels, in particular those with internal gears. A multi-stage internal gear pump is marketed by one manufacturer under the name *Q Pump*, 'Q' signifying quiet.

Size and weight of pump

Generally, the overall size and weight of the hydraulic system is only important in mobile installations. The size and weight of the pump is only part of the whole system and it is the whole system that is important. In the mobile hydraulic field the trend is to reduce the weight of the hydraulic system by:

(a) Increasing the operating pressure.
(b) Reducing the size of the reservoir and using efficient oil coolers.

The best power : weight ratios can usually be achieved in the 200–300 bar operating pressure range.

The actual size and weight of a pump depends upon the particular manufacturer's design. Very light compact units have been developed for use in the aerospace industry but these tend to be extremely expensive.

Efficiency

Reciprocating pumps tend to have higher efficiencies than rotary pumps. The actual efficiency depends on design, operating pressure, speed and fluid viscosity pumped. Table 2.3 gives an indication of the range of efficiencies of various types of pump.

Table 2.3 Efficiency ranges of pumps.

Pump type	Volumetric efficiency (%)	Overall efficiency (%)
Piston		
plunger in-line	$\leqslant 99\%$	$\leqslant 95\%$
radial	$> 95\%$	$> 90\%$
axial	$> 95\%$	$> 90\%$
Precision gear pump	$\leqslant 95\%$	$\leqslant 90\%$
Vane pumps	$\leqslant 90\%$	$\leqslant 80\%$

Cost

The initial cost of a pump is usually of secondary importance to running and maintenance costs except for the equipment manufacturers involved in quantity products.

The lower cost units are gear and vane pumps, the piston types being much dearer, with seated valve in-line plunger pumps probably being the most expensive.

Availability and interchangeability

A number of gear pump manufacturers produce units to CETOP and SAE standards so far as the external dimensions are concerned. This gives direct interchangeability between gear pumps of different manufacture. The shafts, mounting flanges and port connections of most other types also comply with various international standards allowing a degree of interchangeability.

Maintenance and spares

In every type of pump, the components involved in pumping will become worn after a time and need replacing. In gear pumps, it is usual to replace the entire pump. With some types of vane pumps all the wear parts are grouped together as a cartridge which can easily be replaced without dismantling the pump drive.

In the case of piston pumps, it may be advisable to ensure that the manufacturers offer a fast overhaul service, and for 'critical' applications to carry a spare pump in stock.

2.2 PUMPING CIRCUITS

There are a number of basic pumping circuits which form the heart of most hydraulic systems.

Section 2.2 Pumping circuits

Single fixed-displacement pump
Single fixed-displacement pump with accumulator } Open loop systems
Multi-pump
Variable-displacement pump
Hydrostatic transmission } Closed loop system

In an open loop system, fluid is taken from the reservoir by the pump or pumps, used to drive the actuators and then returned to tank. Specialist circuits found in hydrostatic transmissions employ a principle where the fluid returned from the actuator (hydraulic motor) is fed directly to the pump suction forming a closed loop system. Section 4.4 in Chapter 4 contains a detailed explanation of this type of circuit.

2.2.1 Single fixed-displacement pump

A large majority of simple hydraulic circuits employ a single fixed-displacement pump driven at a constant speed by an electric motor (Figure 2.18). This produces an almost constant flow rate whenever the motor is running. The system pressure is limited to the setting of the relief valve. Any flow from the pump in excess of the circuit demand will discharge across the relief valve at full system pressure, generating heat energy which raises the fluid temperature. Excessive heat generation leads to overheating and fluid degradation.

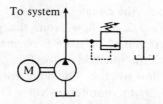

Figure 2.18 Single fixed-displacement pump circuit.

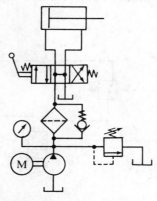

Figure 2.19 Double-acting cylinder controlled by a four-port, three-position, spring-centered, hand-operated directional control valve.

The type of circuit in Figure 2.18 is best used where:

(i) Full pump flow is needed at all times.
(ii) Pump output can be unloaded to tank when not required.
(iii) Power loss due to relief valve action is small and the heat generated easily dissipated.
(iv) There is short and infrequent operation with the drive motor switched off when not required.

A typical application which satisfies (ii) above is shown in Figure 2.19 – a simple circuit where a double-acting cylinder is controlled by a four-port, three-position, spring-centered, hand-operated directional control valve. Movement of the actuator is effected by operating the hand lever. When the lever is released the spring-centering device ensures that the pump output is diverted to tank via an easy flowpath. Resistance to flow being low, there will be little heat created.

2.2.2 Single fixed-displacement pump with accumulator

Hydraulic accumulators are devices for storing energy in the form of hydraulic fluid under pressure. They are capable of providing a high fluid flow rate over a short time period.

Details of various types and applications of accumulators are given in Section 6.4 in Chapter 6. The most widely used accumulator comprises a forged steel bottle containing a flexible bag charged with nitrogen under pressure. The pump is used to force fluid into the bottle and the gas is further compressed. A reduction in input pressure causes the gas to expand forcing fluid back out into the circuit.

In the typical circuit, Figure 2.20, flow from the pump is used to charge the accumulator A. When a pre-determined pressure is reached, pressure switch B operates and breaks the electrical circuit to solenoid valve C which reverts to its unoperated condition unloading the pump flow to tank. The fluid stored under pressure is locked in the upstream part of the circuit by the non-return valve D.

The electrical control circuit in Figure 2.20 is shown as 'failsafe' – the pump flow is dumped when the solenoid valve C is de-energized (i.e. when the control circuit is broken).

This particular circuit is shown in a simplified and hopefully more-easily understood

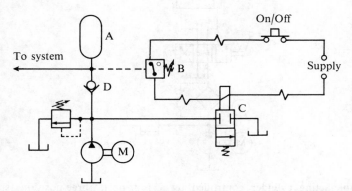

Figure 2.20 Pump and accumulator circuit with electrical controls.

Section 2.2 Pumping circuits

form. In practice, additional valve-work is recommended to assist in the safe operation and maintenance of the accumulator which is a pressure vessel and potentially dangerous. Legislation in some countries makes the use of accumulator safety-valve blocks mandatory.

Accumulator circuits are especially useful for systems where the actuator is at rest for a considerable proportion of the operating cycle time – particularly where pressure has to be maintained at the actuator.

EXAMPLE 2.3

A cylinder has to operate with the following time cycle: extend in 5 seconds at 25 bar, flow rate 12 l/min; remain extended for 25 seconds at 200 bar, no flow; retract in 4 seconds at 35 bar, flow rate 12 l/min; remain retracted for 26 seconds at 200 bar, no flow.

Pressure/flow requirements are shown graphically in Figure 2.21. Flow is required for only 15% of the cycle. With a single fixed-displacement pump circuit (Figure 2.22) the pump output of 12 liters/min will discharge over the relief valve at 200 bar for 85% of the cycle time.

Theoretical input power is

$$\text{Flow} \times \text{Pressure}, = \frac{12}{60} \times 10^{-3} \times 200 \times 10^5 \qquad (\text{m}^3/\text{s} \times \text{N/m}^2)$$

$$= 4000 \text{ Nm/s}$$
$$= 4 \text{ kW}$$

a major portion of which will be wasted as heat energy across the relief valve.

Considering the flow requirement curve in Figure 2.21 the flow needed during a one-minute cycle is:

To extend the cylinder = 12 × 5/60 = 1 liter
To retract the cylinder = 12 × 4/60 = 0.8 liter
Total oil required per minute = 1.8 liters

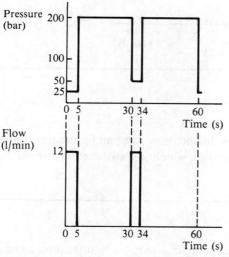

Figure 2.21 Pressure flow requirements.

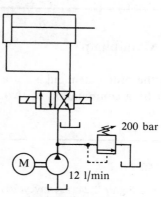

Figure 2.22

Thus by storing the flow from the pump in an accumulator when the cylinder is at rest (see Figure 2.20) a pump with a delivery rate of 1.8 liters/min will be sufficient. However since a minimum pressure of 200 bar is required during the cylinder rest periods, it will be necessary to operate above this pressure, to say 250 bar, and the variations in circuit pressure (see Figure 2.23) may be disadvantageous. The pump output is continuously charging the accumulator up to the new maximum circuit pressure setting. Pressure falls as the accumulator discharges to the circuit and flow controls will have to be included to limit the discharge rate.

It is difficult to make the average pump supply exactly match the time average circuit demand so a larger pump would be chosen with the excess flow discharging over the relief valve.

If pump output greatly exceeds circuit demand so that the accumulator remains at maximum pressure for a large proportion of the operating cycle, a pump unloading system must be incorporated (pressure relief/unloader valves are described in Section 3.1 of Chapter 3).

Theoretical power requirements assuming a pump delivery of 2 l/min with excess flow discharging over the relief valve at 250 bar is:

$$\frac{2 \times 10^{-3}}{60} \times 250 \times 10^5 \times 10^{-3}$$

$$= 0.83 \text{ kW}$$

Detailed examples of accumulator calculations are included in Sections 6.4 and 6.6 of Chapter 6.

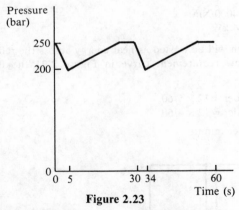

Figure 2.23

2.2.3 Multi-pump

Where the system demand is for a number of distinct flow rates and pressures it may best be met by a combination of two or more pumps which are switched on or off load as required.

EXAMPLE 2.4

Refer to Figure 2.24. A conveyor is driven by a hydraulic motor, and by using three pumps of different volumetric displacements, seven stepped speeds are attainable, in addition to zero.

Section 2.2 Pumping circuits

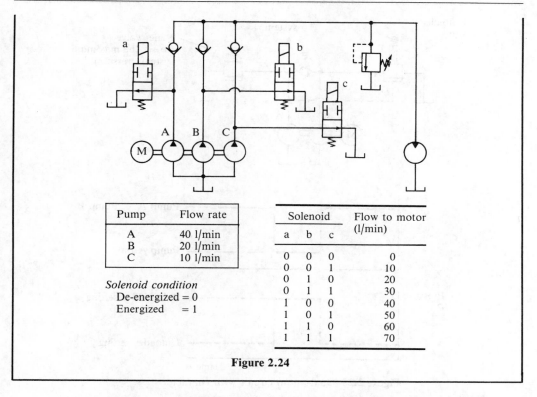

Pump	Flow rate
A	40 l/min
B	20 l/min
C	10 l/min

Solenoid condition
De-energized = 0
Energized   = 1

Solenoid			Flow to motor (l/min)
a	b	c	
0	0	0	0
0	0	1	10
0	1	0	20
0	1	1	30
1	0	0	40
1	0	1	50
1	1	0	60
1	1	1	70

Figure 2.24

A standard, much-used dual-pump arrangement is shown in Figure 2.25 where the pump outputs are automatically switched in and out of circuit according to the system demand pressure. Pump A operates against the relief valve C which sets the maximum circuit pressure. Pump B usually has a higher flow rate than pump A and will discharge to tank with little resistance to flow if the circuit pressure exceeds the setting of unloader valve D. Whenever the circuit demand pressure is below the setting of valve D, both pumps feed the system. Check valve E prevents the output of pump A discharging over valve D.

It is sometimes called a 'Hi–Lo' circuit referring to the two usual characteristics which it can provide of high volume at low pressure or low volume at high pressure.

The circuit will operate equally as well for pumps of equal displacement or with the high-pressure pump having a larger displacement than the low-pressure pump. It is essential, however, that the operating pressures are sufficiently different to give correct operation of the unloader valve. (Unloader valves are described in detail in Section 3.1 of Chapter 3.) Typical pressure and flow characteristics for a dual pump system are shown in Figure 2.25.

A typical application of a dual or multi-pump circuit is a long stroke hydraulic press, the combined outputs of the high-volume low-pressure pump and low-volume high-pressure pump being used to rapidly close the press tools onto the work piece. At this point load resistance suddenly increases and the consequential increase in circuit pressure operates the unloader valve which dumps the flow from the high-volume pump. Flow from the low-volume high-pressure pump completes the stroke at a much reduced speed

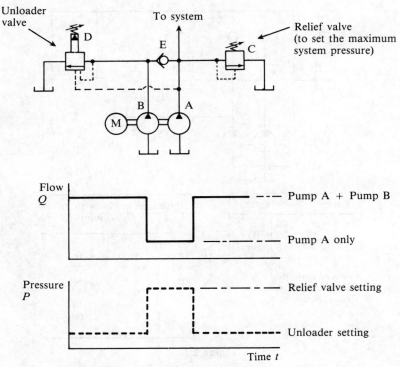

Figure 2.25 Dual-pump circuit with characteristics.

but at pressures up to the relief valve setting. Both pumps are used to open the dies at high speed and low pressure.

This circuit will provide a more constant pressure than an accumulator circuit and achieve considerable power savings compared with a single fixed-displacement pump system. By selection of suitable pump capacities and operating pressures it can be arranged for the same power to be required to drive the pumps for both the low-pressure and high-pressure conditions.

EXAMPLE 2.5

A press requires a flow rate of 200 l/min for high-speed opening and closing of the dies at a maximum pressure of 30 bar. The work stroke needs a maximum pressure of 400 bar but a flow rate between 12 and 20 l/min will be acceptable.

Theoretical power required to open or close the dies is

$$\frac{200 \times 10^{-3}}{60} \times 30 \times 10^5 \, (\text{Nm/s})$$

$$= 10\,000 \text{ Nm/s}$$
$$= 10 \text{ kW}$$

To utilize this power for the pressing process: if q is the available flow at 400 bar, then

$$q \times 400 = 200 \times 30$$

and

$q = 15 \text{ l/min}$

which is acceptable.

Required pump deliveries are:

High-pressure, low-volume pump = 15 l/min
High-volume, low-pressure pump = (200 − 15) = 185 l/min

An equivalent single fixed-displacement pump having a flow rate of 200 l/min and working at a pressure of 400 bar requires a theoretical input power of 133.3 kW.

2.2.4 Variable-delivery pump

A number of variable displacement pumps were described in Section 2.1. They are almost inevitably more complex and expensive than fixed-displacement units, but their use often facilitates the most efficient pumping circuit. According to the method of control adopted, flow and pressure can be varied to match the system demand.

2.3 PUMP DRIVES

Pumps may be driven by an electric motor, an internal combustion engine such as a diesel engine, a pneumatic motor or even by a hydraulic motor.

The driving torques available at low speeds from an electric motor and a diesel engine are very low. It is therefore advisable to unload the pump when it is being started; this can be achieved by venting the relief valve (Figure 2.26). (The principle of venting a two stage relief valve is explained in Section 3.1 of Chapter 3.)

In Figure 2.26 the solenoid vent valve could be connected to an auxiliary 'delta' contact on the 'star-delta' starter of the electric motor. In the case of a diesel-engine drive, a separate unloader valve may be used or the relief valve vented.

Another method of reducing the power requirements during starting against a

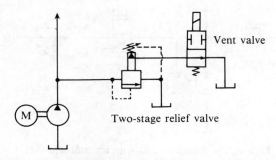

Figure 2.26 Pump unloading by venting the relief valve.

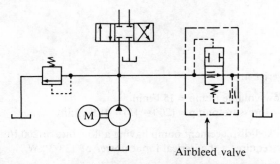

Figure 2.27 Pump circuit with air bleed valve.

blocked circuit is to use an 'airbleed' or 'start-up' valve (Figure 2.27). This not only facilitates pump priming and purges air from the system, but allows the pump to accelerate at low load; pumping is directly to tank before it closes and establishes system pressure. Initially there is an easy flow path to tank through the valve which is then slowly closed by a build-up of pressure caused by the flow forces.

When an air motor is used as the drive (Figure 2.28) no unloading is needed because air motors have a relatively high starting torque. The air motor will also stall off and stop when the load torque is too great. Hence by regulating the air supply pressure to the motor, the pump can be stalled when a predetermined system pressure is reached. The setting of the relief valve must be approximately 15% above the system pressure at which the air motor stalls.

Air motor drives tend to be limited to low-power applications owing to the constraints imposed by the motor power. Among the applications of air-driven pumps are units for pumping water at low pressure but high flow rate, and at the other extreme, pumps working up to 700 bar but at very low flow rates.

The main advantage of an air-driven pump is that the air pressure can be adjusted to set the pressure of the hydraulic fluid at which the pump stalls. This means a circuit can be kept pressurized with the air motor stalled for long periods of time without loss of energy and hence no heat generation in the fluid.

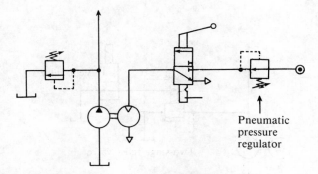

Figure 2.28 Pump circuit with air motor drive.

2.4 PUMP CIRCUIT DESIGN STUDY

To select the type of pumping circuit for a particular application the pressure/flow requirements must be analyzed. In practice the most appropriate solution may well be obvious from the data; or at least some of the possibilities can be eliminated without in-depth calculation. However, in the following design study Example 2.6, for the sake of completeness several basic pumping circuits will be considered showing the relative effectiveness of the solutions.

EXAMPLE 2.6

Design data

The hydraulic system to be supplied by the pump has a circuit demand characteristic for flow and pressure as shown in Figure 2.29. The complete cycle time is 30 seconds. The system demands fluid for only half its cycle time but requires to be pressurized for two-thirds of the cycle. Flow controls may have to be used to set the fluid rate to the values required. The fluid to be used is mineral oil, and there are no other special requirements. Four alternative designs

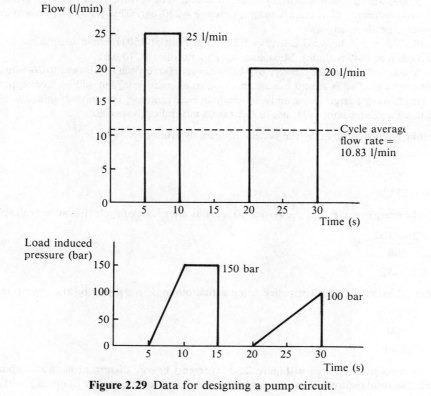

Figure 2.29 Data for designing a pump circuit.

will be considered:

1. Using a single fixed-displacement pump
2. Using two fixed-displacement pumps
3. Using an accumulator system
4. Using a pressure-compensated pump.

1. *Using a single fixed-displacement pump*

Consider a fixed-displacement pump circuit as shown in Figure 2.19.

Theoretical pump delivery = 25 l/min (allow an additional 10% approximately)
Therefore, pump delivery required = 27.5 l/min

System pressure maximum = 150 bar. (Set relief valve at 10% above system pressure.)
Therefore, relief valve setting = 165 bar.

These flow rates and pressures are within the range available for gear pumps (see Table 2.4 giving details of single Dowty gear units).
 Assume direct drive from 1440 rev/min motor. Calculate the equivalent pump delivery at 1500 rev/min. Therefore required pump delivery at 1500 rev/min is

$$27.5 \times \frac{1500}{1440} = 28.7 \text{ l/min}$$

From Table 2.4 the nearest standard gear pumps are:

(a) 1 PL 060 with a nominal delivery of 28.1 l/min at 1500 rev/min (equivalent to 27.0 l/min at 1440 rev/min). Maximum working pressure = 250 bar. This pump is just within the system specification.
(b) 1 PL 072 with a nominal delivery of 33.6 liters/min at 1500 rev/min (equivalent to 33.2 liters/min at 1440 rev/min). Maximum working pressure = 210 bar.
(c) 2 PL 090 with a nominal delivery of 41.5 l/min at 1500 rev/min (equivalent to 26.6 l/min at 960 rev/min). This is almost exactly the same as alternative (a) but will be more expensive owing to using a larger pump and a 960-rev/min electric motor. The only advantage would be if using a fire-resistant fluid, but in this case a mineral oil is specified.

Hydraulic energy required by the system 10 seconds after start of cycle is

$$\frac{25 \times 150}{600} \text{ kW}$$

$$= 6.25 \text{ kW}$$

Hydraulic energy required by the system 30 seconds after start of cycle (i.e. at end of cycle) is

$$\frac{20 \times 100}{600} \text{ kW}$$

$$= 3.3 \text{ kW}$$

Theoretical hydraulic power supplied using actual pump delivery, pump (a) or pump (c) is

$$\frac{27 \times 165}{600}$$

$$= 7.4 \text{ kW}$$

The cross-hatched area in Figure 2.30 represents energy dissipated as heat within the system. The total theoretical energy supplied to the system by the electric motor is 7.4 kW for

Section 2.4 Pump circuit design study

Table 2.4 Dowty Powerline series of gear pump/motors.

Pump type	Motor type	Theoretical displacement (cm³/rev)	Maximum continuous pressure P1(bar)	Speed at pressure P1				Typical pump delivery at 1500 rev/min (l/min)
				Min.pump (rev/min)	Max.pump (rev/min)	Min.motor (rev/min)	Max.motor (rev/min)	
0PL 003		1.22	280	500	4000			1.50
0PL 004		1.63	280	500	4000			1.95
0PL 006		2.18	280	500	4000			2.91
0PL 008		2.87	280	500	4000			3.95
0PL 011	0ML 011	3.81	280	500	4000	500	4000	5.32
0PL 013	0ML 013	4.46	280	500	4000	500	4000	6.27
0PL 015	0ML 015	5.14	280	500	4000	500	4000	7.27
0PL 019	0ML 019	6.26	280	500	4000	500	4000	8.95
0PL 025	0ML 025	8.08	225	500	4000	500	4000	11.73
1PL 020	1ML 020	7.02	250	500	3000	500	3000	10.13
1PL 028	1ML 028	9.46	250	500	3000	500	3000	13.72
1PL 036	1ML 036	11.89	250	500	3000	500	3000	17.32
1PL 044	1ML 044	14.33	250	500	3000	500	3000	20.95
1PL 052	1ML 052	16.76	250	500	3000	500	3000	24.50
1PL 060	1ML 060	19.20	250	500	3000	500	3000	28.10
1PL 072	1ML 072	22.84	210	500	3000	500	3000	33.60
1PL 090	1ML 090	28.12	175	500	2500	500	3000	41.50
2PL 050	2ML 050	16.66	250	500	2500	500	3000	24.36
2PL 070	2ML 070	22.71	250	500	2500	500	3000	33.45
2PL 090	2ML 090	28.77	250	500	2500	500	3000	42.45
2PL 105	2ML 105	33.23	250	500	2500	500	3000	49.10
2PL 120	2ML 120	37.85	250	500	2500	500	3000	55.91
2PL 146	2ML 146	45.50	210	500	2500	500	3000	67.32
2PL 158	2ML 158	49.35	210	500	2500	500	3000	73.05
3PL 150		47.08	250	500	2500			68.9
3PL 180		56.20	250	500	2500			82.5
3PL 210		65.26	250	500	2500			96.1
3PL 250		77.19	210	500	2500			114.1
3PL 300		92.08	175	500	2250			136.6
3PL 330		101.77	160	500	2250			150.9
3PL 380		116.85	140	500	2150			173.5

Dowty Hydraulic Units Ltd

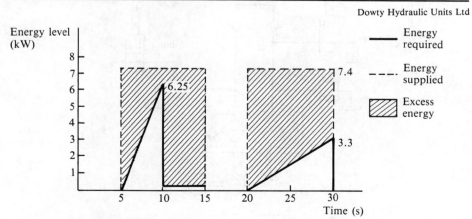

Figure 2.30 Energy supplied and energy dissipated as heat.

20 seconds in a 30-second cycle. The total energy usefully consumed in each cycle is that used between 5 and 10 seconds and between 20 and 30 seconds. Therefore

Total theoretical energy
supplied $= 7.4 \times 20$ kW (joules $\times 10^3$)
                $= 140$ kJ

Total energy usefully
used $= (6.25 \times 5/2) + (3.3 \times 10/2)$ kJ
                $= 32.12$ kJ

The system overall efficiency based on energy usefully used in the system divided by energy supplied is

$$= \frac{32.12 \times 100}{148}$$

$$= 21.7\%$$

2. *Using two fixed-displacement pumps*

When using two fixed-displacement pumps (circuit as shown in Figure 2.31), both pumps are used together to give the higher flow, and one pump only to give the lower flow. Thus the theoretical pump deliveries required are 20 l/min and 5 l/min. As before, allow an additional 10% on theoretical pump deliveries. This gives 22 l/min and 5.5 l/min.

Tandem or double pumps are available from some gear pump manufacturers but use a limited range of units. Table 2.5 shows a selection of units which can be obtained in any combination.

The system is time-based and therefore a control timer can be used to switch the pumps on and off load.

Actual pump deliveries taken from data sheet Table 2.5 are 22.9 l/min and 5.7 l/min for the size 16 and size 4, respectively, at 1440 revs/min and 175 bar; the deliveries will be almost the same at 165 bar. (A reduction in system pressure improves the pump delivery by reducing leakage.)

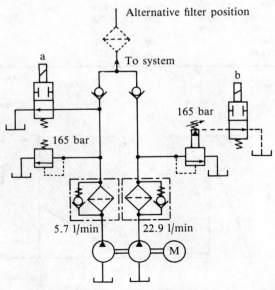

Figure 2.31 Double pump circuit.

Section 2.4 Pump circuit design study

Table 2.5 Units for use as tandem (double) pumps. Any two of the pumps may be combined as a double unit.

Pump size	Theoretical displacement (cm³/rev)	Nominal delivery at 1440 rev/min (l/min)	Maximum pressure (bar)	Maximum speed (rev/min)	Volume efficiency at 175 bar (%)	Actual delivery (l/min)
4	4.4	6.34	250	3000	90	5.7
8	8.5	12.24	250	3000	91	11.1
12	11.9	17.14	240	2500	92	15.8
16	17.3	24.9	220	2000	92	22.9

With the filters in the position shown in Figure 2.31, there will be a constant flow through each filter irrespective of whether the system is on or off load and all the oil pumped will be filtered. If a single filter is placed in the alternative position only the oil used by the system will be filtered, and the filter element will be subjected to flow surges as solenoids (a) and (b) (Figure 2.31) are energized. Flow to the circuit is controlled by these solenoid valves a and b,

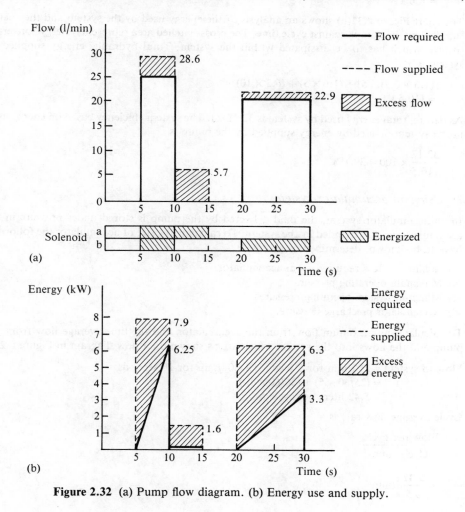

Figure 2.32 (a) Pump flow diagram. (b) Energy use and supply.

actuated from a timer. (Solenoid valves are described in Chapter 3.) With neither solenoid energized, flow from both pumps is returned to tank at low pressure and hence little waste of energy. When a solenoid is energized, the route to tank is blocked and the appropriate pump feeds the circuit.

Figure 2.32(a) shows the quantity of oil delivered by the pump circuit (as the relative solenoids are energized or de-energized) and the oil demanded by the system, both to a base of cycle time. The cross-hatched area denotes the excess pump delivery which will flow over the relief valves. This assumes flow-control valves are used in the circuit to regulate the actuator speeds.

Hydraulic energy supplied by the 22.9 l/min pump to the system at 165 bar is given by:

$$22.9 \times 165 \times 1/600$$
$$= 6.3 \text{ kW}$$

Similarly the hydraulic energy supplied to the system by the 5.7 l/min pump at 165 bar is given by:

$$5.7 \times 165 \times 1/600 \text{ kW}$$
$$= 1.6 \text{ kW}$$

The chart Figure 2.32(b) shows an analysis of the energy used by the system and the energy supplied by the pumps against cycle time. The cross-hatched area represents energy converted to heat which has to be dissipated within the system. Total hydraulic energy supplied to system is

$$[(1.6 + 6.3) \times 5] + (1.6 \times 5) + (6.3 \times 10)$$
$$= 110.5 \text{ kJ}$$

As before, total energy used by system is 32.12 kJ. The system efficiency based on energy used in the system divided by energy supplied by the pump is

$$\frac{32.12}{110.5} \times 100 = 29.1\%$$

3. *Using an accumulator system*

In an accumulator system, the fluid delivered by the pump is stored under pressure in the accumulator until demanded by the system. To calculate the size of accumulator the following have to be known, determined or assumed:

(a) Maximum flow required from accumulator.
(b) Maximum operating pressure.
(c) Minimum system operating pressure.
(d) Accumulator precharge pressure.

To calculate the maximum flow from the accumulator find the time-average flow from the pump and the flows into the system which are as shown on a flow diagram in Figure 2.29.

Flow to system = 25 l/min for 5 seconds + 20 l/min for 10 seconds
$$= (25/60 \times 5) + (20/60 \times 10)$$
$$= 5.42 \text{ liters per cycle}$$

Cycle average flow rate is

$$\frac{\text{Flow per cycle}}{\text{Cycle time}}$$

$$= \frac{5.42}{0.5} \text{ l/min}$$

$= 10.84 \text{ l/min}$
$= 0.18 \text{ l/s}$

The flow of fluid into or out of the accumulator can be calculated by multiplying the flow rate by the flow time.

(i) Between 0 and 5 seconds the flow rates are:
Pump delivery = 0.18 l/s
System demand = 0
Flow rate into accumulator is 0.18 l/s
Flow into accumulator between 0 and 5 seconds is 0.18×5 liters = 0.9 liter.

(ii) Similarly between 10 and 20 seconds the pump output flows into the accumulator:
Flow into accumulator between 10 and 20 seconds is $0.18 \times 10 = 1.8$ liters.

(iii) During period 5 to 10 seconds:
Pump delivery $= 0.18 \text{ l/s}$
Circuit demand $= 25 \text{ l/min}$
$= 0.417 \text{ l/s}$
Flow rate from accumulator $= 0.417 - 0.18$
$= 0.237 \text{ l/s}$
Flow from accumulator between 5 and 10 seconds is 0.237×5 liters = 1.185 liters.

(iv) During period 20 to 30 seconds:
Pump delivery $= 0.18 \text{ l/s}$
Circuit demand $= 20 \text{ l/min}$
$= 0.333 \text{ l/s}$
Flow rate from accumulator $= 0.333 - 0.18$
$= 0.153 \text{ l/s}$
Flow from accumulator between 20 and 30 seconds is 0.153×10 liters = 1.53 liters.

The flow of oil to and from the accumulator is shown in Figure 2.33. The volume of oil to be stored in the accumulator is the maximum amplitude of Figure 2.33, i.e. $1.53 + 0.285 = 1.815$ liters.

The maximum working pressure of the system is the maximum safe working pressure of the lowest rated component. In this case assume a gear pump has been selected with a maximum continuous working pressure of 207 bar and an intermittent rating above this value. The minimum system pressure is set by the design criteria, i.e. 150 bar. The gas precharge pressure for the accumulator is usually 90% of minimum system pressure, i.e. $0.9 \times 150 = 135$ bar.

In order to calculate the actual size of the accumulator, the various conditions of the gas charge in the accumulator will be considered. These are shown in Figure 2.34. It should be noted that values of pressure and temperature must be in absolute units for all gas calculations.

The precharge pressure, $P_1 = 135$ bar gauge = 136 bar absolute.
The maximum system pressure, $P_2 = 207$ bar gauge = 208 bar absolute.
The minimum system pressure, $P_3 = 150$ bar gauge = 151 bar absolute.

The minimum volume of oil to be stored in the accumulator is $V_3 - V_2 = 1.815$ liters. Assume isothermal compression between conditions (a) and (b), the charging period of the accumulator, then

$$P_1 V_1 = P_2 V_2$$

$$\frac{V_1}{V_2} = \frac{P_2}{P_1} = 208/136 = 1.529$$

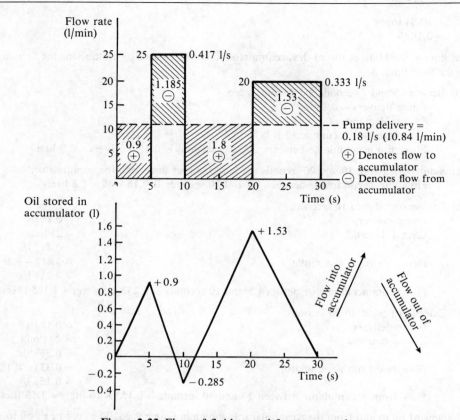

Figure 2.33 Flow of fluid to and from accumulator.

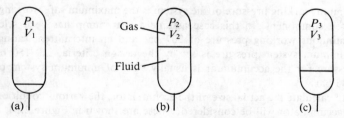

Figure 2.34 Gas charge in the accumulator. (a) Pre-charged with gas. (b) Fully charged with fluid. (c) Fully discharged of usable fluid.

Assume isentropic discharge between conditions (b) and (c) then

$P_2 V_2^\gamma = P_3 V_3^\gamma$ where γ is the adiabatic index which may be taken as 1.4.

$(V_3/V_2)^\gamma = P_2/P_3 = 208/151$

$V_3/V_2 = (208/151)^{1/\gamma} = 1.257$

Thus

$$V_3 - V_2 = 1.815 \tag{2.5}$$
$$V_1 = 1.529\, V_2 \tag{2.6}$$
$$V_3 = 1.257\, V_2 \tag{2.7}$$

Section 2.4 Pump circuit design study

From equations (2.5) and (2.7),

$$0.257\ V_2 = 1.815$$
$$V_2 = 7.062$$

From equation (2.6)

$$V_1 = 1.529 \times 7.062$$
$$= 10.8 \text{ liters}$$

An accumulator with a minimum capacity of 10.8 liters precharged to 135 bar is required with a maximum working pressure of 207 bar. From accumulator manufacturers' data sheets there is a choice of a 10- or 20-liter nominal capacity unit. If the 10-liter accumulator is used it will result in a slightly longer cycle time. This can be compensated for by using a slightly larger delivery pump. If the 20-liter capacity accumulator is used, the maximum working pressure can be reduced resulting in a more efficient system.

The pump has to deliver 10.84 l/min at a maximum pressure of 207 bar. From the pump data sheet (Table 2.4) an OPL 025 has a delivery of 11.73 l/min at 1500 rev/min and a maximum working pressure of 225 bar. A 1PL 028 has a nominal delivery of 13.72 l/min at 1500 rev/min and a working pressure of 250 bar. Because of the higher working pressure, select the 1PL 028 which will deliver 13.17 l/min at 1440 rev/min. Redraw the system demand and accumulator diagrams using a pump delivery of 13.17 l/min, i.e. 0.219 l/s. (See Figure 2.35.)

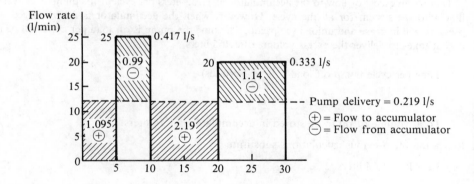

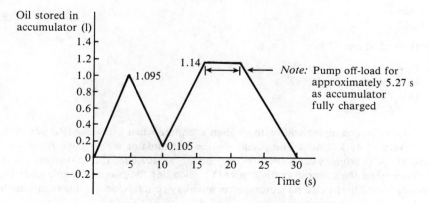

Figure 2.35 Flow of oil to and from accumulator.

(i) From 0 to 5 seconds flow into accumulator is $0.219 \times 5 = 1.095$ liters.

(ii) From 5 to 10 seconds:
Flow from pump $= 0.219 \times 5$
$= 1.095$ liters
Circuit demand $= 0.417 \times 5 = 2.085$ liters
Therefore net flow from accumulator $= 2.085 - 1.095$
$= 0.99$ liter

(iii) From 10 to 20 seconds flow into accumulator $= 0.219 \times 10 = 2.19$ liters.

(iv) From 20 to 30 seconds:
Flow from pump $= 0.219 \times 10$
$= 2.19$ liters
Circuit demand $= 0.333 \times 10$
$= 3.33$ liters
Therefore net flow from accumulator $= 3.33 - 2.19$
$= 1.14$ liters

Using these values,

total flow into accumulator per cycle is $1.095 + 2.19 = 3.285$ liters.

And,

total flow from accumulator per cycle is $0.99 + 1.14 = 2.13$ liters.

There is an excess of flow to the accumulator of 1.155 liters per cycle if the pump is delivering fluid into the system for all the cycle. However, when the accumulator is fully charged, the pressure will increase and unload the pump. The time the pump is off load per cycle will be the time it takes to deliver the excess volume of 1.155 liters.

$$\text{Time per cycle pump off load} = \frac{1.155}{0.219} \text{ (seconds)}$$

$$= 5.27 \text{ s}$$

Total volume of oil to be stored in accumulator is 1.14 liters.

Repeating the previous calculations substitute

$$V_3 - V_2 = 1.14 \text{ liters} \quad (2.8)$$

Now, assuming the pressures used are the same,

$$V_1 = 1.529 \; V_2$$
$$V_3 = 1.257 \; V_2 \quad (2.9)$$

From equations (2.8) and (2.9)

$0.257 \; V_2 = 1.14$
$V_2 = 4.436$
$V_1 = 1.529 \; V_2$
$= 6.78$ liters

A 10-liter capacity accumulator will be more than adequate when using the IPL 028 pump having a delivery of 13.17 l/min. The circuit for the accumulator power pack is shown in Figure 2.36(a). A pressure switch (PS) set to operate at 207 bar, the maximum system pressure, de-energizes the solenoid venting valve (V) unloading the pump. The solenoid valve will also be connected to the electric motor starter auxiliary contacts, so that the pump can be started under no-load conditions.

Section 2.4 Pump circuit design study

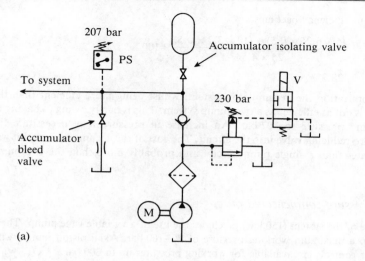

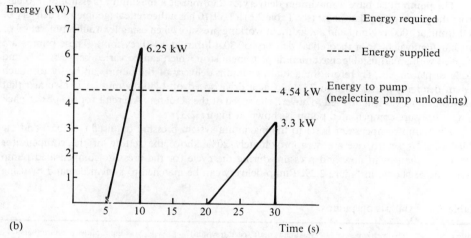

Figure 2.36 (a) Accumulator circuit. (b) Energy analysis.

$$\text{Hydraulic energy to pump} = 13.17 \text{ l/min} \times 207 \text{ bar} \times \frac{10^2}{60 \times 10^3} \text{ (kW)}$$

$$= 4.54 \text{ kW}$$

Figure 2.36(b) shows an analysis of the energy used by the system and the energy supplied neglecting the unloading period.

$$\text{System efficiency} = \frac{\text{Energy used in system}}{\text{Energy supplied to system}} \times 100$$

$$= \frac{(5 \times 6.25 \times 0.5) + (10 \times 3.3 \times 0.5)}{30 \times 4.54} \times 100$$

$$= 23.5\%$$

Taking into account the time for which the pump is off load of approximately 5 s as shown in

Figure 2.35 the system efficiency becomes

$$\frac{(5 \times 6.25 \times 0.5) + (10 \times 3.3 \times 0.5)}{25 \times 4.54} \times 100$$

$$= 28.3\%$$

In this particular application the accumulator system, whilst being more efficient than the single pump system, is not as efficient as a two-pump system. This is partly owing to having to increase the operating pressure to 207 bar. An increase in pressure may necessitate the inclusion of a pressure-reducing valve into the circuit. The cost of the accumulator circuit will be considerably greater than a single pump circuit and probably more than the two-pump circuit.

4. *The use of a pressure-compensated pump*

The working pressure of the system (150 bar) precludes the use of a variable vane pump. These are usually limited to a maximum working pressure of 70–100 bar. Axial piston pumps with pressure-compensator controls are available for working pressures up to 300 bar.

The pump must have a maximum delivery of 25 l/min at a maximum pressure of 150 bar. From the piston pump data sheet (see Table 2.6) a PVB10 has a theoretical (geometric) delivery of 21.1 l/min at 1000 rev/min and a maximum working pressure when using hydraulic mineral oil of 210 bar (equivalent to a theoretical delivery of 30.4 l/min at 1440 rev/min). These pumps are supplied with an adjustable maximum displacement stop which can be varied between 25% and 100% displacement. Therefore the actual maximum delivery of the pump can be set to match system demand, in this case to 25 l/min at a drive speed of 1440 rev/min. It is assumed that flow-control valves will be used to govern the speed of the actuator. A circuit for the power pack using a pressure-compensated pump is shown in Figure 2.37.

The pump compensator is set to the maximum system pressure required (150 bar) and the relief valve is set to operate at approximately 20% above the setting of the compensator (180 bar). Changes in flow and pressure during the cycle for the pressure-compensated pump circuit are as shown in Figure 2.29. Pump delivery can be matched to system demand by using

Table 2.6 Axial piston pumps.

Basic model designation	Geometric displacement (cm^3/rev) or geometric flow rate (liters per 1000 rev/min)	Maximum shaft speed (rev/min)			Maximum outlet pressure (bar)		
		Anti-wear hydraulic oil	Water–glycol	Water-in-oil emulsion (40%/60%)	Anti-wear hydraulic oil	water–glycol	Water-in-oil emulsion (40%/60%)
PFB5	10.55	3600			210		
PFB10	21.10	3200	1800	1800	210	175	175
PFB20	42.80	2400			175		
PVB5	10.55				210	140	140
PVB6	13.81				140	105	105
PVB10	21.10				210	140	140
PVB15	33.00	1800	1800	1800	140	105	105
PVB20	42.80				210	140	140
PVB29	61.60				140	105	105
PVB45	94.50				210	140	140
PVB90	197.50	1800	1200	1200	210	140	140

Vickers Systems Ltd

Section 2.4 Pump circuit design study

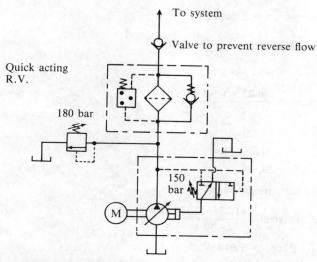

Figure 2.37 Circuit of power pack using a pressure-compensated pump.

flow-control valves. System pressure is set at 150 bar by the pump compensator with excess pressure energy being dissipated as heat across the flow control valves.

Figure 2.38 shows the energy supplied by the pump and the energy used by the system.

$$\text{System efficiency} = \frac{\text{Energy required by system}}{\text{Energy supplied to system}}$$

The pump delivery will equal the circuit demand but the pressure at the pump-delivery port will equal the compensator setting of the pump.

Hydraulic energy supplied during

$$\text{cycle period 5 to 10 s} = \frac{25 \times 150}{600} \text{ (kW)}$$

$$= 6.25 \text{ kW}$$

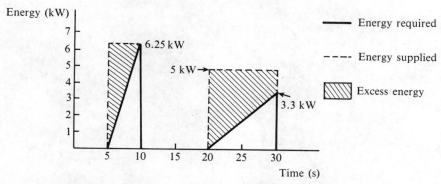

Figure 2.38 Energy supplied and used in pressure-compensated pump circuit.

Hydraulic energy supplied during

$$\text{cycle period 20 to 30 s} = \frac{20 \times 150}{600} \text{ (kW)}$$

$$= 5 \text{ kW}$$

$$\text{System efficiency} = \left(\frac{6.25 \times 5}{2} + \frac{3.3 \times 10}{2}\right) \Big/ (6.25 \times 5) + (5 \times 10)$$

$$= \frac{32.125}{81.25} \times 100$$

$$= 39.5\%$$

Time-average heat energy

$$\text{dissipated to system} = \frac{81.25 - 32.125}{30} \left(\frac{\text{kW}}{\text{s}}\right)$$

$$= 1.64 \text{ kW}$$

Although the pump delivery is exactly matched to system demand, the pump-operating pressure is fixed. If the pump-delivery pressure can also be matched to system pressure, the hydraulic efficiency will be 100%. However, to match the pressures the flow-control valves being used to set the system actuator speeds must be eliminated. This can be done by using a servo-control pump and driving the pump swash-control piston by a profile cam, the profile being cut to suit system flow demand. The operating pressure would be the load-induced pressure. However, this type of system is very inflexible as any required alterations in speed involve making a new profile cam. It is an ideal solution for automatic machines which operate on continuous or very long runs. Alternatively a more flexible arrangement can be obtained by using a pump controlled by a microprocessor via proportional valves to exactly match the system demand (see Chapter 8).

CHAPTER THREE
HYDRAULIC VALVES

Hydraulic valves provide the interface between the hydraulic fluid, the control signal and the hydraulic actuators. They are used to control the flow rate, the direction of flow and the pressure of the fluid. The control signals may be mechanical, manual, hydraulic, pneumatic or electrical. The action of the control valve may be digital (in which the valve changes from one set position to another) or analogue (in which the movement of the valve control element is dependent upon the strength or value of the control signal). A two- or three-position lever-operated directional-control valve is an example of a digital valve whereas a relief valve is an example of an analogue control valve.

In general a valve influences just one of these functions:

1. A relief valve is used to regulate maximum *pressure* in a circuit or part of a circuit.
2. A four-port spool valve may be used to change the *direction* of rotation of a hydraulic motor.
3. A flow-control valve can alter the speed of an actuator by changing the fluid *flow rate* to or from the actuator.

In practice, two or more valves may be combined in the same envelope to produce a composite valve which has more than one function. A typical example of this is where a flow control and check valve are combined producing a unidirectional speed-control valve with virtually unrestricted flow in the reverse direction.

Again, altering one particular parameter of a circuit may affect another. Adjusting a needle valve to restrict flow in a circuit will cause a change in pressure difference over the control orifice. Similarly, if the cracking pressure of a relief valve is set too close to the load-induced pressure at an actuator, adjustment of the valve pressure setting will affect the speed of the actuator. Some of the flow bleeds over the relief valve instead of feeding the actuator.

These points will be amplified later in this section which covers various types and applications of valves under the headings of their main characteristics.

3.1 PRESSURE-CONTROL VALVES

A pressure-control valve can be used to limit the maximum pressure (a relief valve), to set a back pressure (a counterbalance valve), or to pass a signal when a certain pressure has been reached (sequence valve). The principal feature of most pressure controls is that the hydraulic forces are resisted by a spring. The action of a simple pressure-control valve is shown in Figure 3.1.

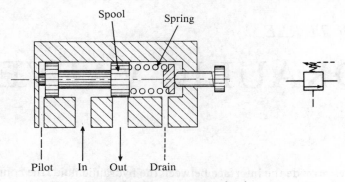

Figure 3.1 Pressure-control valve.

When the force arising from the pilot pressure is greater than the spring force, the valve spool will move towards the spring until an equilibrium position is obtained where the pilot pressure is just equivalent to the spring force. As the pilot pressure varies, the spool position will alter to try to maintain the force equilibrium. The valve spool may be of the normally closed type as shown or normally open. In the case of the latter, the valve closes as the pilot pressure increases.

3.1.1 Relief valves

The function of a relief valve is to set the maximum pressure in a hydraulic system. Although there are many designs and varieties, they can all be denoted by the general symbol (a) in Figure 3.2. It is a normally closed valve which partially opens permitting flow to the tank port when the pressure at the inlet port overcomes the spring force. Symbol (a) does however more accurately represent a direct acting valve. If there is no arrow through the spring the valve is pre-set, i.e. non-adjustable. A two-stage relief valve may be shown as Figure 3.2(b) as this more closely illustrates its operation which will be described later.

When the pressure control valve shown in Figure 3.1 is closed the inlet and outlet ports are isolated by the valve spool. An adequate hydraulic seal is obtained owing to the minute clearance between the spool and its housing. This seal becomes less efficient as the working pressure increases. Many simple direct acting relief valves employ either a conical poppet or a ball to seal against a mating valve seat. Being a contact type seal this is more effective at high pressure. In the poppet type relief valve (Figure 3.3) pressure at port P acts on the exposed surface of the poppet to apply a force which is resisted by the spring

Figure 3.2 Relief valve symbols. (a) General or direct-acting. (b) Two-stage.

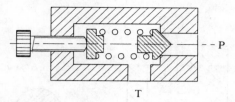

Figure 3.3 Poppet type direct acting relief valve.

force. When the pressure at port P has risen sufficiently to overcome the spring force, the poppet is lifted off its seat permitting fluid to flow to the tank port T, relieving the pressure in the system.

Relief valves of the ball or poppet type have a rapid response to pressure surges, typically 25 ms, but the pressure flow characteristic is not constant. The poppet or ball tends to hammer on the seat giving rise to relief valve whine; seat damage can occur with resultant leakage and they are best suited for infrequent duty. A variation is the guided poppet type relief valve which has the advantage of a direct poppet valve but is more suitable for continuous duty.

The guided piston relief valve (Figure 3.4) is of much quieter operation but is best suited for low-pressure applications (up to 100 bar) under constant flow conditions. The response time is still fast although slightly slower than the direct poppet-type relief valve. In common with the preceding direct-acting relief valves, it has a high pressure over-ride characteristic. The pressure override is the difference between the cracking pressure or opening pressure and the pressure drop across the valve when it is passing the maximum rated flow at the same valve setting.

The differential piston/poppet type relief valve (Figure 3.5) is suitable for pressures up to 350 bar. The pressure acts on the differential areas between the poppet and the seat. When the valve operates, a large flow area opens for a relatively small poppet movement. This results in a low-pressure over-ride, but the reset pressure may be appreciably lower than the opening pressure.

The pilot-operated relief valve (Figure 3.6) is a two-stage valve which gives good regulation of pressure over a wide range of flow. It consists of a main spool controlled by a small built-in direct-acting relief valve. Pressure is sensed at the pilot relief valve via a small hole or jet in the spool or through the housing. When the control valve is closed the main spool is in hydraulic balance; it is however held onto its seat by a light spring. Any increase in pressure sufficient to open the control valve throws the main spool out of balance owing to the pressure drop across the jet, and the spool lifts against the spring

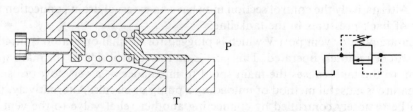

Figure 3.4 Guided piston relief valve.

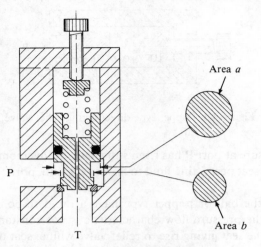

Figure 3.5 Differential poppet relief valve: force to overcome spring = pressure $\times (a - b)$.

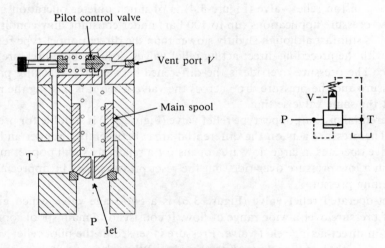

Figure 3.6 Pilot-operated relief valve.

relieving the major flow from the pressure to tank port. The small amount of flow which passes through the control section is also returned to the tank port (i.e. it is internally drained). Alternatively the control section may have an external drain connection to avoid the effect of back pressures in the tank line.

A separate pilot or vent port V which is plugged for normal operation is fitted so that the valve can be remotely operated. This port is on the control side of the main spool and connection to the tank causes the main spool to imbalance at a very low pressure. This venting feature is a useful method of unloading a pump or circuit. Alternatively, the main valve can be remotely controlled by connecting another relief valve to the vent port V. This will regulate the pressure from its minimum up to the limit set by the main valve pilot

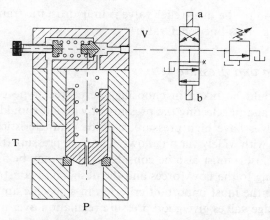

Figure 3.7 Solenoid-controlled relief valve.

section. Both these features are demonstrated in Figure 3.7 in which a three-position solenoid actuated directional-control valve enables the relief valve to be remotely operated by an electrical signal to give three different pressure settings, one of which is nominally zero, the relief valve then being vented. With solenoid *a* energized, internal pressure control is achieved; with solenoid *b* energized, remote pressure control is achieved; and with both solenoids *a* and *b* de-energized, the valve is vented. The directional control valve may be integral with the relief valve or a separate valve connected to the vent port.

Dual-relief valves

Frequently, relief valves are required in pairs to relieve the pressure on either side of an actuator. These usually take the form of a sandwich block which can be built into a valve stack as shown symbolically in Figure 3.8. The pressure in the service lines (A and B) may be relieved directly to the tank line T (port relief) or to the opposite service line (cross-line relief).

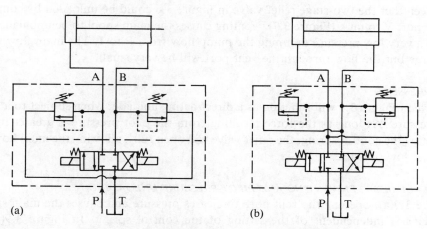

Figure 3.8 Dual valves. (a) Port relief. (b) Cross-line relief.

Another application of the dual-relief valve is in hydrostatic transmission where it is usually referred to as a cross-line relief valve.

Relief valve selection and pressure setting

Considerable care should be taken in choosing the correct type of relief valve for a particular application and in selecting the pressure at which it should just open or crack. Most direct acting valves have high pressure override characteristics which make them unsuitable for systems with widely varying flows. The reset pressure (that is the pressure at which an open valve closes) must also be considered. This may be as low as 50% of the opening pressure, owing to the flow forces and the design and construction of the valve. Response time may be the most important criterion in a specific application.

In general, two stage valves give good pressure regulation over a wide range of flows with low pressure override and close tolerance between the opening (or cracking) and resetting pressures. Direct acting valves have rapid response times. Poppet types are the most tolerant to fluid contamination and also tend to have less internal leakage than spool valves, which makes them suited to high pressure working.

A frequently used rule of thumb is for the main relief valve in a circuit to be set at 10–20% above the maximum required working pressure, taking into account the type of valve, its position relative to the actuator and the pressure losses in the system. Where there is more than one pressure valve in a circuit or when used in conjunction with pressure compensated pumps, the controls must not be set at pressures which are too close together as interaction or hunting may result. It is usual to set secondary relief valves such as port or cross-line reliefs at a pressure higher than the main relief valve.

Unloader valves

A relief valve can be unloaded in two ways: by pressure release, i.e. venting, or by a pilot pressure.

PRESSURE RELEASE (VENTING)
It was seen that the two-stage relief valve in Figure 3.6 could be unloaded by connecting the vent port V to tank (Figure 3.7). Venting causes the main spool to be unbalanced and open at a very low pressure dumping the pump flow from P to T. The main flow may be quite large but the flow through the vent port will be very small.

PILOT PRESSURE
The valve in Figure 3.1 will function as a direct-acting unloader when subject to a remote pilot pressure. As long as the force resulting from the pilot pressure is greater than the force set by the control spring, the relief valve will open fully, allowing the main flow to go to tank at low pressure.

DIFFERENCE BETWEEN VENTING AND PILOT-PRESSURE UNLOADING
In Figure 3.9(a), opening the vent port V releases pressure and causes the main spool to open. This is independent of the setting of the control spring. In Figure 3.9(b), the pressure signal at X from a remote source pilots the valve open against the spring setting.

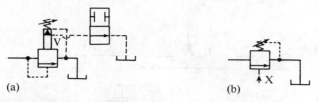

Figure 3.9 Relief valve unloading. (a) By venting. (b) By pressure signal.

In a two-stage unloader valve pilot pressure from a remote source causes a piston to unseat the control poppet of the relief valve. The main spool is imbalanced and opens, dumping the main pump flow from P to T at a very low pressure. Although it is the high pressure on the pilot-port piston which causes the valve to unload, the action of pushing the control poppet off its seat vents the main spool. Generally, the valve will still respond to pressure at port P and function as a normal relief valve. A typical application is in a twin pump circuit (sometimes called a 'high–low' or 'hi–lo' circuit), as described in Section 2.2.3 of Chapter 2. This is frequently employed on presses where both pumps supply fluid to move the tooling with just the small pump carrying out the pressing operation. Considerable savings in input power can be achieved. These circuits employ a check valve to isolate the high and low pressure halves of the circuit and sometimes the check valve is incorporated into the unloader valve (Figure 3.10). Another established application is with an accumulator and Figure 3.11 depicts the valve from Figure 3.10 in such a circuit.

It can be seen that whereas the direct-acting valve (Figure 3.1) can only be opened by a pressure at the pilot port, the two-stage valve (Figure 3.10) may still function as a normal relief valve responding to an internal pressure. However, since the pilot piston has a slightly larger area than the control-valve poppet, the external pilot pressure needed to open the valve is less than the direct pressure setting of the relief-valve spring.

A particular type of direct-acting unloader valve (not illustrated) specifically used in dual-pump systems unloads the secondary pump when the main pump circuit reaches a predetermined, non-adjustable pressure (say, 20 bar) below the relief-valve setting.

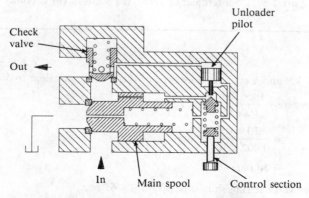

Figure 3.10 Two-stage unloader valve with integral check valve.

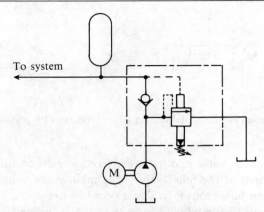

Figure 3.11 Accumulator circuit using the valve illustrated in Figure 3.10.

3.1.2 Counterbalance valves

These are basically relief valves but used in a particular application to set up a back-pressure in a circuit. They are frequently employed to 'counterbalance' a load as shown in the circuit in Figure 3.12. Here the valve creates a back-pressure to prevent the load running away when the cylinder is retracting. The usual pressure setting is 1.3 times the load-induced pressure.

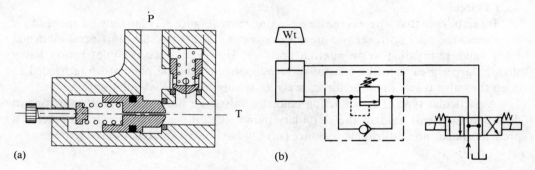

Figure 3.12 Counterbalance valve. (a) Section. (b) Circuit.

EXAMPLE 3.1

With a load of 10 kN and a cylinder bore area of 0.002 m² (equivalent to 50 mm diameter).

$$\text{Load-induced pressure} = \frac{10 \times 10^3}{0.002} \ (N/m^2)$$

$$= \frac{10 \times 10^3}{0.002 \times 10^5} \ (bar)$$

$$= 50 \ bar$$

The counterbalance valve setting should be $50 \times 1.3 = 65$ bar.

Section 3.1 Pressure-control valves

A check valve is incorporated in the circuit in Figure 3.12 to allow free-flow in the reverse direction (i.e. to bypass the counterbalance valve when raising the load). Care must be taken if using a conventional relief valve for this application as at some stage of operation the tank port will be subject to maximum circuit pressure. This is not permissible with many relief valves. The counterbalance valve shown has an integral check valve. A separate drain connection from the spring chamber is unnecessary because the pressure section of the valve is inoperative when the T port is pressurized (flow is through the check valve). When it is counterbalancing the back pressure at T should be kept to a minimum.

Over-center valve

A disadvantage of a counterbalance valve is that it reduces the available force. Consider the press circuit in part (a) of Figure 3.13 where the valve is used to counteract the weight of the press tools whilst they are closing. During the forming operation, part of the possible pressing force will be lost in overcoming the back-pressure set up by the counterbalance valve.

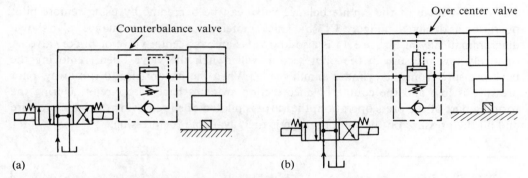

Figure 3.13 Press circuit. (a) With counterbalance valve. (b) With over-center valve.

EXAMPLE 3.2

Consider a 100-kN press where the tools weigh 5 kN:

Cylinder bore = 80 mm
Cylinder rod = 60 mm
Full bore area = $0.08^2 \pi/4 = 0.005$ m^2
Annulus area = $(0.08^2 - 0.06^2) \pi/4 = 0.0028$ m^2

Pressure at annulus side to balance tools = $\dfrac{5 \times 10^3}{0.0028} \times 10^{-5} = 17.8$ bar

Suggested counterbalance valve setting $17.8 \times 1.3 = 23$ bar
Pressure at full bore side of cylinder to overcome counterbalance = $23 \times 0.0028/0.005 = 13$ bar

Pressure to achieve 100 kN pressing force = $\dfrac{100 \times 10^3 \times 10^{-5}}{0.005} + 13 = 213$ bar.

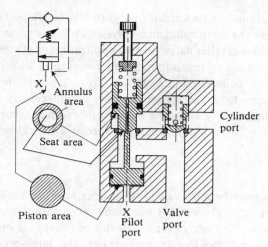

Figure 3.14 Over-center valve (pilot-operated counterbalance valve or brake valve).

The disadvantage of the counterbalance valve can be overcome by using remote pilot operation as depicted in Figure 3.13(b). This remotely piloted counterbalance valve shown diagrammatically in Figure 3.14 is also known as an over-center valve or brake valve. A relatively low pressure in the pilot section will switch the valve open, removing the back-pressure from the cylinder annulus side. When the piston tries to run away, pilot pressure is lost and the counterbalance section switches back into circuit. During the forming part of the press operation, the valve is piloted open removing the back pressure and all the pressure on the full bore side is then available for pressing.

EXAMPLE 3.3

Consider the application in Example 3.2 but using an over-center valve with a 2:1 pilot input ratio, set at 23 bar to balance the tools, instead of the counterbalance valve.

Pressure on the pilot required to open the valve = 23/2 = 11.5 bar, i.e. pressure at full bore side to drive down the tooling = 11.5 bar.

Pressure required to achieve 100 kN pressing force is

$$\frac{(100-5) \times 10^3 \times 10^{-5}}{0.005} = 190 \text{ bar}$$

This is greater than the 11.5 bar pressure needed to pilot the over-center valve open. Therefore there will be no back-pressure set up on the annulus side of the piston during the pressing operation.

It was seen in Example 3.2 using a conventional counterbalance valve that a pressure of 213 bar was necessary to achieve the same pressing force. The over-center valve also functions as a brake valve decelerating the load when the directional control valve is moved to its center position.

The over-center valve is frequently used in motor circuits (hydrostatic transmissions) as a brake valve. In Figure 3.15, the circuit shows a simple winch driven by a hydraulic

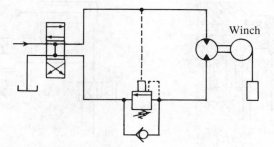

Figure 3.15 Over-center valve used in a winch circuit.

motor; the over-center valve will:

(a) Hold the load in the neutral position.
(b) Prevent over-run during lowering.
(c) Gently brake the motor to a halt on switching from 'lowering' to 'neutral'.

The ratio between the pilot pressure to the direct pressure necessary to open the valve is generally from 2:1 to 10:1 according to the application.

Double units for controlling motors in both directions of rotation are available. A particular variation incorporates a series of check valves and is known as a 'motion control and lock valve'. It has a port for the make-up oil input for closed-circuit transmissions and in the event of the motor stalling it functions as a cross-line relief valve. Its application in a hydrostatic transmission circuit is shown in Figure 4.38 in Section 4.4.2 of Chapter 4.

3.1.3 Pressure sequence valves

Sequence valves sense a change in pressure in a system and transmit a hydraulic signal when the set pressure has been reached. The valve may be normally open or normally closed, changing state when the system reaches the set pressure. They may be used to assure priority hydraulic pressure in one system before another can operate.

An important feature of all sequence valves is a separate drain connection from the spring chamber. This is because, unlike a conventional relief valve, a high pressure can occur in the output port during the normal course of operation. Should it be internally drained, any pressure in the output port will be reflected back into the spring chamber causing a malfunction. In fact a sequence valve may be used as a relief valve in any circuit where excessive back pressures are encountered in the return line. The independently drained pilot makes sequence valves insensitive to downstream back pressure.

A normally closed sequence valve with integral reverse-flow check valve is shown in Figure 3.16 together with an established application which is to sense that a component has been clamped before initiating the next stage in a 'sequence' of operations. When the component is unclamped, the pressure falls and the sequence valve closes. The check valve prevents the signal being trapped and allows it to decay back past the sequence valve poppet.

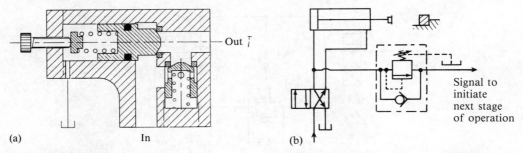

Figure 3.16 (a) Normally closed sequence valve with integral reverse-flow check valve. (b) Clamping application.

Two-stage sequence valves suitable for high flow rates are available and one specialized form known as a 'Circuit Breaker' or 'Kickdown' sequence valve is shown in Figure 3.17. The valve is normally closed until the pressure setting of the control section is reached, when the main spool opens fully with very little resistance to flow. It remains open even if conditions in the downstream circuit cause the circuit pressure to fall below the control setting.

The function of the valve is similar to that of the two-stage relief valve in Figure 3.6, except that once the main spool has lifted, the 'kickdown' jet is connected to the output port. In this condition, the input pressure necessary to hold the valve fully open has only to overcome the resistance caused by the secondary circuit pressure and the light spring situated behind the main spool. It remains open even when the secondary circuit pressure is less than the valve set pressure only resetting at a very low value.

Direct-acting sequence valves are employed in low flow applications such as providing signals to operate directional control valves or to positively release a brake before a machine can function. Where the output flow is used to drive cylinders directly, two-stage valves are usually more appropriate.

As the name might imply, the sequencing of cylinder movements is a common application. In Figure 3.18 when the directional control valve is switched to the 'tramline' condition, cylinder A will extend followed by cylinder B. The flow to cylinder B is through sequence valve S1 which will open when the pressure at the full bore end of cylinder A has attained a certain value, probably owing to it having been stopped by some external object

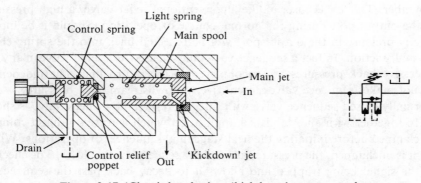

Figure 3.17 'Circuit breaker' or 'kickdown' sequence valve.

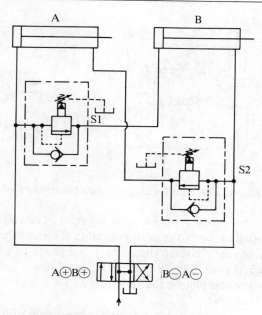

Figure 3.18 Cylinder sequence circuit.

or at the extremity of its stroke. With the control valve in its 'crossover' condition, cylinder B will retract before cylinder A, with change-over initiated by S2.

In circuits where pressure-sensing is used to control cylinder movements, it must be borne in mind that sequence valves operate when a specific pressure has been achieved and do not guarantee that the cylinders have completed or reached a particular point in their strokes.

3.1.4 Pressure-reducing valves

These are used to limit the pressure in part of the circuit to a value lower than that required in the rest of the circuit. The pressure-reducing valve is a normally open valve which throttles or closes to maintain constant pressure in the regulated line. Direct-acting pressure-reducing valves are available for low flow rates up to about 45 l/m and pressures up to 210 bar; they can be supplied with or without a reverse-flow check valve.

Pressure-reducing valves may be:

(a) Non-relieving, i.e. they do not limit any pressure increase downstream of the valve set up by an external force.

(b) Relieving type. This limits the pressure downstream of the valve even when it is increased by an external force.

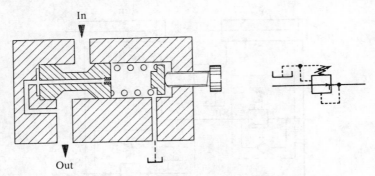

Figure 3.19 Direct-acting pressure-reducing valve.

Figure 3.19 shows a direct-acting pressure-reducing valve. The valve is biassed open by the spring. Pressure is sensed at the outlet port and fed to the end of the spring-loaded spool. As pressure in the secondary circuit rises, the valve tends to close against the spring pressure. Flow through the small bleed hole in the spool to the spring chamber and drain prevents the valve closing completely, thus averting a pressure build up in the downstream circuit.

Pilot-operated (two-stage) pressure-reducing valves are used for higher flow rates and in general give better regulation of pressure with flow.

The action of pressure-reducing valves always generates heat energy because of the throttling effect. This heat generation must be taken into account when considering their application. Where two separate pressures are continuously required in a circuit, a two-pump system may prove a better solution than one using pressure-reducing valves. This will depend upon the flow and pressures required.

EXAMPLE 3.4

The primary part of a circuit is operating at 180 bar. A secondary circuit supplied from the primary circuit via a pressure-reducing valve requires a constant flow of 30 l/min at 100 bar. The power loss over the pressure-reducing valve will be:

$$\frac{(180 - 100) \text{ (bar)} \times 30 \text{ (l/min)}}{600} \text{ (kW)} = 4 \text{ kW}$$

This may well be more than can be dissipated by natural cooling. In practice, the cost of fitting a heat exchanger and operating costs should be weighed against alternative circuitry such as a two-pump system.

3.2 FLOW-CONTROL VALVES

These are used to regulate the fluid rate to actuators and so give speed control. This is primarily achieved by varying the area of an orifice and flow characteristics of orifices play a major part in the design of hydraulic control devices.

Flow through the control orifice is usually considered to be turbulent and the quantity

Section 3.2 Flow-control valves

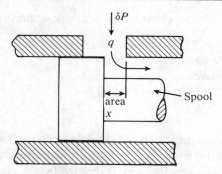

Figure 3.20 Flow through a control orifice.

of a fluid flowing can be given by

$$q = Kx\,(\delta P)^{1/2}$$

where q is the quantity flowing, x is the orifice area, δP is the pressure drop over the orifice, and K is a constant which may include such functions as the orifice characteristics, viscosity of the fluid and the Reynolds Number.

An orifice is a sudden restriction in the flow path and may be fixed but is generally variable. Ideally it should be of zero length and sharp-edged in which case it will be insensitive to temperature (i.e. viscosity) changes in the fluid flowing.

The flow through the orifice shown in Figure 3.20 will vary as the square root of the pressure drop and will be sensitive to viscosity changes. This type of orifice can be used to control flow rates if the pressure drop and fluid temperature is reasonably constant and minor variations in flow rate are acceptable.

When precise speed control is required under varying load conditions it is necessary to maintain a constant pressure drop over the orifice.

The relationship between flow and position of the adjusting device can be linear, logarithmic or specially contoured to follow a particular curve.

The characteristics of a simple needle valve are shown in Figure 3.21. Generally, a non-return valve is incorporated enabling regulated flow in one direction with free flow in the reverse direction.

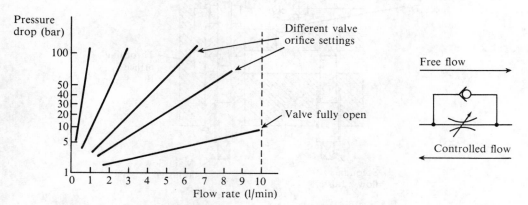

Figure 3.21 Characteristics of a simple needle valve.

Three specialist forms of flow-control valves are now considered:

1. Deceleration valves
2. Viscosity or temperature-compensated valves
3. Pressure-compensated valves.

Deceleration valves

These are a throttle-type valve in which the throttle opening is controlled by a roller or roller lever. The valve may be either normally open or normally closed so that the flow and hence acceleration or deceleration may be controlled. A check valve and secondary throttle valve can be fitted. The former for free reverse flow; the latter to provide an adjustable minimum flow when the main throttle is closed. A sectional view through a deceleration valve is shown in Figure 3.22.

In the circuit in Figure 3.23 a deceleration valve is used to retard a cylinder towards the end of its stroke. During the initial part of the stroke, speed is largely controlled by the restrictor A metering the flow leaving the cylinder with a small amount of flow through restrictor C. As the cam depresses the operating roller, the main spool B gradually closes the main flow path. Control over the final part of the stroke is by restrictor C. When the cylinder retracts, flow by-passes the deceleration valve through the check valve D.

Deceleration valves are most suitable for high-flow application and are generally not recommended for flows below 15 l/min.

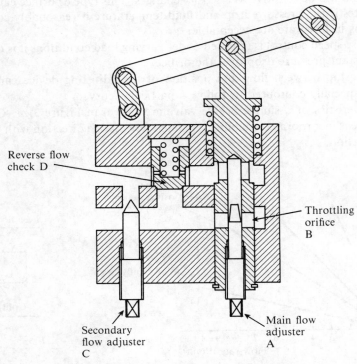

Figure 3.22 Deceleration valve.

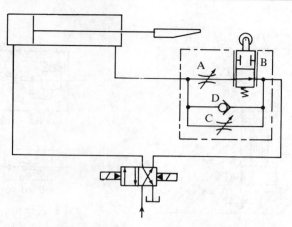

Figure 3.23 Deceleration valve circuit.

Viscosity or temperature-compensated flow-control valves

The viscosity of a hydraulic oil is dependent on the oil temperature; hence some valve manufacturers refer to temperature compensation and others to viscosity compensation.

The simplest way to eliminate the effect of viscosity is to use a sharp-edged orifice, the flow through which is independent of viscosity.

In some designs of viscosity/temperature-compensated throttle valve the orifice aperture over which throttling of flow takes place consists of two adjacent flat plates — one fixed and one movable. A V-shaped notch in one of the plates is masked or unmasked as the movable plate is rotated relative to the fixed plate. The design of the throttle gives a sharp-edged orifice which makes the flow practically independent of viscosity and hence temperature, particularly at the higher flow rates. Problems can still occur at low flows (< 0.5 l/min) in which case a valve will function better with a low viscosity oil. Flow through these valves is load-dependent but this can be remedied by the addition of a pressure-compensating spool. A check valve is frequently incorporated to allow relatively unrestricted reverse flow.

An alternative method of temperature compensation favoured by some manufacturers is to have part of the orifice adjusting mechanism made of a material with a high coefficient of thermal expansion. When the temperature of the fluid increases, a spindle in the mechanism lengthens thus reducing the control orifice opening.

Pressure-compensated flow-control valve

A pressure-compensating spool built into a flow control valve maintains a constant pressure drop across the metering orifice independent of changes in supply and load pressure.

Figure 3.24 shows diagrammatically a two-port pressure-compensated flow-control valve together with its symbols. Flow rate is set by an adjustable metering orifice (1) which may also be viscosity-compensated. In the unoperated condition, the compensating spool (2) is biased fully open by the compensator spring (3). As soon as flows occurs, there will

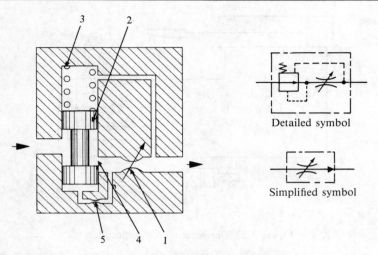

Figure 3.24 Two-port pressure-compensated flow-control valve (with symbols), see text for explanation.

be a pressure drop across the valve and pressure upstream of the metering orifice tends to close the valve but this is opposed by the spring assisted by pressure from downstream of the metering orifice. The compensator spool adopts a balanced position with a consequential pressure drop over the compensating orifice (4) formed by the partially closed spool. A rise in supply pressure tends to close the spool and the increased pressure drop across the compensating orifice balances the increase in supply pressure. If the load pressure rises, the compensating orifice opens, again maintaining the pressure drop over the metering orifice at a set value. This pressure drop is usually 3–6 bar, dependent upon the size of the metering orifice. The total pressure drop across the valve is dependent upon the difference between supply and load pressure, but a minimum total pressure loss across the valve of 5–12 bar is normally required for the valve to function correctly. (Typical curves are shown in Figure 3.25.) The damping orifice (5) stabilizes the compensator and prevents hunting as pressure fluctuates. A stroke limiter or anti-lunge device is sometimes fitted to the compensator spool to eliminate a flow surge which occurs when the circuit

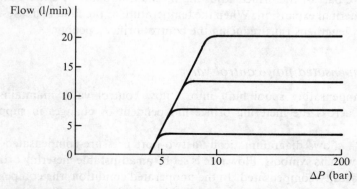

Figure 3.25 Two-port pressure-compensated flow-control valve curves.

Section 3.2 Flow-control valves

starts up. When there is no flow through the metering orifice, the pressure-compensating spool will be fully open and as soon as flow commences, there will be a pressure drop through the valve causing the compensator to lunge or jump. The stroke limiter is a movable end-stop which limits the travel of the compensating spool. This device, which has to be adjusted every time the setting of the flow control valve is changed, is used to position the compensating spool somewhere near its expected final location. However large variations in pressure can no longer be corrected.

Pressure-compensated flow controls must be used when accurate speed control at varying supply or load pressures is required. The minimum regulated stable flow from a good-quality flow-control valve will be in the region of 0.1 l/min. In any precision flow-control valve application it is essential to have well-filtered fluid (better than 10 μm absolute) to promote efficient control and long life of the valve. The smaller the flow to be controlled, the finer the filtration necessary.

Various types of valve adjusting mechanism are available – hand knob, lockable hand knob, lever, DC motor control etc.

It must be remembered that whenever a flow-control valve is used in a system, there will always be some pressure drop and associated heat generation.

3.2.1 Speed control of a cylinder

In a simple cylinder circuit there are three positions in which the flow control valve can be placed relative to the cylinder: meter in, meter out, and bleed off.

Meter in

The quantity of oil entering the cylinder is controlled as shown in Figure 3.26(a). (Note the directional control valve is not shown.) The pump must deliver more oil than is required to drive the cylinder at the selected speed, excess oil passing to tank at the relief valve setting. The circuit pressure has to be at a higher value than that required to overcome the load owing to the requirements of the flow-control valve (a drop of approximately 10 bar, as previously stated).

When the circuit is initially started, the compensator spool will be fully open causing a flow surge before the compensator adjusts to take correct control. In many machine-tool

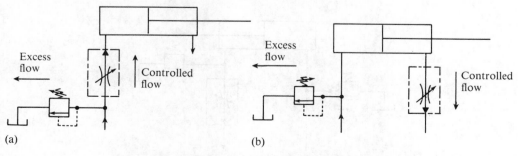

Figure 3.26 Flow control. (a) 'Meter-in' (b) 'Meter-out'.

applications, an initial flow surge would cause the tool to dig into the work piece. In these circumstances flow-control valves with anti-lunge devices must be used. An alternative is to design the circuit so that there is always flow through the flow-control valve – this keeps the compensating spool 'active', preventing flow surges or kicks.

The fluid in the cylinder has to be pressurized before the piston begins to move; this requires a flow of fluid to cause compression. The force or pressure needed to start the cylinder moving is generally greater than the pressure needed to maintain movement (owing to static friction and load inertia). Once the load has started to move, the resistance to movement reduces and the pressure on the piston falls with an expansion of the fluid causing a sudden acceleration. Some degree of instability exists, initially caused by the action of the pressure compensator in the flow control valve.

There may be a tendency for the direction of the load to reverse, that is, act in the direction of motion, or to over-run. The 'meter-in' system loses control. To overcome this problem, a back-pressure has to be introduced by using a counterbalance valve or over-center valve in the tank line, which in turn means increasing the system pressure.

'Meter in' provides accurate control providing the load always opposes actuator movement. If a fixed-displacement pump is used over a wide range of piston speeds, a large percentage of the fluid flows over the relief valve, resulting in a 'hot' system.

Meter out

The flow-control valve is installed in the return line metering the fluid discharged as shown in Figure 3.26(b). As in the case of 'meter in', the pump must deliver more oil than is required by the cylinder. The circuit pressure has to overcome the cylinder load resistance and the pressure drop across the flow-control valve. However as the flow-control valve is on the annulus side of the piston, a somewhat reduced pressure is required at the full bore end (owing to differential areas) to overcome the pressure drop across the flow-control valve. This makes it marginally more efficient on the extend stroke.

Initially, the compensator spool is fully open, and full pump flow is passed into the cylinder until the piston moves forward building up pressure at the flow-control valve. The compensator spool will now come into operation and restrict the flow to its correct value. There is an initial flow surge before the compensator spool adjusts as in the case of 'meter in'.

When using a 'meter out' system, the pressure in the annulus end of the cylinder must be carefully considered, e.g. if the ratio of piston area to piston rod areas is 2:1 and the

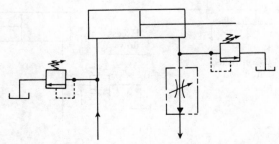

Figure 3.27 Relief valve preventing over-pressurization owing to 'meter-out' control.

Section 3.2 Flow-control valves

system pressure is 150 bar. Then with no external load on the piston, the pressure in the annulus end will be 300 bar. If this condition is likely to occur, a separate relief valve may be fitted to the annulus side of the cylinder to prevent over-pressurization, as shown in Figure 3.27. (**Note:** If the secondary relief valve 'blows', speed control will be lost.)

With 'meter-out' speed control, the quantity of oil leaving the cylinder is controlled. When the cylinder is extending, the fluid from the annulus end is metered which is a smaller quantity than that flowing into the full bore end. Consequently under extend conditions, 'meter-out' flow control is not as sensitive as 'meter-in' control. When the cylinder is retracting, the reverse is true.

'Meter out' provides accurate speed control even with reversing loads. However, as with the 'meter-in' system, considerable heat will be generated when used with a fixed-delivery pump and a wide range of piston speeds.

Bleed off

The flow-control valve is arranged to bypass part of the pump output directly to tank as shown in Figure 3.28. The system pressure only reaches the setting of the relief valve when the piston is stationary. Thus the excess oil bled off over the flow control valve is at a pressure induced by the cylinder load. This results in a cooler and more efficient system.

The accuracy of this method depends upon the delivery from the pump, as this varies with pressure. 'Bleed off' is used where the pressure is reasonably constant or where precise speed control is not critical. (It is the unwanted flow which is being accurately controlled.)

In general, 'bleed-off' speed control is best employed when the vast majority of the pump output is utilized by the cylinder and only a small percentage is bypassed.

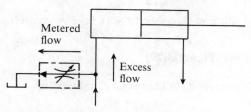

Figure 3.28 'Bleed-off' flow control.

EXAMPLE 3.5: RELATIVE EFFICIENCY OF 'METER-IN' AND 'METER-OUT' FLOW CONTROL

A cylinder has to exert a forward thrust of 100 kN and a reverse thrust of 10 kN. The effects of using various methods of regulating the extend speed will be considered. In all cases the retract speed should be approximately 5 m/min utilizing full pump flow. Assume that the maximum pump pressure is 160 bar and the pressure drops over the following components and their associated pipework (where they are used):

Filter = 3 bar
Directional valve (each flow path) = 2 bar
Flow control valve (controlled flow) = 10 bar
Flow control valve (check valve) = 3 bar

Determine:

(a) the cylinder size (assume 2:1 ratio piston area to piston rod area),
(b) pump size, and
(c) circuit efficiency

when using:

Case 1: No flow controls (Figure 3.29) (calculate extend speed)
Case 2: 'Meter-in' flow control for extend speed 0.5 m/min
Case 3: 'Meter-out' flow control for extend speed 0.05 m/min.

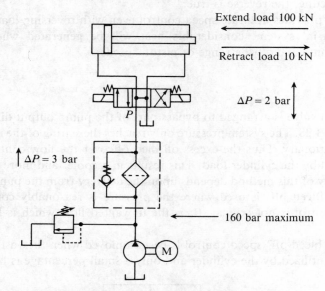

Figure 3.29 Example 3.5 with no flow controls.

Case 1: No flow controls (Figure 3.29)

(a) Maximum available pressure at full bore end of cylinder is

$$160 - 3 - 2$$
$$= 155 \text{ bar}$$

Back-pressure at annulus side of cylinder = 2 bar. This is equivalent to 1 bar at the full-bore end because of the 2:1 area ratio. Therefore, maximum available pressure to overcome load at full bore end is

$$155 - 1$$
$$= 154 \text{ bar}$$

Full bore area = Load/Pressure

$$= \frac{100 \times 10^3}{154 \times 10^5} \text{ (Nm}^2\text{/N)} = 0.00649 \text{ m}^2$$

Piston diameter $= \left(\frac{4}{\pi} \times 0.00649\right)^{1/2} = 0.0909 \text{ m}$

$$= 90.9 \text{ mm}$$

Select a standard cylinder (refer to Table 4.1) say with 100-mm bore × 70-mm rod diameter.

Full bore area $= 7.85 \times 10^{-3}$ m^2
Annulus area $= 4.00 \times 10^{-3}$ m^2

This is approximately a 2:1 ratio.

(b) Flow rate required for a retract speed of 5 m/min is

$$\text{Area} \times \text{velocity}$$
$$= 4.00 \times 10^{-3} \times 5 \text{ m}^3/\text{min}$$
$$= 20 \text{ l/m}$$

Extend speed $= \dfrac{20 \times 10^{-3}}{7.85 \times 10^{-3}}$

$= 2.55$ m/min

Pressure to overcome load on extend is

$$\dfrac{100 \times 10^3}{7.85 \times 10^{-3}}$$
$$= 12.7 \times 10^6 \text{ N/m}^2$$
$$= 127 \text{ bar}$$

Pressure to overcome loan on retract is

$$\dfrac{10 \times 10^3}{4.00 \times 10^{-3}}$$
$$= 2.5 \times 10^6 \text{ N/m}^2$$
$$= 25 \text{ bar}$$

(i) Pressure at pump on extend (working back from directional valve tank port):

Pressure drop over directional control valve B to T is 2 bar × ½ (piston area ratio)	= 1
Load-induced pressure	= 127
Pressure drop over directional control valve P to A	= 2
Pressure drop over filter	= 3
Therefore pressure required at pump during extend stroke	= 133 bar

Relief valve setting = 133 + 10% = 146 bar.

(ii) Pressure required at pump on retract (working from directional valve tank port as before) is (2 × 2) + 25 + 2 + 3   = 34 bar

Note The relief will not be working other than at the extremities of the cylinder stroke.

Also when movement is not required, pump flow can be discharged to tank at low pressure through the center condition of the directional control valve.

(c) System efficiency

This is $\dfrac{\text{Energy to overcome load on cylinder}}{\text{Total energy into fluid}}$

$= \dfrac{\text{Flow to cylinder} \times \text{pressure owing to load}}{\text{Flow from pump} \times \text{pressure at pump}}$

$$\text{Efficiency on extend stroke} = \frac{20 \times 127}{20 \times 133} \times 100$$
$$= 95.5\%$$
$$\text{Efficiency on retract stroke} = \frac{20 \times 25}{20 \times 34} \times 100$$
$$= 73.5\%$$

Case 2: 'Meter-in' flow control for extend speed of 0.5 m/min (Figure 3.30)

From Case 1,

Cylinder 100 mm bore diameter × 70 mm rod diameter
    Full bore area $= 7.85 \times 10^{-3} \text{m}^2$
    Annulus area $= 4.00 \times 10^{-3} \text{m}^2$
    Load-induced pressure on extend $= 127$ bar
    Load-induced pressure on retract $= 25$ bar
    Pump flow rate $= 20$ l/min

Flow rate required for extend speed of 0.5 m/min is
$$7.85 \times 10^{-3} \times 0.5$$
$$= 3.93 \times 10^{-3} \text{ m}^3/\text{min}$$
$$= 3.93 \text{ l/min}$$

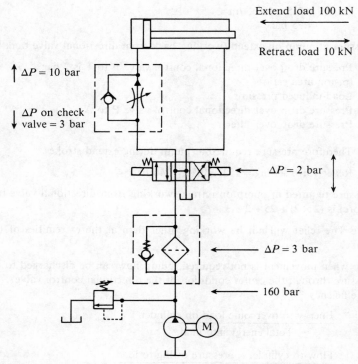

Figure 3.30 Example 3.5 with 'meter-in' flow control.

Section 3.2 Flow-control valves

Working back from directional control-valve tank port:

Pressure required at pump on retract is

$(2 \times 2) + (2 \times 3) + 25 + 2 + 3$
$= 40$ bar

Pressure required at pump on extend is

$(2 \times \frac{1}{2}) + 127 + 10 + 2 + 3$
$= 143$ bar

Relief valve setting is

$143 + 10\%$
$= 157$ bar

This is close to the maximum working pressure of the pump (160 bar). In practice, it would probably be advisable to select either a pump with a higher working pressure (210 bar) or use the next standard size of cylinder. In the latter case, the working pressure would be lower but a higher flow-rate pump would be necessary to meet the speed requirements.

Now that a flow-control valve has been introduced when the cylinder is on the extend stroke, the excess fluid will be discharged over the relief valve.

System efficiency on extend is

$$\frac{3.93 \times 127}{20 \times 157} \times 100$$
$= 15.9\%$

System efficiency on retract is

$$\frac{20 \times 25}{20 \times 40} \times 100$$
$= 62.5\%$

Case 3: 'Meter-out' flow control for extend speed of 0.5 m/min (Figure 3.31)

Cylinder, load, flow rate and pump details are as before. Working back from directional control valve tank port:

Pressure required at pump on retract is

$(2 \times 2) + 25 + 3 + 2 + 3$
$= 37$ bar

Pressure required at pump on extend is

$(2 \times \frac{1}{2}) + (10 \times \frac{1}{2}) + 127 + 2 + 3$
$= 138$ bar

Relief valve setting is

$138 + 10\%$
$= 152$ bar

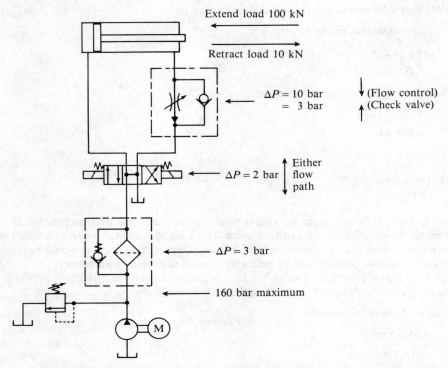

Figure 3.31 Example 3.5 with 'meter-out' flow control.

System efficiency on extend is

$$\frac{3.93 \times 127}{20 \times 152} \times 100$$

$$= 16.4\%$$

System efficiency on retract is

$$\frac{20 \times 25}{20 \times 37} \times 100$$

$$= 67.6\%$$

As can be seen, 'meter out' is marginally more efficient than 'meter in' owing to the ratio of piston area to piston rod area. Both systems are equally efficient when used with through rod cylinders or hydraulic motors. It must be remembered that 'meter out' should prevent any tendency of the load to run away.

In both cases if the system is running light, i.e. extending against a low load, excessive heat will be generated over the flow controls in addition to the heat generated over the relief valve. Consequently there will be further reductions in the efficiency. Also in these circumstances, with 'meter-out' flow control, very high-pressure intensification can occur in the annulus side of the cylinder and within the pipework between the cylinder and the flow-control valve.

Take a situation where in the 'meter out' circuit just considered (Figure 3.31) the load on extend is reduced to 5 kN without any corresponding reduction in the relief valve settings:

Flow into full bore end is 3.93 l/min.
Therefore, excess flow from pump is

$20 - 3.93$
$= 16.07$ l/min,

which will pass over the relief valve at 152 bar.

The pressure at full bore end of cylinder is

$152 - 3 - 2$
$= 147$ bar

This exerts a force which is resisted by the load and the reactive back-pressure on the annulus side,

$$147 - \left(\frac{5 \times 10^3}{7.85 \times 10^{-3} \times 10^5}\right) = (2 + 10 + P) \times 4.00/7.85$$

where P = pressure within the annulus side of the cylinder and between the cylinder and the flow control valve, and

$P = [(147 - 6.4) \times 7.85/4.00] - 12$
$= 264$ bar

The system efficiency on extend is

$$\frac{3.93 \times 6.4}{20 \times 152} \times 100$$

$= 0.83\%$

Almost all of the input power is wasted and dissipates as heat into the fluid, mainly across the relief and flow-control valves.

3.2.2 Three-port or bypass-type flow-control valve

This is basically a pressure-compensated flow-control valve with a 'built-in' relief valve, so that any excess flow is bypassed to tank at a pressure just above the load pressure. It can only be used as a 'meter-in' control. It is shown diagrammatically in part (a) and symbolically in parts (b) and (c) of Figure 3.32.

Let ΔP be the relief-valve spring setting. (It is also the pressure drop across the control orifice.) P_L is the load pressure. Then the system pressure is

$P_s = P_L + \Delta P$

An accepted value for ΔP is 7 bar. Therefore, the system pressure P_s would be at 7 bar above the load-induced pressure P_L.

The spring-loaded spool sets up a constant pressure drop across the control orifice independent of load or supply pressure. Once the regulated flow circuit is supplied, the excess flow is bypassed to tank. In this design, the tank line must go directly to the reservoir and not to a line which may be pressurized.

Bypass flow control can accurately regulate the speed of an actuator which operates against a wide range of loads and reduce the heat generated in the circuit.

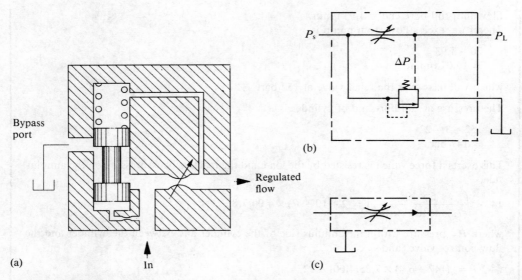

Figure 3.32 Bypass-type pressure-compensated flow control valve. (a) Section. (b) Detailed symbol. (c) Simplified symbol.

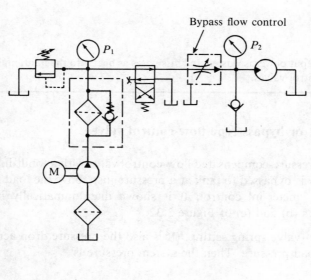

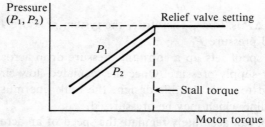

Figure 3.33 Motor circuit using bypass flow control with pressure/torque characteristics.

Section 3.2 Flow-control valves

Consider the circuit which is shown in Figure 3.33 together with its pressure/torque characteristics. Excess flow will be bypassed to tank from a pressure (P_1) slightly higher than that induced by the load (P_2). It should be noted that if the circuit cannot accept the whole of the regulated flow as in the case of the motor stalling, the internal relief valve will close, blocking the bypass outlet. Therefore, a separate relief valve must always be included in the circuit.

3.2.3 Priority flow control

This is a valve similar to the bypass flow control, except that the valve construction has been modified to allow any excess flow to be fed to a secondary circuit. It is shown diagrammatically in Figure 3.34.

The priority flow rate is set by a needle valve (1). This flow is pressure-compensated by the spool (2) which is biassed against the needle by a light spring (3). After regulated flow demand is satisfied, the balance of inlet flow is directed via the bypass spool (4) to the secondary circuit. The spools automatically adopt settings such that primary circuit flow requirements are accurately met irrespective of pressure changes in either circuit or the supply. The bypass flow can be used at any pressure up to the maximum operating pressure of the valve. Various symbolic representations of the valve are shown in Figure 3.35.

A typical application of a priority flow valve is in a system where one pump is used to supply two or more circuits, the requirements of one of the circuits having to be fully met before any fluid is fed to the other circuits. The primary circuit may be a cooling pump motor, a brake system, a steering circuit or some form of safety circuit. Figure 3.36 shows an application using two priority flow-control valves so that motors A and B will always receive the correct amount of fluid provided the pump is driven. As the pump drive speed increases, the excess flow goes to motor C.

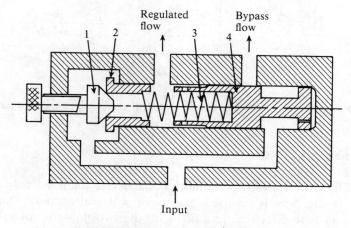

Figure 3.34 Priority flow control.

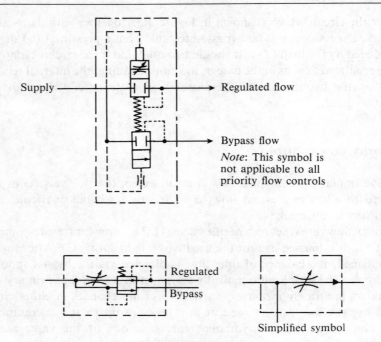

Figure 3.35 Priority flow control: symbolic representations.

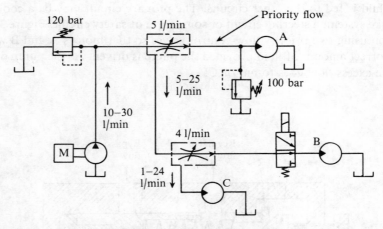

Figure 3.36 Application of two priority flow-control valves with varying input flow.

3.2.4 Bridge networks

A compensated flow-control valve is a unidirectional device, i.e. it will only accept flow in one direction; if the flow is reversed, the valve will malfunction. One method of controlling flow in both directions in a line is to use two flow-control valves with two check valves as shown in Figure 3.37. This system is ideal if different flows are required in

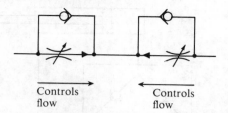

Figure 3.37 Accurate flow control in either direction using two flow control valves.

each direction but if the same flow is required there will be considerable difficulty in balancing the settings of the flow-control valves.

A simpler alternative is to use a bridge network and a single flow-control valve as shown in Figure 3.38. The flowpaths through the check valves are as shown. The path of fluid through the flow-control valve is in one direction only, independent of the direction of flow through the main line. The flow rate which will be the same in each direction of flow, is easily adjusted, and can be used for 'meter-in' or 'meter-out' flow control.

A modification to the bridge network can cause it to function as a lock valve provided that leak-free check valves are used. This is accomplished by applying a pilot signal at point C through a check valve. A pressure slightly higher than that at A and B will shut off check valves V_a and V_b in the bridge and prevent any flow from A or B. As fluid continues to flow through the flow control, the pressure-compensating spool remains active, reducing the possibility of surge when the actuator restarts. (**Note** that the flow from the locking signal source must exceed that permitted by the flow control valve.)

The principle is demonstrated in the circuit in Figure 3.39 where when neither solenoid is energized the piston is locked in position. Energizing solenoid A causes the piston to retract under 'meter-out' control by releasing the locking pressure. Solenoid B enables the extend stroke and the control is 'meter-in'. Speed will be the same in both directions providing the load cannot over-run on the extend stroke. Should solenoids A and B be energized together, the extend stroke occurs. When the piston is stationary it is under pressure and this in conjunction with the 'active' compensator spool gives the best possible starting behaviour for the 'meter-out' retract stroke.

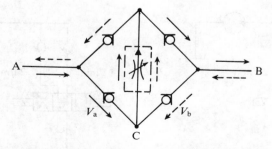

Figure 3.38 Accurate flow control in either direction using a bridge network and single flow-control valve.

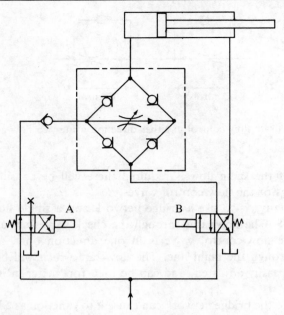

Figure 3.39 Application of bridge network as a lock valve.

3.2.5 Multi-speed systems using flow-control valves

Choice of different speed ranges for an actuator may be desirable. One method of achieving this is to use a number of flow-control valves and select the ones required.

In the condition shown in Figure 3.40(a) the flow to the motor will be 1 l/min. If the solenoid is energized, the second flow-control valve is switched into circuit and the flow becomes 2 l/min. This gives two selectable motor speeds. If a three-position valve is used as in Figure 3.40(b), it is possible to obtain three flow rates and hence three motor speeds dependent upon the position of the directional control valve. The flow rates will be 3 l/min with the valve centered and 1 or 2 l/min depending upon whether the left- or right-hand solenoid is energized.

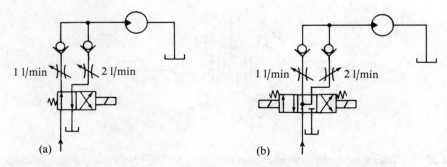

Figure 3.40 Selectable motor speeds. (a) Two speeds. (b) Three speeds.

3.2.6 Flow dividers

There are two distinct types of flow divider:

1. Valve type

2. Motor type

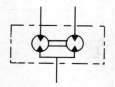

Valve type

The valve-type flow divider consists of a pair of matched orifices and interlocking spools. Usually the flow splits equally between the outputs, but it is possible to obtain other flow-dividing ratios by special order. A valve-type flow divider is shown in Figure 3.41.

If the flow through one orifice increases, an increased pressure drop results and a spool reduces the orifice, equalizing the flows. Should one of the output ports be blocked, the pressure drop becomes zero and a spool moves to completely close the other outlet port. When flow is reversed, the unit becomes a flow combiner allowing equal flows into the valve. Thus one valve can be used to synchronize two actuators in both directions.

Valve-type flow dividers will only divide the flow into two parts and are limited to fairly low flow applications (below 200 l/min). Each basic size is designed around an optimum flow and they become less accurate when working at other flow rates.

Although these valves are very accurate they do not exactly split the flow and cannot be used to synchronize two cylinders over a number of cycles unless a synchronizing valve is used at the end of each stroke.

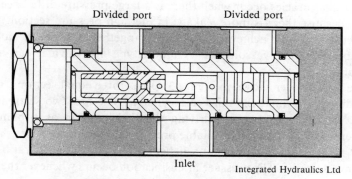

Figure 3.41 Valve-type flow divider.

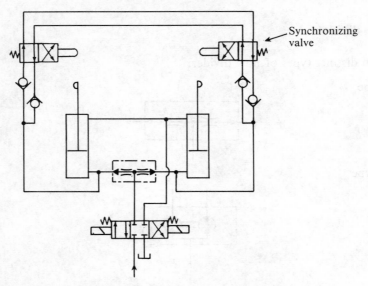

Figure 3.42 Flow divider circuit with synchronization at the stroke end.

In Figure 3.42, if one cylinder reaches the end of its stroke before the other, a synchronizing valve opens and connects the full bore ends of the cylinders. This allows the second cylinder to reach the end of its stroke. Alternatively, manual-synchronizing valves or electrical limit switches can be used.

Motor type

Motor-type flow dividers consist of a number of precision gear motors built on to a common shaft. This shaft ensures that the gear motors are synchronized. The gear motors can have the same volumetric displacement giving an equal division of the flow, or they can have other volumetric displacements giving whatever flow ratios are required. It must be appreciated that although these flow dividers are precision units, they do not exactly split the flow as leakage occurs across the gears in each motor section and the leakage flow will not be the same for each section.

The greatest inaccuracies occur when there is a large pressure difference between the output circuits, because there is more leakage in the high-pressure section.

Greater accuracy can be achieved by coupling together two piston motors but this may prove to be an expensive solution. A motor-type flow divider can also act as a pressure intensifier.

Assume the flow divider shown in Figure 3.43 has three equal sections, two of which are connected directly to tank, the third to a cylinder. The first two sections of the flow divider will act as hydraulic motors driving the third section as a pump. This will theoretically increase the maximum available pressure at P_3 to three times the main relief valve setting P_1. The actual intensification depends on the volumetric ratios, efficiencies and the number of sections. The pressure intensification occurs whenever the output from one section of a flow divider is returned to tank or has a low resistance circuit. Particular

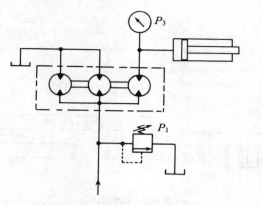

Figure 3.43 Motor-type flow divider.

attention must be paid to the possibility of pressure intensification wherever a motor-type flow divider is used in a circuit.

Some systems take advantage of this feature and employ a gear-type flow divider to enable part of a circuit to work above the pressure set by the main relief valve.

EXAMPLE 3.6

A press circuit is as shown in Figure 3.44. Determine the speeds and maximum thrusts:

1. During rapid closing.
2. During compaction.
3. During final forming.

Neglect any pressure drops in the circuit. The capacities of the motor sections are 20, 5 and 5 cm^3/rev as shown.

(i) Consider rapid closure (Solenoid C energized):

Flow to cylinder = 10 l/min

$$\text{Rapid closure speed} = \frac{10 \times 10^{-3}}{0.04} \left(\frac{\text{m}^3/\text{min}}{\text{m}^2} \right)$$

$$= 0.25 \text{ m/min}$$

Maximum rapid closure thrust is
70 (bar) × 0.04 (m^2)
= 70 × 10^5 (N/m^2) × 0.04 (m^2)
= 280 kN

(ii) Consider initial compacting (Solenoids A and C energized):

$$\text{Flow to cylinder} = \frac{(5+5)}{(20+5+5)} \times 10 \text{ l/min}$$

$$= 3.3 \text{ l/min}$$

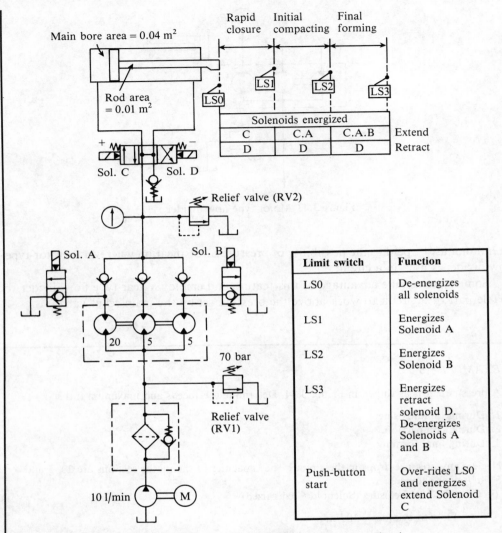

Figure 3.44 Motor-type flow divider used in a press circuit.

Initial compacting speed is

$$\frac{3.3 \times 10^{-3}}{0.04} \left(\frac{m^3/min}{m^2}\right)$$

$$= 0.083 \, m/min$$

Theoretical maximum pressure during initial compacting is

$$70 \times \frac{(20 + 5 + 5)}{(5 + 5)} \, (bar)$$

$$= 210 \, bar$$

Theoretical maximum thrust during initial compacting is

$$210 \text{ (bar)} \times 0.04 \text{ (m}^2\text{)}$$
$$= 840 \text{ kN}$$

Consider final forming (Solenoids A, B and C energized):

Flow to cylinder $= 5/(20 + 5 + 5) \times 10$ l/min
$= 1.67$ l/min

Final forming speed $= 0.0416$ m/min

Theoretical maximum pressure during final forming is

$$70 \times \frac{(20 + 5 + 5)}{5} \text{ (bar)}$$
$$= 420 \text{ bar}$$

Therefore, theoretical maximum final forming thrust is

$$420 \times 10^5 \times 0.04 \times 10^{-3}$$
$$= 1680 \text{ kN}$$

The intensified pressures and consequently the thrusts are theoretical values. In practice, these figures will be lower because of the inefficiencies of the flow divider.

Care must be taken not to exceed the pressure limitations of the components. Relief valve RV2 should be set to limit the maximum pressure in this circuit.

3.3 DIRECTIONAL CONTROL VALVES

These valves are used to direct the flow of fluid to the required line.

3.3.1 Check valves

The simplest type of directional control valve is the non-return or check valve which allows flow in one direction and prevents flow in the reverse direction. Such a valve, its symbols and characteristic curve is shown in Figure 3.45.

Check valves are available with different spring rates to give particular cracking pressures. The cracking pressure is that at which the check valve just opens. If a specific cracking pressure is essential to the functioning of a circuit, it is usual to show a spring on the check valve symbol. The pressure drop over the check valve depends upon the flow rate; the higher the flow rate, the further the ball or poppet has to move off its seat and so the higher the spring force.

Ball-type check valves have the least expensive form of construction, but as the ball is not guided there is a tendency for leakage to occur. Although manufacturers claim their check valves are leak-free in one direction of flow and allow free flow in the reverse direction, a tiny scratch, wear mark, or imperfection on the poppet or seat will permit some leakage. Soft-seated check valves use Delrin or similar polymer material for the seating and 100% sealing is possible but at the expense of valve life. However, they are not

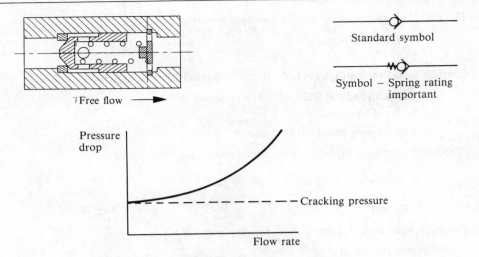

Figure 3.45 Ball-type check valve with symbols and curves.

generally suitable for pressures above 200 bar or temperatures above 35°C. Valves which seal satisfactorily at high pressure may leak at lower pressures. At high pressure, the poppet is forced onto the seat hydraulically giving a good seal; at low pressures, the sealing force is less and the valve may leak.

Pilot-operated check valves

These are normally closed check valves which may be opened by a pilot signal or less commonly held closed by a pilot signal. The pilot pressure needed to open the check valve against a load pressure depends upon the ratio of the areas of the pilot piston and check valve. A pilot-operated check valve is shown in Figure 3.46(a). Most manufacturers offer a range of pilot ratios, i.e. if the pilot ratio is 4:1, the pilot pressure required to open the valve is 25% of the load pressure. A typical application is shown in the circuit in Figure 3.46(b) where a pilot-operated check valve is used to lock in pressure to prevent a load from falling. With a long stroke cylinder the lowering motion of the load may be jerky. If the load overruns, the pressure in the full bore end of the cylinder drops, the check valve closes and the cylinder jerks to a stop. The pressure at the full bore end increases, the check opens, the cylinder lowers the load, the load overruns and so on. This problem can

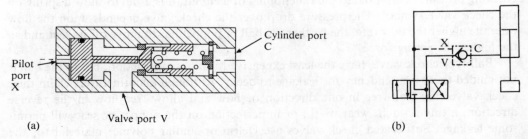

Figure 3.46 Pilot-operated check valve. (a) Section. (b) Application.

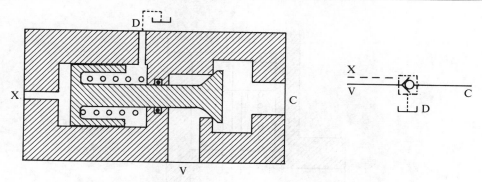

Figure 3.47 Vented pilot-operated check valve.

be overcome by using:

(a) a meter-out flow control valve to limit the cylinder speed;
(b) a counterbalance valve to prevent overrun; or
(c) an over-center valve.

With the directional control valve in the mid-position and the load raised, there will be a tendency for the pilot-operated check valve to leak at low loads, since the hydraulic sealing force on the check valve poppet is also reduced. Zero leakage is possible by using soft-seated versions of the valve.

Pressure on the pilot port X of the pilot-operated check valve shown in Figure 3.46 has not only to overcome the closing pressure which is present at cylinder port C but is also sensitive to any back-pressure at the valve port V. This can be overcome as shown in Figure 3.47 by incorporating a seal on the pilot stem and a separate vent or drain connection D for the spring chamber. Any back-pressure at port V will assist the pilot to open the valve.

Prefill valves

Prefill valves are basically large pilot-operated check valves. They are used in hydraulic press circuits to prefill the main cylinder with fluid whilst the press dies are being closed. The valve shown diagrammatically in Figure 3.48 is similar to a large pilot-operated check valve both in construction and operation but incorporates a decompression feature.

Hydraulic fluids are compressible to varying degrees and the volume of free fluid compressed into a cylinder is greater than its internal capacity. For example, in a cylinder having an internal volume of 0.3 m^3, approximately 0.31 m^3 of a typical mineral oil hydraulic fluid at atmospheric pressure will compress into the cylinder at 400 bar. (This quantity will be much greater if the oil is aerated.) Special valves have to be employed to control the decompression of large cylinders because the additional fluid (10 liters in this case) will attempt to discharge instantaneously resulting in extremely high shock forces.

The decompression feature incorporated in the prefill valve (Figure 3.48) is composed of a small poppet built within the main poppet. When the valve is piloted open by a pressure at port X, the main poppet is initially held firmly on its seat by pressure within the cylinder. The first part of the movement operates the pilot poppet opening up a small

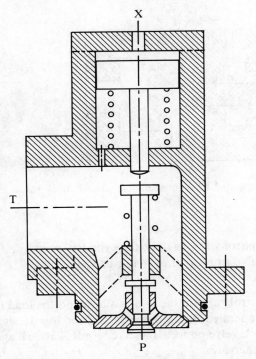

Figure 3.48 Prefill valve with decompression feature.

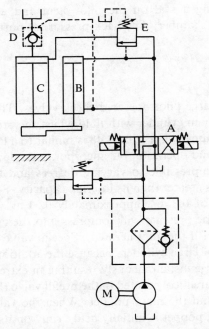

Figure 3.49 Press circuit utilizing a prefill valve.

flow path facilitating a controlled decompression. Further movement opens up the main poppet and the valve functions as a normal pilot-operated check.

Figure 3.49 is a press circuit which utilizes a prefill valve. Operating directional control valve A to the crossover condition initiates the closing of the dies. The side cylinder B drives down the main ram C and fluid from the reservoir which is mounted above the press, is sucked in through the prefill valve D to charge the full bore end of C. As the dies close onto the workpiece, pressure builds up opening the sequence valve E and flow from the pump pressurizes the full bore end of the main cylinder. During the pressing operation the prefill valve isolates the cylinder from the reservoir. On the retract stroke, (valve A in tramline condition) the prefill valve is piloted open and as the side cylinder pulls back the main ram, fluid from the full bore end is pushed into the reservoir. Using a prefill valve in this manner enables rapid movement of a large bore cylinder from a small delivery circuit.

Pilot to close check valves

In the valve shown in Figure 3.50(a), application of sufficient pilot pressure at port X prevents flow through the check in either direction. At other times the valve performs as a normal check valve with free flow one way (B to A) and blocked flow in the opposite direction (A to B). A representative application could be as a safety valve. In Figure 3.50(b) if pressure is lost in circuit number 1, circuit number 2 exhausts immediately.

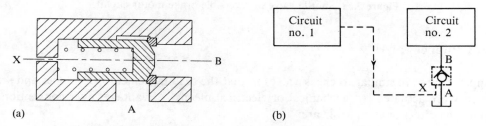

Figure 3.50 Pilot to close check valve. (a) Section. (b) Application.

Sandwich plates

Both check valves and pilot-operated check valves are manufactured as single or double units in sandwich plates for mounting in a valve stack between the directional control valve and base plate. (See section 5.4.6 in Chapter 5.)

Restrictor checks

Check valves are available with small holes through or bypassing the poppet to give a controlled leakage rate in the normally blocked direction. Such a valve may be pilot operated and is used as a safety feature in some circuits, or to give a pilot supply through the closed check to the downstream circuit.

Shuttle valves

The shuttle valve is a single-ball check valve with two alternative inputs A and B, and one output C. It is used for load-sensing and will accept a signal from the higher of two pressure inputs.

A typical application is in a reversible brake motor circuit (Figure 3.51) where it is used to release the brake when the motor is driven in either direction.

A double-ball shuttle valve or back-to-back check valve is able to sense signals from different inputs but prevents pressure feedback or interaction from other circuits. Care has to be taken in its use as it is possible to 'lock in' a pressure signal on the output side.

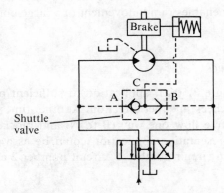

Figure 3.51 Shuttle valve in reversible brake motor circuit.

3.3.2 Poppet valves

Poppet valves are similar to check valves in that the sealing element is a poppet and seal but they are actuated by mechanical or electrical means. Advantages over conventional spool-type directional controls are:

1. Virtually zero leakage in closed position.
2. Poppet elements do not stick even when left under pressure for long periods.
3. Fast, consistent response times (down to 15 ms).

Disadvantages are:

1. Axial pressure balance is almost impossible and considerable force may be needed to open the poppet against the flow at high pressure. This limits valves which have direct mechanical actuation to low flow duties.
2. Generally individual poppets are required for each flow path which significantly increases the complexity of multiport valves.

Mechanical operation

Mechanical actuated valves are predominantly employed on presses and machine tool applications where they often form part of a dedicated control mechanism.

Electrical control

TWO-PORT SOLENOID OPERATED POPPET VALVES

These are available either as normally open or normally closed devices. In certain small models primarily intended for use as pilot valves the poppet is directly actuated by the solenoid. They are used as the control section of two-stage valves which are basically pilot operated check valves.

In one type of normally closed two-port solenoid controlled poppet valve shown in Figure 3.52, pressure at port A is applied to the back of the poppet via a small hole (orifice X) in the side-wall. Pressure keeps the poppet closed in the manner of a check valve. Energizing the solenoid lifts a plunger unblocking a hole (Y) in the center of the poppet. Imbalance occurs because of the pressure differential across orifice (X) and the poppet lifts permitting flow through the valve from A to B. Note that the solenoid armature (plunger) is permanently surrounded by the hydraulic fluid and hence balanced.

Relatively unrestricted reverse flow from B to A is possible when the solenoid is de-energized, the valve behaving as a normal check valve with only a small pressure differential needed to overcome the bias spring. If the solenoid is energized restricted reverse flow characteristics may be exhibited depending upon valve geometry.

Cartridge type construction is usually adopted, the valve fitting into a cavity similar to that in Figure 3.68.

Possible applications are similar to those for the pilot operated check valve (e.g. cylinder locking). A variation suitable for use as fast response, high flow rate, pump dump valves is the solenoid controlled logic element which is discussed in depth in Section 3.4 of this chapter.

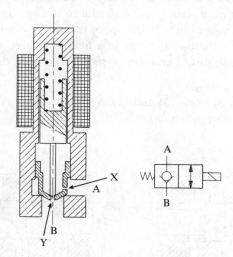

Figure 3.52 Two port solenoid-controlled normally closed poppet valve.

THREE- AND FOUR-PORT SOLENOID OPERATED POPPET VALVES

In one design of three-port valve a solenoid assisted by an internally piloted piston is used to switch a double-coned poppet from one seat to another. Various configurations are

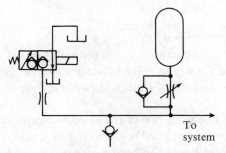

Figure 3.53 Three-port poppet valve as safety unloader valve in accumulator circuit.

available including four-port versions which incorporate a second double seated poppet section.

An established application for a three-port solenoid-poppet valve is as a safety unloader valve in an accumulator circuit (Figure 3.53). The valve symbol shows the solenoid energized which would be the normal operating condition and the circuit fails 'safe'. When the solenoid is de-energized, fluid in the accumulator depressurizes slowly through the fixed restrictor. The solenoid may be built into an accumulator safety block. In this situation a spool valve would be a constant source of leakage.

3.3.3 Sliding spool-type directional control valves

The vast majority of directional control valves fall into this category. The construction consists of a valve body with a sliding spool. Reduced sections in the spool interconnect ports and passages in the valve body as the spool is moved axially.

Figure 3.54 shows how the passageways in a five-port valve are connected when the spool is in the extreme positions. The valve ports are normally designated:

P   Supply on pressure port (1)
T   Return or tank port or ports (3) and (5). May also be marked T1 and T2
A, B   Cylinder or service ports (2) and (4)

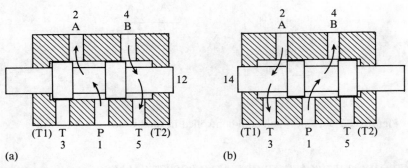

Figure 3.54 Five-port valve passageways: connections in extreme positions. (a) Spool moved over to left-hand position: P(1) to A(2), B(2) to T(5). (b) Spool in right-hand position: P(1) to B(4), A(2) to T(3).

Section 3.3 *Directional control valves* 101

The associated numbers bracketed in the list are the CETOP port identification numbers. With the spool in its extreme right-hand position, there are connections from P to B and from A to T1. When the spool is moved over into its left-hand position P connects to A, and B to T2.

In the majority of directional control valves the two tank ports T1 and T2 are permanently interconnected within the valve body or base plate to produce a four-port valve with just one common tank port T.

Components are machined to very close tolerances and fine surface finishes. The seal depends upon the closeness of fit between the lands (spool outside diameter) and valve bore. Elastomer seals and 'O'- rings cannot be used for this purpose because of the high flow forces and pressures involved. Sometimes spools are selectively fitted to give precisely the required clearances. A certain amount of leakage over the spool is necessary for lubrication to overcome friction.

Actuation can be by a variety of means: hand lever, mechanical trips, cam, solenoid, pilot pressure or pneumatic pressure. Some of these will be discussed in more detail later.

A part section through a four-port, double solenoid operated spool-type valve is shown in Figure 3.55. In this direct-acting valve the spool is switched by the solenoid force transmitted through push pins. When the solenoid is de-energized the spool is returned to its original position by centralizing springs.

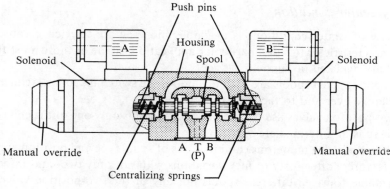

Mannesmann Rexroth

Figure 3.55 Four-port double solenoid-operated spool-type valve.

Spool transition states

It was seen in Figure 3.54 that in a two-position valve the normal connections made by the spool are P to A, B to T in one position and P to B, A to T in the other position. There is however a transition state between the 'tramline' and 'crossover' position and this transition state may be included in the symbol using dotted lines.

A three-position valve is basically the same valve but it is possible to select the mid-state as a definite operating position. Here again transition conditions exist between the end and center positions. The spool in the valve illustrated in Figure 3.55 is centralized by springs and when in its mid-position there is no connection between the ports. This is

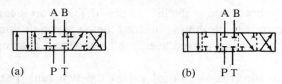

Figure 3.56 Spool transition states switching from center to end position. (a) Pressure port opening. (b) Tank port opening.

known as a blocked center and two such spools are shown symbolically in Figure 3.56. Switching from the closed-center to an end condition the spool passes through a transition state which depends upon the flow path first opened.

PRESSURE PORT OPENING (FIGURE 3.56a)
Port P is connected to the appropriate service port before the other service port is opened to tank.

TANK PORT OPENING (FIGURE 3.56b)
One service port is opened to tank before the pressure port and other service ports are interconnected.

These variations may be extremely critical to circuit performance.

Selection of center condition

A wide variety of center conditions is available and they are selected because of their switching characteristic or to suit a particular application. Some of those most frequently encountered are depicted in Figure 3.57.

Notched and tapered spools are used to smooth the switching action, reduce pressure shocks on change-over and to facilitate decompression.

Some two-position valves use the center condition plus only one of the end conditions and again there is an intermediate state.

For each size of valve numerous combinations of spool type and position are possible. Some manufacturers offer forty to fifty variations with not only two- and three-position valves using some fourteen different spools but any of the two-position valves may be offset to a particular end condition. Two- and three- port versions are also included.

If a three-port valve is required, this can often be achieved by plugging one port on a four-port valve. However, in many designs the tank port T is limited to a lower pressure than the other ports on the valve. Care must be taken not to block the tank port if system pressure can rise above the tank port rating.

The pressure drop through the valve depends on the flow rate, the spool type, the flow path, the fluid viscosity and temperature. Figure 3.58 shows for one size of valve, typical performance curves giving pressure drops through various flow paths with the spool types shown in Figure 5.57 and with a fluid viscosity $v = 36$ cSt.

Performance curves are measured at a specific viscosity and temperature and when used under other circumstances the figures must be adjusted accordingly. Under most conditions pressure drop will vary directly with the viscosity. The flow can be affected by silting, therefore the data are also dependent upon fluid cleanliness and the maximum

Section 3.3 Directional control valves

Spool reference	Center condition	Switching characteristic or typical application
a	⊏⊐	Prevents collapse of pressure during changeover (may cause pressure shocks)
b	⊢⊣	Pump unloading (In two-position valve pressure collapses momentarily during change-over)
c	⊏⊐	Unloading pump circuit but blocking ports A and B giving a degree of locking. (NOTE This spool causes a higher pressure drop through the valve than most other spools)
d	⊢⊐	Pilot-operated check valve circuits. Hydrostatic transmission to give free-wheeling effect and reduce pressure surges. Used when a second directional valve has to be supplied with fluid
e	⊏⊣	Single acting cylinder circuits
f	⋈	To gradually relieve pressure in the service lines on change-over to mid-position
g	⊢⊣	To maintain pressure on both service ports in mid-position, e.g. clamping Regeneration in mid-position

Figure 3.57 Spool valve center conditions.

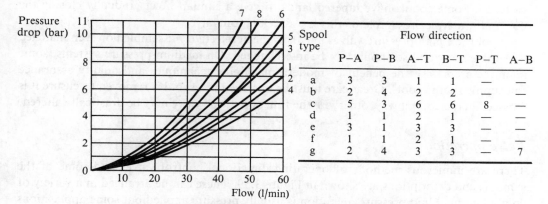

Spool type	P–A	P–B	A–T	B–T	P–T	A–B
a	3	3	1	1	—	—
b	2	4	2	2	—	—
c	5	3	6	6	8	—
d	1	1	2	1	—	—
e	3	1	3	3	—	—
f	1	1	2	1	—	—
g	2	4	3	3	—	7

Figure 3.58 Typical performance curves for various spool types.

values will not be achieved unless a high level of filtration is used. Data also refer to normal operation such as for a four-port valve where flow is simultaneously from P to A and B to T. When used as a three-port valve with port A or B blocked, flow is in one direction only. In this case, the performance is influenced by the flow forces acting within the valve and optimum results may not be achieved.

Pressure drops both ways across a valve should be taken into account. When driving a double-acting cylinder there will be a pressure drop from P to A in the fluid going to the actuator plus a pressure drop from B to T in the return oil from the actuator causing back-pressure. They will both reduce the effective force at the actuator. It should also be remembered that when cylinders are retracting, the return oil flow rate may be much higher than the input rate owing to the differential areas of the piston. The individual pressure drops must be taken into account in calculations.

When selecting a valve, bear in mind that performance curves issued by a manufacturer are designed to show the valve in the best light and will have been obtained under ideal conditions. Generally, they refer specifically to the valve itself and may not take into account pressure drops in the valve sub-plate or mounting block.

Spool shapes

The valve spools have annular grooves machined in the lands to form a hydrostatic bearing and prevent hydraulic locking of the spool. This can occur if spools with plain lands are fitted.

There is always some clearance to allow for temperature changes in the valve and ensure free movement of the spool. Although this clearance is only in the order of 5 to 15 μm there will be a slight fluid leakage dependent upon the fluid viscosity, temperature, the difference in pressure between ports and actual spool clearance. The effective clearance is largely the radial clearance between the spool and housing but is also dependent on the length of the gap, i.e. the amount by which the spool land overlaps the port slot in the housing. The leakage will increase with spool wear and is an indication of the valve condition. Good filtration will reduce wear and prevent silting up of the valve clearances.

Notches may be cut in the lands to give a gradual change in flow rates as the valve is switched. Some spools have tapered lands to give a similar effect gradually closing the flow path; a typical example of this is a deceleration valve.

Spools on high-pressure valves may have been selectively assembled or even matched to a valve body and so should not be interchanged. On medium-pressure systems, some manufacturers with the benefit of modern machining techniques and quality assurance can supply spare spools. Great care must be observed if a spool is removed to ensure it is replaced in the correct way, otherwise the function of the valve may be drastically altered.

Valve operation

There are numerous methods of operating directional control valves and some of the symbols and descriptions are shown in Figure 3.59. These can be arranged in a variety of combinations. Fluid pressure operation is usually pressure applied but some applications use pressure release, i.e. pressure is normally applied to both ends of the spool and movement is achieved by removing or reducing the pressure at one end.

Section 3.3 Directional control valves

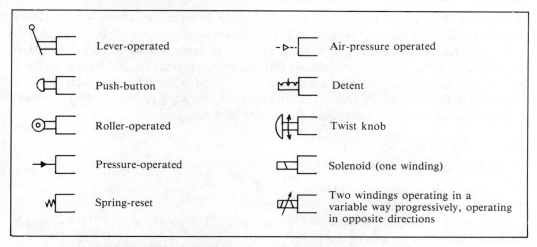

Figure 3.59 Symbols for directional control valve operators.

Actual force necessary to move a spool varies with the operating conditions and is affected by fluid flow forces, fluid acceleration forces and the geometry of the spool and housing. Sometimes the resultant forces assist and at other times oppose the movement.

Spool position may be controlled by:

(a) Springs which can be arranged to return the spool to any of the positions when the external operating force is removed.
(b) Mechanical detent where spring-loaded balls or catches hold the spool in its selected position when the external operating force is removed.
(c) Hydraulic detent which locks the spool in the required position by applying hydraulic pressure to the appropriate end of the spool.

Solenoid operation

When the coil is energized, the armature is pulled into the coil by electromagnetic forces. The armature moves the push rod and hence the spool, switching the flow paths through the valve (Figure 3.60). The coil may be AC or DC, air-gap or oil-immersed (sometimes called 'wet pin').

Control circuits for AC solenoid systems are generally less sophisticated and consequently less expensive than those for DC systems. However, great care must be taken to ensure that two opposing AC solenoids cannot be energized at the same time. In

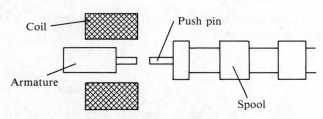

Figure 3.60 Solenoid operation.

AC circuits when the coil is energized it is subjected to a high inrush current which reduces rapidly as the armature pulls into the coil. The current necessary to keep it energized (the holding current) is only about one-seventh of the inrush current. Solenoids are designed to withstand the holding current indefinitely but the inrush current can only be applied for a very short time. Consequently, if anything prevents an AC solenoid pulling fully home, a burnt-out coil will result. In DC circuits the solenoid operating current is fairly constant and coils are designed to withstand this for indefinite periods.

TYPES OF SOLENOID
Air-gap AC solenoids have the following characteristics:

- Very short switching times (30 ms).
- Simple electrical control.
- High inrush current (7–10 times the holding current). They can only withstand inrush current for short periods of time.
- Maximum switching frequency of about 7000 per hour.

Air-gap DC solenoids have the following characteristics:

- Switching time approximately 60 ms.
- Smooth switching.
- Stopping armature in an intermediate position will not damage the coil.
- High maximum switching frequency (15 000 per hour—approximately twice that of AC solenoids).

Oil immersed or 'wet pin' solenoids have the armature encased within a non-magnetic tube which is not sealed from the low-pressure ports of the valve. Consequently, the armature is immersed in the fluid. The actuating coil is mounted on the outside of the tube. Many manufacturers are adopting this method of construction.

The AC wet-pin solenoids have similar characteristics to the air-gap AC solenoids plus the advantages owing to oil immersion of the armature:

- Less wear.
- Good heat dissipation.
- Cushioned armature end stops.

The DC wet-pin solenoids have similar characteristics to the air-gap DC solenoids plus the advantages owing to oil immersion of the armature:

- Less wear.
- Better heat dissipation.
- Cushioned armature end stops.

DC solenoids can be supplied with a built-in rectifier bridge so that they can be connected directly to an AC supply.

Soft switching

The change-over time of a DC solenoid-operated directional control valve can be controlled by means of a throttle in a connection between the ends of the spool as shown

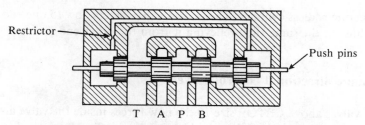

Figure 3.61 Throttle connection for soft switching.

in Figure 3.61. Fluid from the chamber at one end of the spool is 'metered' by the restrictor as it transfers to the corresponding chamber at the opposite end of the spool. In some designs the connection between the two end chambers is via a nozzle and drilling through the center of the spool.

Neon indicators are often an optional extra on solenoid valves; these will show when an electrical supply is applied to the solenoid but not that the solenoid has energized. Valves with devices which detect spool end position are available for use in specialist applications where it is essential to know that the valve has switched (e.g. mine winding gear and lifts). Some proportional valves (see Chapter 8) have devices which monitor actual spool position. Manual over-rides can be fitted to solenoid valves: these over-riders are useful when setting up a system and for emergency use.

Pressure-operated directional control valves

To operate a valve hydraulically requires another valve (or valves) to direct the flow to move the spool (see Figure 3.62). In this instance the slave spool may be considered as a

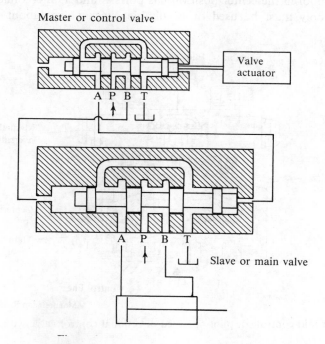

small double-acting rodless cylinder. The master valve directs flow to one end of the slave valve spool whilst at the same time relieving it from the other end.

3.3.4 Two-stage directional control valves

In the case of valves above CETOP size 10 the flow forces inside the valve are usually too great for direct operation of the main spool by a solenoid. To overcome this a pilot stage is introduced; the pilot stage is solenoid-operated, directing pressure fluid to move the main spool.

The speed of change-over of the spool can easily be controlled in a two-stage valve by restricting the flow of pilot fluid from the main spool. This control is achieved by inserting a choke pack between the solenoid pilot valve and main valve as shown in Figure 3.63. The choke pack is often a standard flow-control module of the same size range as the pilot valve. Two-stage valves without choke packs are frequently used when fast switching is needed.

The pilot supply to the solenoid valve can be taken internally from the main pressure port, or from a separate pilot source. If an internal pilot is used, the X (pilot) port in the base must be blocked. The connection between the P port and X port may contain a check valve or a removable plug. Similarly, the control-valve tank line may be internally drained into the main-valve tank port or externally drained. If the main-valve tank line is subjected to high pressure or pressure surges, the control valve must be externally drained.

Connecting valves for internal/external, pilot/drain combinations is achieved by moving plugs to block, or unblock passageways. The various combinations are shown in the circuit and table in Figure 3.64. Plugs a and d may be in the base plate.

If the main spool in the center position has ports P and T interconnected, either an external pilot supply must be used or a suitable check valve (about 4-bar cracking

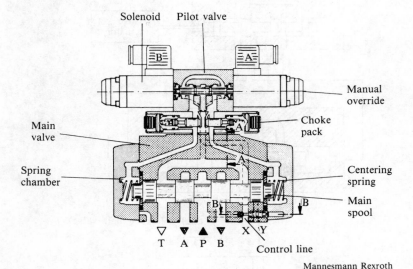

Mannesmann Rexroth

Figure 3.63 Solenoid-controlled, pilot-operated directional control valve with choke pack.

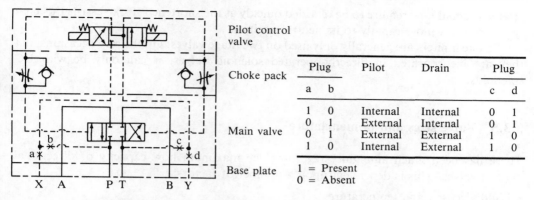

Figure 3.64 Position of plugs for internal or external pilots and drains.

pressure) must be incorporated downstream of the pilot pressure take-off, to provide sufficient pilot pressure. The check valve may be fitted in the pressure line after the pressure take-off or in the tank line. It can be integral with the directional valve (see Figure 3.65)

Stroke limiters are available on some makes of valve. The stroke limiter is fitted into the end cap of the main valve (see Figure 3.66) and acts as an adjustable stop to limit the spool movement in one or both directions. The main spool acts as a metering orifice providing a crude method of flow control through the valve. It can only satisfactorily be used where there is little variation in operating temperature, i.e. when the fluid viscosity remains reasonably constant. It may have an application where loads driven by a

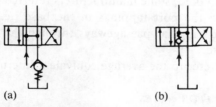

Figure 3.65 Incorporation of check valve to provide pilot pressure. (a) In tank line. (b) In pressure port.

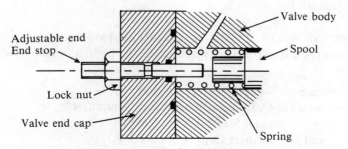

Figure 3.66 Stroke limiter.

relatively small flow require to be retarded quickly as the shortened stroke assists the main spool to revert more rapidly to its neutral position.

Stroke limiters are generally only used on two-stage valves. On the rare occasions that they are incorporated in directly-operated solenoid valves, it can only be with DC solenoids.

3.3.5 Valve sizes and nomenclature

There has been much ambiguity regarding the maximum flow capacity of directional control valves. This is dependent upon a number of factors:
- Fluid viscosity and temperature
- Area of flow passageways
- Port connections and base plates
- Type of spool
- Symmetrical flow forces
- Fluid cleanliness level

The size may be quoted with reference to:
- Average equivalent diameter of flow passageways
- Manufacturer's recommended maximum flow rate
- Standards organizations' reference numbers

These are discussed below.

AVERAGE SIZE OF FLOW PASSAGEWAYS

When Imperial sizes were quoted, some manufacturers referred to the valve passage sizes whereas others referred to the port tappings in the base i.e. $\frac{3}{8}$ inch could mean the equivalent diameter of the irregular passageways through the valve or $\frac{3}{8}$ inch BSP (or NPT) ports in the sub-plate.

Metric sizes generally refer to the average equivalent internal diameter.

RECOMMENDED MAXIMUM FLOW RATES

These are very dependent upon fluid viscosity, acceptable pressure drop and choice of spool type. Artificially-high figures are often quoted by referring to a valve without a sub-plate and may be much reduced when considered in association with the necessary mounting plate or manifold.

VALVE INTERFACES AND INTERNATIONAL REFERENCE NUMBERS

The following organizations are largely responsible for hydraulic component standard specifications:

       ISO = International Standards Organization
   CETOP = European Oil-hydraulic & Pneumatic Committee
     ANSI = American National Standards Institute
     NFPA = National Fluid Power Organization (USA)
      DIN = Deutsche Industrie-Norm (West Germany)

Section 3.3 Directional control valves

Standardization of the mounting interface facilitates interchangeability of valves from different manufacturers. Although valves which comply with these standards are interchangeable, they will not necessarily have the same flow capacity. Precise dimensions of the port and mounting configurations for directional control valves are specified in:

ISO 4401
CETOP R35H
DIN 24340 (Shape A)

Interfaces for pressure and flow control valves and non-return valves are specified in:

ISO 6264
CETOP R69H (Form P)
DIN 24340 (Shape D and Shape E)

The NG reference number approximates to the diameter in millimeters of the valve ports. Directional control valve standard sizes are: NG4, 6, 8, 10, 16, 25 and 32 (sizes 4 and 8 are rarely encountered).

Standard sizes for the pressure, flow and check valve two-port interfaces are: NG8, 10, 25 and 32. However, where these types of valve are employed in valve stacks (see Section 5.4.6 of Chapter 5) the configuration naturally has to suit the appropriate directional control sub-plate.

Figure 3.67 illustrates mounting details applicable to a popular size of directional control valve: in one form suitable for direct acting valves and two other forms for pilot operated valves to US and International Standards.

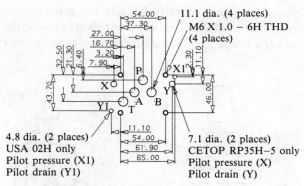

Figure 3.67 Size 10 valve interface: without pilots to CETOP-5, ISO-05, NFPA-DO2, NG10 specifications; with pilots X and Y to CETOP RP35H-5 specifications; with pilots X1 and Y1 to US-02H specifications.

Identification of valve actuator positions

On a circuit diagram the valve actuator (e.g. solenoid) is shown against the symbol box to which it refers. For example in Figures 3.65 (a) and (b), energizing a solenoid shown on the left-hand side of the symbol would switch the spool to the tramline condition. In practice, identifying the appropriate actuator is difficult as there are two conflicting systems in common use.

The US ANSI B93.9 standard designates solenoid A as the solenoid which selects the condition where P is connected to A irrespective of solenoid location and spool type. With two-stage valves the ports refer to those in the main stage.

The German DIN system defines solenoid A as the solenoid which is mounted at the end of the valve nearest to port A irrespective of spool type. With two-stage valves the reference is to the pilot valve and is independent of main stage valve port locations.

Any concurrence between the two is purely coincidental and because of the construction adopted by most manufacturers the systems conflict for the majority of spools – only with spool c in Figure 3.57 do they more frequently correspond.

3.4 CARTRIDGE VALVES

These consist of a valve shell which can be mounted in a standard recess in a valve block or manifold. This form of construction has been used for many years particularly for pressure controls, flow controls and check valves, but both poppet- and spool-type solenoid valves are available. Form drills and reamers are used to machine the standard cavities which are then tapped to accept the cartridges (Figure 3.68). The machine manufacturer does not have to worry about tolerances for moving spools and poppets, spring ratings, etc. because these are taken care of by the hydraulic valve manufacturer. The system is particularly advantageous for batch production and the modularized packages or integrated circuits eliminate expensive and potentially leaking pipework and connectors.

The term 'cartridge valve' has recently become the generic name for what are in principle a range of pilot-operated check valves, controlled to give directional, check, flow and pressure functions (*cartridge logic valves*). They are sometimes referred to as 'logic

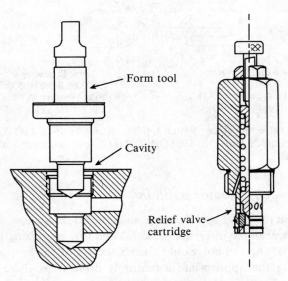

Figure 3.68 Cartridge valves.

Section 3.4 Cartridge valves

elements' and a particular characteristic is high-flow capacity relative to physical size. A range of standard cavities which accepts these valves is detailed in specification DIN 24342.

The valve shell or body has two main ports (A and B), which are connected or separated by a poppet or spool. The *poppet-type cartridge valve* is basically a check valve which can be pilot-operated in a number of ways whereas the *spool-type cartridge valve* is used as a variable restrictor which is either normally fully open and closed by action of the control or vice versa. The action of the two types of cartridge valve is completely different and will be considered separately.

3.4.1 Poppet-type cartridge valves

In some designs the poppet fits into a cavity and is held in position by a cover or top plate which contains all the pilot connections. Others are designed to fit the standard cavities used by some conventional cartridge valves (Figure 3.68). Logic elements which have balanced poppets or spools can be modulated and are largely used as pressure controls. Those with unbalanced poppets are primarily used for switching functions such as directional controls or where the poppet movement can be limited as flow controls.

The principal advantages of poppet-type valves are:

- Very high flow rates for relatively small physical size.
- A positive seal can be obtained.
- May be extremely rapid acting but may also be easily adapted for soft switching.
- The shape of the poppet or spool together with its seat can be varied to give different operating characteristics to the valve assembly.

The major disadvantage is that the unbalanced poppets, being responsive to pressure change on all ports, may malfunction owing to pressure surges. Particular care has to be taken in the circuit design to ensure safe operation.

The opening and closing movements of the poppet in a cartridge valve are pressure-dependent and a function of the forces on three areas.

If:

A_A is effective area of poppet at Port A
A_B is effective area of poppet at Port B
A_X is effective area of poppet at Port X

$A_X = A_A + A_B$

In the balanced poppet-type valve shown diagrammatically and symbolically in Figure 3.69:

$A_B = 0$

and

The areas A_A and A_X are equal.

The pilot X controls the function of the valve. If X is connected to port B the valve will

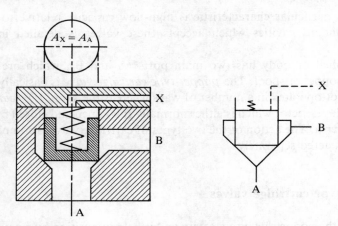

Figure 3.69 Balance poppet cartridge valve: area ratio $A_X = A_A$.

operate as a check valve allowing flow from A to B by opening the poppet but preventing flow from B to A by closing the poppet. If port X is connected to an external pressure, the valve can be used to control pressure.

In the unbalanced poppet-type valve shown diagrammatically and symbolically in Figure 3.70 it is possible to obtain different area ratios, typically:

$A_A : A_X = 1 : 1.1$ where $A_B = 0.1 A_A$
$A_A : A_X = 1 : 1.05$ where $A_B = 0.05 A_A$
$A_A : A_X = 1 : 2$ where $A_B = A_A$

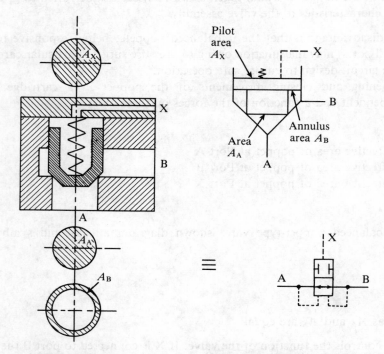

Figure 3.70 Unbalanced poppet-type valve: area ratios $A_X = A_A + A_B$.

Section 3.4 Cartridge valves

With the pilot X vented, pressure at port A or B only has to overcome the bias spring force for flow in either direction. However, it can be held closed by a pressure on the pilot port which is dependent upon the poppet area ratios.

EXAMPLE 3.7

Consider the valve shown in Figure 3.70 where the ratio $A_A : A_X$ is 1:1.1. If the force exerted by the control spring is equivalent to 3 bar and the pilot pressure is 7 bar,

$$A_X = A_A + A_B$$

If
$$A_A : A_X = 1 : 1.1$$
$$A_B : A_X = (1.1 - 1) : 1.1$$
$$= 0.1 : 1.1$$

When the flow is from A to B the pressure required at A to just open the valve is calculated by equating forces on the poppet.

$$P_A \times A_A = (P_X + \text{spring}) A_X$$
$$P_A = (P_X + \text{spring}) A_X / A_A$$
$$= (7 + 3)(1.1/1) = 11 \text{ bar}$$

If the flow is from B to A, the pressure required at B to just open the valve is again obtained by equating forces on the poppet.

$$P_B \times A_B = (P_X + \text{spring}) A_X$$
$$P_B = (P_X + \text{spring}) (A_X / A_B)$$

Thus
$$P_B = (7 + 3) \times (1.1/0.1) = 110 \text{ bar}$$

Hence a very low pressure on the pilot port X can balance a high pressure on port B.

In the absence of pressure at port X, the valve will open to flow in either direction provided the pressure at port A or B is sufficient to overcome the spring biassing force. The actual pressure will be also dependent on the ratio of the poppet areas. In this example (where $A_X : A_A = 1.1 : 1$) when P_X is zero the valve will open if $P_A = 3.3$ bar or $P_B = 33$ bar.

Normally closed cartridge valve

Valves which are closed unless a control signal is present are useful for blocking circuits or preventing movement of actuators. (Sun Hydraulics Corporation have a patent pending for a normally closed unbalanced poppet-type valve.)

The logic element shown in Figure 3.71 is normally closed and will not respond to pressures at either A or B while the pilot is vented. The pilot pressure at port X necessary to open the valve is dependent upon the poppet effective area ratios, the biassing spring and the back-pressures on ports A and B.

$$A_X (P_X - \text{spring}) = (A_A \times P_A) + (A_B \times P_B)$$

In its basic form the valve functions as a pilot-operated check valve. Variations are available which have an internal pilot supply from port A or B through an orifice. These are discussed in more detail later in this section.

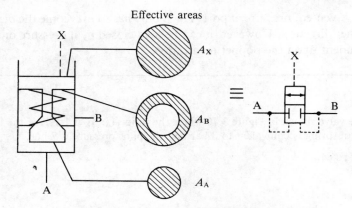

Figure 3.71 Normally closed cartridge valve.

Restrictor cartridge poppets

Poppet valves are extremely rapid acting and at times this may be undesirable. Restrictor poppets enable soft switching and may even be used as flow controls. In the basic valve (Figure 3.72) the restrictor is formed by a notched projection on the nose of the poppet. As the poppet lifts, the tapered notches are uncovered, gradually opening the flow path between ports A and B; smooth closing is similarly achieved. The controlled action is called 'soft switching'.

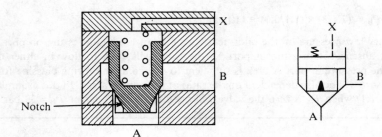

Figure 3.72 Restrictor poppet cartridge valve.

When used as a flow control, adjustment is by a mechanical device which limits the opening of the poppet. Flow is restricted equally in both directions but the valve can still be closed by a pilot pressure at port X (Figure 3.73).

A unidirectional flow control can be arranged by self-piloting the valve from one of the work ports. i.e. piloting from port B prevents flow from B to A whilst permitting controlled flow from A to B.

Stroke-limiting devices on cartridge valves are useful aids to maintenance. They may be screwed fully home to isolate a circuit or actuator. In the case of the normally closed cartridge (Figure 3.71) a stroke limiter can be used to 'crack open' the valve to depressurize a circuit or unlock an actuator to allow manual movement.

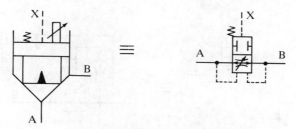

Figure 3.73 Flow-control (bidirectional).

Solenoid-operated cartridge valves

By combining with a conventional solenoid-operated directional control, the logic element can be used as a two-position, two-port internally-piloted solenoid valve (Figure 3.74). If the flow is from A to B then with the solenoid de-energized, the poppet opens against the spring and there is flow through the valve. When the solenoid is energized, the poppet is held closed by the input pressure on the pilot. For flow from B to A and with the solenoid de-energized, the valve is piloted closed. Energizing the solenoid releases the pilot signal allowing the poppet to open against the spring.

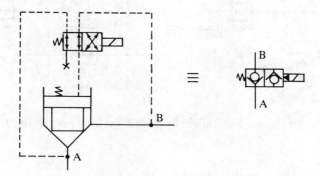

Figure 3.74 Two-position, two-port internally-piloted solenoid valve.

Orifice cartridge valves

An orifice in the poppet provides a pilot flow to port X from one of the work ports. Operation of the valve is controlled by blocking or venting port X. When X is blocked, pressure from the appropriate work port builds up on the full poppet area and closes the valve. This is shown symbolically in Figure 3.75 where (a) is internally-piloted from port A, (b) from port B and (c) from either A or B. In all three arrangements when the solenoid is de-energized and port X is vented, the valve is open to flow in either direction. When the solenoid is energized, blocking port X in Figure 3.75(a) only permits flow through the valve from B to A. The arrangement in Figure 3.75(b) permits flow from A to B and that in Figure 3.75(c) prevents flow in either direction.

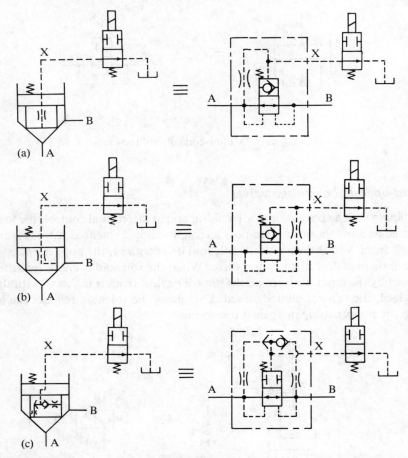

Figure 3.75 Orifice cartridges valves. Internally piloted: (a) from port A; (b) from port B; (c) from either A or B.

Remote switching of cartridge elements

Figure 3.76 shows a two-port, two-position cartridge valve which is externally piloted from a remote pressure source P. By energizing the solenoid, the flow path through the cartridge valve is blocked; with the solenoid de-energized, there is free flow from A to B and B to A. The solenoid valve can be connected in the opposite manner so that energizing the solenoid opens the flow path through the cartridge valve. They can be used as safety valves to dump the pump flow in case of control circuit failure. The action of the valve will be very fast resulting in pressure surges caused by rapid opening or closing. This effect can be reduced by using a restrictor-type poppet to give a gradual opening or closing of the flow path. A further refinement is to fit a restrictor – normally a fixed orifice – into the pilot line to slow down the movement of the poppet. This gives a soft start to a system. By using an adjustable mechanical stop to limit the movement of the poppet, the valve becomes a remote solenoid-operated variable-flow control valve (Figure 3.77). The solenoid does not vary the flow rate. This is preset by the mechanical control and will be

Section 3.4 Cartridge valves

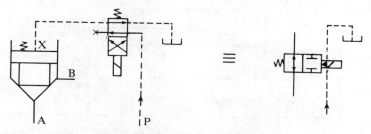

Figure 3.76 Externally-piloted, two-port, two-position cartridge valve.

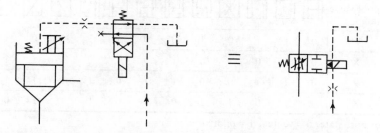

Figure 3.77 Remote solenoid-operated variable flow-control valve.

the same in either direction and it is not pressure- or viscosity-compensated. The solenoid is used to switch the valve on or off as required.

Pilot control sources

Cartridge logic elements may be:

(a) Self-piloted by using either, or both, of the work ports (e.g. Figure 3.74)
(b) Internally-piloted from a work port through an orifice in the poppet (e.g. Figure 3.75), control being by venting or blocking the pilot port
(c) Externally-piloted from a remote pressure source (e.g. Figure 3.76).

Elements which are self- or internally-piloted are passive devices. Pilot pressure presence is dependent upon there being pressure at the workport. Externally-piloted elements are active devices. Pilot pressure may be available at all times independent of the conditions at the work ports.

Selection of the correct method of piloting is important in circuit design. The use of internally-piloted components simplifies circuit and manifold design but it is necessary to consider the whole cycle of operation when checking that a pilot source is available. External piloting sometimes leads to a more complex circuit but it is possible for several different functions in a system to be performed by a single element.

Switching of multiple elements

Four cartridge logic elements are required to control a double-acting hydraulic cylinder. Manipulation of the pilot signals to independently open or close four logic elements

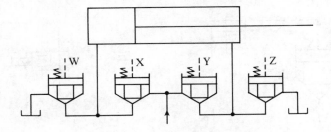

Pilot port state: 0 = vented; 1 = piloted

Figure 3.78 Twelve equivalent four-port spool valve conditions.

enables 12 equivalent four-port spool-valve conditions to be achieved. These are tabulated in Figure 3.78. Each of the pilot signals has to be controlled by a separate directional control valve. This variety of conditions enables the actuator to be:

- Driven in either direction.
- Locked in an intermediate position.
- Allowed to free-wheel in either direction.
- Extended regeneratively.
- Prevented from extending but permitted to retract and vice versa.

If the pilot connections are coupled in pairs, i.e. W + Y and X + Z and are controlled by a two-position, four-port valve, the network will function as a four-port, two-position pilot operated valve, having the third and fourth equivalent spool valve states tabulated in Figure 3.78. The cylinder will only stop at the ends of its stroke and if the control valve is a single solenoid spring return type the cylinder will always reset to one stroke extremity. Using a three-position pilot valve enables either the first or second equivalent spool valve state to also be obtained, depending upon the center condition of the valve selected.

Pressure-control cartridge valves

These consist of a balanced poppet (i.e. with 1:1 ratio) with the pilot connected to a pressure-control valve which may be integral or remote to the cartridge valve. In Figure 3.79, port A is connected to the pressure supply which is to be controlled and port B is to tank. The orifices in the pilot lines damp out any pressure surges and prevent valve flutter. Port V is a vent port which can be used to remotely vent the valve or cause it to operate at a lower pressure than the setting of the main pilot relief valve. This is shown

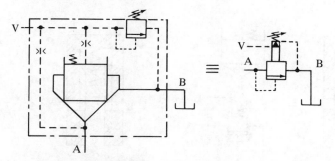

Figure 3.79 Pressure-control cartridge valve.

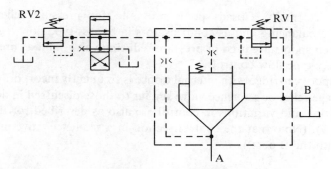

Figure 3.80 Remote pressure control.

diagrammatically in Figure 3.80. In this circuit, the state of the directional control valve in the vent line determines the control source on the valve pilot. In the center condition, the line is vented and the pressure at port A only has to overcome the cartridge valve spring. With the valve in the 'tramline' condition, control is by the remote relief whilst in 'crossover' the internal relief sets the operating pressure.

The pressure can be electrically modulated by substituting a proportional pressure-control valve for the external relief valve RV2 in Figure 3.80. If this type of control is adopted, it is recommended that the internal manually-operated pilot pressure-relief valve RV1 (set at the maximum circuit pressure) is still incorporated in case of malfunction or the failure of the electrical supply to the proportional valve. (Proportional valves are discussed further in Chapter 8.)

3.4.2 Spool-type cartridge valves

Spool-type cartridge valves are used for pressure regulation and pressure compensation. They employ balanced spools (1:1 area ratio) in the valve body to gradually open or close the flow path according to the valve configuration, which may be normally closed or normally open (Figure 3.81). The function of the valve is determined by the manner in which the spool is controlled. Most pressure-regulating functions can be performed by

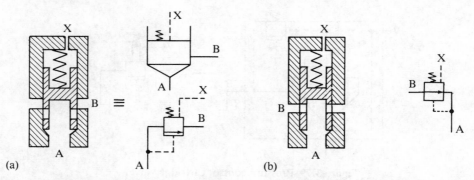

Figure 3.81 Spool-type cartridge valves. (a) Normally closed. (b) Normally open.

cartridge valves. Normally, closed spools are used for pressure relief, sequencing, unloading, counterbalancing and as compensators in bypass-type flow regulators.

Normally open spools are used in pressure-reducing valves and as pressure compensators in restrictive-type flow controls.

A balanced spool cartridge with internal orifice is frequently incorporated as the main stage of a two-stage relief or sequence valve similar to those discussed in detail in Section 3.1.1. The operation and variations in control are also as described for the poppet-type relief (Figure 3.79). (Note that the control section in a sequence valve must be drained externally, not into the B port.)

Pressure compensators

For pressure-compensating applications, cartridges work in conjunction with an external orifice which is generally adjustable.

The normally closed balanced piston element is used in bypass-type flow controls (Figure 3.82(a)). Excess flow is bypassed to tank at slightly in excess of the pressure at the 'Out' port. Restrictive-type pressure-compensated flow controls use a normally-open

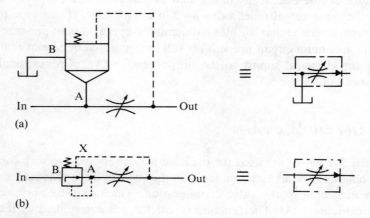

Figure 3.82 Pressure-compensated flow control. (a) Bypass-type. (b) Restrictive type.

balanced piston element (Figure 3.82(b)). The compensator maintains a constant pressure drop across the metering orifice regardless of variations in the upstream and downstream pressures.

Pressure-reducing valve

By connecting the remote pilot connection X to a pressure-relief valve, a normally-open spool type will act as a pressure-reducing valve (Figure 3.83). A restrictor in the supply line to the pilot relief valve acts as a damper and prevents the spool in the cartridge valve from oscillating. The maximum pressure acting on port X is set by the relief valve. If the downstream pressure at port A is greater than that at port X, the spool valve will close causing a pressure drop until the pressure at A is equal to the setting of the relief valve.

The check valve in the pilot line will allow any load-induced pressure in line A to be fed to the relief valve, so ensuring that the downstream pressure never exceeds the setting of the valve. A pressure-reducing cartridge valve can be controlled by a proportional relief valve (connected to port Y in Figure 3.83) with a simple relief valve acting as a safety valve in case of malfunction of the proportional valve. The safety valve should be set at the maximum permitted downstream pressure. By incorporating a solenoid-operated directional control valve, between port Y (Figure 3.83) and a secondary simple relief valve, the pressure-reducing valve is able to operate at high or low pressure. If a three-position valve is used the reducing valve can also be vented in the mid-position of the solenoid valve which would limit the downstream pressure to a very low figure. This arrangement is similar to Figure 3.7 in Section 3.1.1 of this chapter which describes the operation of a solenoid-controlled two-stage relief valve.

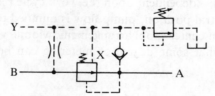

Figure 3.83 Pressure-reducing valve.

Along with proportional control, cartridge logic elements probably represent the most important development to take place in the hydraulics industry over recent years. Their versatility presents the circuit designer with many new and interesting opportunities. Cartridge construction promotes the use of integrated circuits, manifolds and modular packages, with not only cost-benefits, but improved reliability owing to better leakage control.

3.5 MOBILE HYDRAULIC VALVES

A family of special hydraulic valves has been developed largely for use in mobile hydraulic applications: excavators, cranes, fork-lift trucks, etc. The directional control valves are usually six-port spool valves, banked together in groups often with in-built relief and check cartridges.

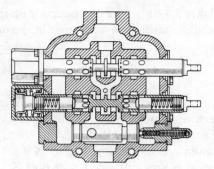

Hydreco Hamworthy Ltd

Figure 3.84 Monoblock valve.

The construction can be either by individual valve sections bolted together, or a one-piece (monoblock) casting containing several spools (Figure 3.84). The former is more versatile, reducing the number of units which have to be stocked and facilitating changes in design. Monoblock valves are, however, neater, smaller for a given flow rate, and less prone to leakage.

3.5.1 Valve arrangements

There are three basic arrangements for the interconnection within the groups − parallel, series or tandem connection − dependent upon the work cycle required. Generally, when all the spools are in the neutral position, pump flow is dumped to tank by either venting the main relief or by an open-center spool arrangement. Mobile valves usually have fairly long spool movements so that some degree of metering can be accomplished by only partially operating the valve.

Parallel connection

With the arrangement shown in Figure 3.85 there is a common pressure feed which is

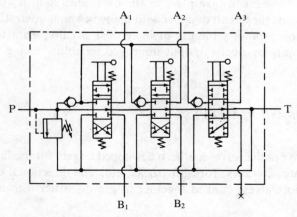

Figure 3.85 Parallel connection.

available simultaneously to all the spools. Two or more spools can be actuated at once but the circuit which requires the lowest pressure will operate first. Where several circuits function together the flow will be shared amongst them and speeds will be lower than when the circuits work independently. Simultaneous operation can be achieved by partially moving a spool, the metering effectively balancing the loads.

Series connection

The arrangement shown in Figure 3.86 also allows simultaneous operation of more than one service, the return oil from one actuator being used to supply the next spool. High speeds are possible but the available pressure is divided between the services. The operating pressure of a later section forms the back-pressure for the preceding section. Care must be taken when using cylinders in series circuits as they may prevent complete movement of the preceding actuator, or limit the amount of fluid available for the next actuator. For this reason, series connection circuits are predominantly used for hydraulic motor applications.

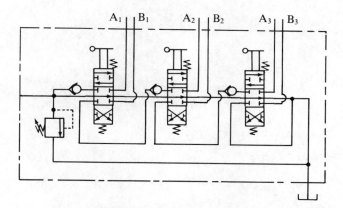

Figure 3.86 Series connection.

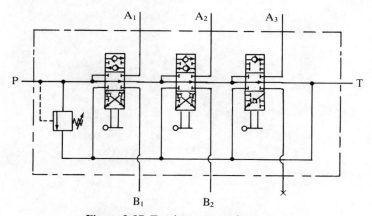

Figure 3.87 Tandem connection.

Tandem connection

With this arrangement (Figure 3.87) it is only possible to operate one service at a time. If two sections are actuated together the spool nearer to the inlet takes priority.

Combinations of the arrangements shown can be made. In addition to a main relief section, built-in port-relief valves, cross-line relief valves, anticavitation check valves, cylinder lock valves, counterbalance valves, regenerative and special spools, etc. may be incorporated. In Figure 3.87 spool 3 is suitable for controlling a single acting cylinder which retracts under gravity.

Electrical microswitches are often incorporated in spool actuators as a safety interlock, to detect spool position or so that on such as an electrically-powered fork-lift truck, the pump is switched on when the valve is actuated. Electrically-controlled proportional valves (see Chapter 8) are being increasingly used on mobile applications. This has the advantages of more sensitive control, facilitates ergonomic layouts and reduces the noise level in the cab by removing the hydraulic valves elsewhere on the vehicle.

CHAPTER FOUR
ACTUATORS

Hydraulic systems are used to control and transmit power. A pump driven by a prime mover such as an electric motor creates a flow of fluid, in which the pressure, direction and rate of flow are controlled by valves. An actuator is used to convert the energy of the fluid back into mechanical power. The amount of output power developed depends upon the flow rate, the pressure drop across the actuator and its overall efficiency.

There are three basic types of hydraulic actuator:

(a) Linear – hydraulic cylinder.
(b) Rotary (continuous rotation) – hydraulic motor.
(c) Rotary (limited angle of movement) – semi-rotary actuator.

Linear actuators, as their name implies, provide motion in a straight line. The total movement is a finite amount determined by the construction of the unit. They are usually referred to as cylinders, rams or jacks. All these terms are synonymous in general use although 'ram' is sometimes intended to mean a single acting cylinder and jack often refers to a cylinder used for lifting.

Continuous angular movement is achieved by rotary actuators, more generally known as 'hydraulic motors'. In basic construction they are similar to hydraulic pumps but whereas a pump shaft is rotated to generate flow, a motor shaft is caused to rotate by fluid being forced into the driving chambers. The driving chambers may be a generated form or a series of pistons and may have either a fixed or variable displacement.

Semi-rotary actuators are capable of a limited angular movement which can be several complete revolutions but 360° or less is more usual.

4.1 *HYDRAULIC CYLINDERS*

Hydraulic cylinders or linear actuators can be divided into three main groups:

Displacement
Single acting
Double acting.

Each is used to convert the pressure energy of a fluid into a linear thrust.

4.1.1 Displacement cylinders

A displacement-type hydraulic cylinder shown diagrammatically in Figure 4.1 consists of a rod which is displaced from inside a tube by pumping hydraulic fluid into the tube. The volume of the rod leaving is equal to the volume of fluid entering the tube, hence the name 'displacement cylinder'.

The rod of the displacement cylinder is guided by bearings in the nose or neck of the cylinder body. A collar on the end of the rod prevents it being ejected and limits the stroke of the cylinder. Elastomer seals in the neck prevent any leakage of fluid along the outside of the rod. This design is a single-acting 'push' or extension cylinder, which has to be retracted by gravity, a spring or some external force. The bore of the cylinder body does not require machining other than for the neck bearing and the inlet port; the manufacturing cost is therefore low when compared with other types of hydraulic cylinder.

The maximum thrust exerted by the displacement cylinder shown in Figure 4.1 is given by

Maximum thrust = pressure × rod area

$$= P \times \frac{\pi d^2}{4}$$

where d is the diameter of the rod. The extend speed of the rod is given by

$$\text{Rod speed} = \frac{\text{Flow rate of fluid entering cylinder}}{\text{Area of cylinder rod}}$$

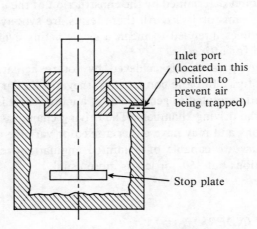

Figure 4.1 Displacement cylinder.

EXAMPLE 4.1

A displacement-type cylinder has a rod of 65 mm diameter and is powered by a hand pump with a displacement of 5 ml per double stroke. The maximum operating pressure of the system is to be limited to 350 bar.

Section 4.1 *Hydraulic cylinders* 129

(a) Draw a suitable circuit diagram showing the cylinder, pump and any additional valving required
(b) Calculate the number of double pumping strokes needed to extend the cylinder rod by 50 mm
(c) Calculate the maximum load which could be raised using this system.

Solutions

(a)

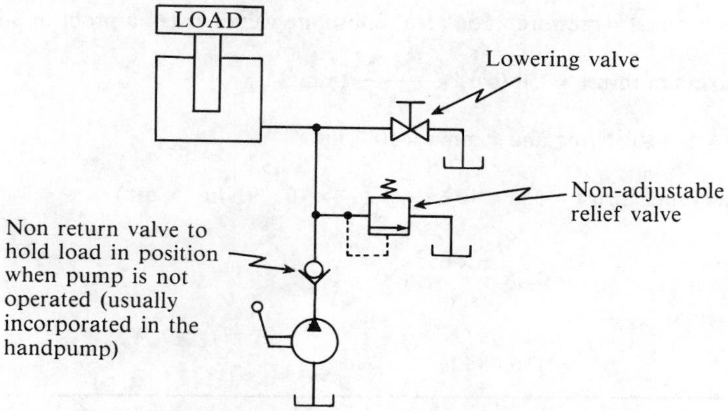

Figure 4.2 Jack circuit.

(b) The volume of rod displaced is equal to the volume of fluid entering the cylinder. Let rod diameter be d, the distance rod extends be L, the displacement per double stroke of pump be V and the number of double pump strokes be S

Rod volume displaced = fluid volume entering

$$\frac{\pi d^2}{4} \times L = V \times S$$

Substituting values given in the problem and showing units for each value

$$\frac{\pi \times 65^2}{4} \text{ (mm}^2\text{)} \times 50 \text{ (mm)} = 5 \text{ (ml)} \times S$$

$$\frac{\pi \times 65^2}{4} \times 50 \text{ (mm}^3\text{)} = 5S \text{ (ml)}$$

The units on both sides of the equation must be the same

Note

1 ml	$= 1 \times 10^{-3}$ liter
1 liter	$= 1 \times 10^{-3}$ m^3
1 ml	$= 1 \times 10^{-6}$ m^3
1 mm^3	$= 1 \times 10^{-9}$ m^3

Thus for dimensional equality

$$\frac{\pi \times 65^2}{4} \times 50 \times 10^{-9} \text{ (m}^3\text{)} = 5 \times 10^{-6} \text{ (m}^3\text{)} \times S$$

or

$$\frac{\pi \times 65^2}{4} \times 50 \text{ (m}^3\text{)} = 5 \times 10^3 \text{ (m}^3\text{)} \times S$$

Therefore

$$S = \frac{\pi \times 65^2 \times 50}{4 \times 5 \times 10^3}$$

$$= 33.17 \text{ double strokes}$$

(c) Maximum thrust = pressure × rod area. Substitute values given in problem and show units

$$\text{Maximum thrust} = 350 \text{ (bar)} \times \frac{\pi \times 65^2}{4} \text{ (mm}^2\text{)}$$

Note 1 bar = 1×10^5 N/m² and 1 mm² = 10^{-6} m²

$$\text{Maximum thrust} = 350 \times 10^5 \times \frac{\pi \times 65^2}{4} \times 10^{-6} \text{ (N/m}^2 \times \text{m}^2\text{)}$$

$$= 35 \times \frac{\pi \times 65^2}{4} \text{ (N)}$$

$$= 116\,080 \text{ N}$$
$$= 116.08 \text{ kN}$$

Telescopic cylinders

Telescopic cylinders are used when a long stroke is required and the length available for installation is limited. A typical application is the tipping gear of a lorry. They generally

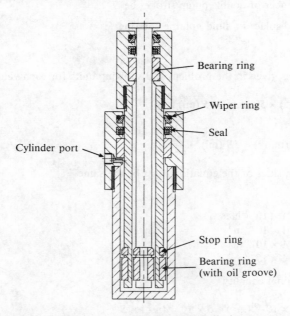

Figure 4.3 Displacement type telescopic cylinder (two-stage).

Section 4.1 Hydraulic cylinders

consist of a nest of tubes as shown diagrammatically in Figure 4.3 and operate on the displacement principle. The tubes are supported by bearing rings, the innermost (rear) set of which have grooves or channels to allow fluid flow. The front bearing assembly on each section includes seals and wiper rings. Stop rings limit the movement of each section preventing separation. When the cylinder extends, all the sections move together until the outer section is prevented from further extension by its stop ring. The remaining sections continue out-stroking until the second outermost section reaches the limit of its stroke; and so on until all sections are extended, the innermost one being last of all.

For a given input flow rate, the speed of operation will increase in steps as each successive section reaches the end of its stroke. Similarly, for a specific pressure the load-lifting capacity reduces for each successive section.

EXAMPLE 4.2

A three-stage displacement type telescopic cylinder (Figure 4.4) is used to tilt the body of a lorry. When the lorry is fully laden the cylinder has to exert a force equivalent to 4000 kg at all points in its stroke. The outside diameters of the tubes forming the three stages are 60, 80 and 100 mm. If the pump powering the cylinder delivers 10 liters per minute, calculate the extend speed and pressure required for each stage of the cylinder when tilting a fully laden lorry.

(i) First-stage

First-stage diameter = 100 mm

First-stage speed = $\dfrac{\text{Quantity flowing}}{\text{Area}}$

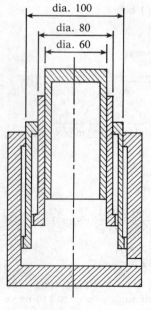

Figure 4.4

$$= \frac{10 \times 10^{-3}}{(\pi/4) \times 0.1^2} \left(\frac{m^3}{min \times m^2}\right)$$

$$= 4/\pi$$
$$= 1.27 \text{ m/min}$$

First-stage pressure $= \dfrac{\text{Load}}{\text{Area}}$

$$= \frac{4000 \times 9.81}{(\pi/4) \times 0.1^2} \left(\frac{N}{m^2}\right)$$

$$= 5 \times 10^6 \text{ (N/m}^2\text{)}$$
$$= 50 \text{ bar}$$

(ii) Second stage

Second-stage diameter $= 80$ mm

Second-stage speed $= \dfrac{\text{Quantity flowing}}{\text{Area}}$

$$= \frac{10 \times 10^{-3}}{(\pi/4) \times 0.08^2} \left(\frac{m^3}{min \times m^2}\right)$$

$$= 1.99 \text{ m/min}$$

Second-stage pressure $= \dfrac{\text{Load}}{\text{Area}}$

$$= \frac{4000 \times 9.81}{(\pi/4) \times 0.08^2} \text{ (N/m}^2\text{)}$$

$$= 7.81 \times 10^6$$
$$= 78.1 \text{ bar}$$

(iii) Third stage

Third-stage diameter $= 60$ mm

Third-stage speed $= \dfrac{\text{Quantity flowing}}{\text{Area}}$

$$= \frac{10 \times 10^{-3}}{(\pi/4) \times 0.06^2} \left(\frac{m^3}{min \times m^2}\right)$$

$$= 3.54 \text{ m/min}$$

Third-stage pressure $= \dfrac{\text{Load}}{\text{Area}}$

$$= \frac{4000 \times 9.81}{(\pi/4) \times 0.06^2} \text{ (N/m}^2\text{)}$$

$$= 13.9 \times 10^6$$
$$= 139 \text{ bar}$$

Telescopic cylinders are made in a standard range for vehicle applications. Although non-standard cylinders can be obtained, they tend to be very expensive if ordered singly.

4.1.2 Single-acting cylinders

These can be powered in one direction only (either extend or retract) by hydraulic forces; the return movement is brought about by either a spring built into the cylinder or an external force (Figure 4.5).

In order to produce a good seal at the piston, the cylinder barrel has to be machined to a high-quality surface finish (usually by honing) and elastomer seals or metal rings fitted to the piston. The face of the piston not acted on by the fluid pressure must be drained to tank or atmosphere to prevent a build up of fluid which has leaked across the piston seal. If the drain port is blocked, the gradual leakage of fluid may reduce the length of cylinder stroke.

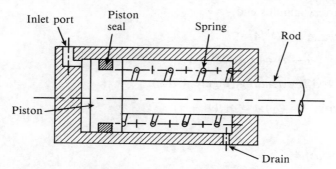

Figure 4.5 Single-acting cylinder.

4.1.3 Double-acting cylinders

These are hydraulically powered in both directions by applying fluid pressure to the appropriate side of the piston. The cylinder consists of a honed barrel or tube with end caps which may be welded to the tube, screwed on, or held in place by tie rods. At least one end cap will contain a bearing, seal and wiper ring to suit the piston rod. A double-acting cylinder is shown diagrammatically in Figure 4.6.

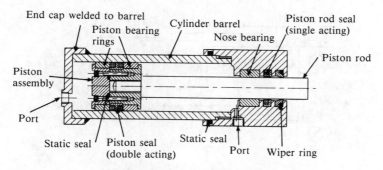

Figure 4.6 Double-acting cylinder.

Speed of a double-acting cylinder

Consider Figure 4.7.
  Let

  D = piston diameter
  d = rod diameter
  A = full bore area
  a = piston rod area
  Q_E = flow into full bore end of cylinder when extending
  q_E = flow from annulus end of cylinder when extending
  V = extend velocity of cylinder
  v = retract velocity of cylinder
  P_1 = pressure at full bore end
  P_2 = pressure at annulus end.

The full bore area $A = \pi D^2/4$

  Annulus area $(A - a) = (\pi/4)(D^2 - d^2)$

(i) When piston rod is extending (Figure 4.7(a)):

  Piston velocity, $V = Q_E/A = q_E/(A - a)$

and thus

$$q_E = Q_E(A - a)/A$$

Thus as the piston rod is extending, the flow rate of fluid leaving the cylinder is less than the flow rate entering.

(ii) When the piston rod is retracting (Figure 4.7(b)):

Let q_R be the flow into the annulus end of the cylinder and Q_R be the flow from the full bore end of the cylinder.
Thus

  Piston velocity, $V = q_R/(A - a)$

$$= \frac{Q_R}{A}$$

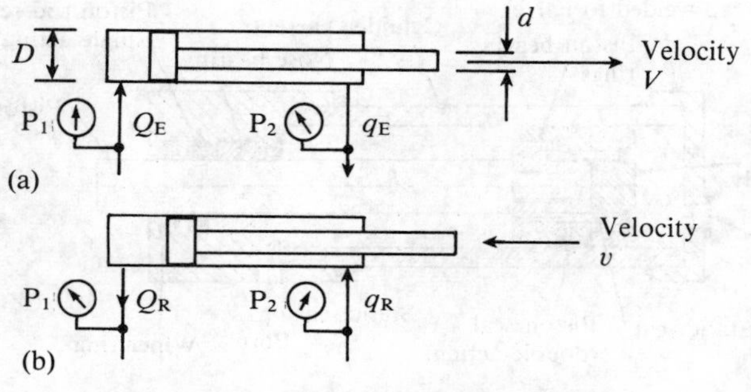

Figure 4.7 Cylinder. (a) Under extend conditions. (b) Under retract conditions.

or

$$Q_R = q_R \times A/(A-a)$$

Thus, when the piston rod is retracting the flow rate of fluid leaving the cylinder is greater than the flow rate entering.

EXAMPLE 4.3

A hydraulic cylinder has a bore of 200 mm and a piston rod diameter of 140 mm. For an extend speed of 5 m/min, calculate:

(a) the supply flow rate (Q_E),
(b) the flow rate from the annulus side on extend (q_E),
(c) the retract speed using Q_E, and
(d) the flow rate from the full bore end on retract (Q_R).

Solutions

(a) Flow rate of oil to extend cylinder at 5 m/min.

Q_E = Area of piston × Velocity
  $= (\pi/4) \times (200/1000)^2 \text{ (m}^2\text{)} \times 5/60 \text{ (m/s)}$
  $= 0.00262 \text{ m}^3/\text{s}$
  $= 0.00262 \times 60 \times 1000 \text{ (l/min)}$
  $= 157 \text{ l/min}$

(b) Flow of oil leaving cylinder q_E is given by

q_E = Annulus area × Velocity
  $= (\pi/4) \times [(200/1000)^2 - (140/1000)^2] \times (5/60) \text{ (m}^3/\text{s)}$
  $= 0.00134 \text{ m}^3/\text{s}$
  $= 80 \text{ l/min}$

(c) The same fluid flow rate used to extend the cylinder (157 liters/min) is used to retract the cylinder. Retract cylinder velocity V is given by

$$V = \frac{Q_E}{(A-a)}$$

where

$(A - a)$ = annulus area
  $= (\pi/4)[(200/1000)^2 - (140/1000)^2]$
  $= 0.01602 \text{ m}^2$

and

$Q_E = 157 \text{ l/min}$
  $= 0.00262 \text{ m}^3/\text{s}$

$$V = \frac{0.00262}{0.01602} \left(\frac{\text{m}^3/\text{s}}{\text{m}^2}\right)$$

  $= 0.164 \text{ m/s}$
  $= 9.8 \text{ m/min}$

(d) Flow from full bore end of cylinder Q_R is given by

$Q_R = A \times V$ where A is the full bore area = 0.03142 m^2

$= 0.03142 \times 0.164$ (m$^2 \times$ m/s)
$= 0.00515$ (m^3/s)
$= 309$ l/min

Note This flow is almost twice the flow rate for the extend stroke and must be allowed for when sizing components. For very fast stroking cylinders it is sometimes necessary to have oversize ports.

Cylinder thrust

STATIC

The static thrust developed by a hydraulic cylinder is the product of the pressure and area. Consider the cylinder as shown in Figures 4.7.

$$\text{Nett forward thrust} = P_1\left(\frac{\pi D^2}{4}\right) - P_2\left(\frac{\pi D^2}{4} - \frac{\pi d^2}{4}\right)$$

$$= \left(\frac{\pi}{4}\right) \times [P_1 D^2 - P_2(D^2 - d^2)]$$

$$\text{Nett retract thrust} = P_2\left(\frac{\pi D^2}{4} - \frac{\pi d^2}{4}\right) - P_1\left(\frac{\pi D^2}{4}\right)$$

$$= \frac{\pi}{4} \times [P_2(D^2 - d^2) - P_1 D^2]$$

DYNAMIC

In dynamic applications the load inertia, seal friction, load friction, etc. must be allowed for in calculating the dynamic thrust.

As a first approximation, the dynamic thrust can be taken as 0.9 times the static thrust. (It must be realized that this is only an approximation and can be considerably in error, dependent on load conditions and associated circuitry.)

Cylinder seal friction varies with seal and cylinder design. The pressure required to overcome seal friction is not readily available from the majority of cylinder manufacturers. The seal friction breakout pressure can be taken as 5 bar for calculation purposes. It will reduce when the piston starts to move. The pressure required to overcome seal friction will reduce as the cylinder bore size increases and will vary according to seal design.

EXAMPLE 4.4

If the maximum pressure applied to the cylinder in Example 4.3 is 100 bar, calculate the

(a) dynamic extend thrust, and
(b) dynamic retract thrust

assuming that dynamic thrust = 0.9 × static thrust.

(a) Full bore area $= \dfrac{\pi \times 0.2^2}{4}$

$= 0.0314 \text{ m}^2$

Dynamic extend thrust $= 0.9 \times \text{Pressure} \times \text{Full bore area}$

$= 0.9 \times 100 \times 10^5 \times 0.0314 \left(\dfrac{\text{N}}{\text{m}^2} \times \text{m}^2\right)$

$= 283 \text{ kN}$

(b) Annulus area $= (\pi/4)(0.2^2 - 0.14^2)$

$= 0.016 \text{ m}^2$

Dynamic retract thrust $= 0.9 \times 100 \times 10^5 \times 0.016 \left(\dfrac{\text{N}}{\text{m}^2} \times \text{m}^2\right)$

$= 144 \text{ kN}$

Regenerative circuits

If the cylinder is connected as shown in Figure 4.8, the pressure on both sides of the piston will be the same. However, since the full bore area is larger than the annulus area, a nett force will cause the piston rod to extend.

The flow q from the annulus end combines with the pump delivery Q and is fed into the full bore of the cylinder. If the extend velocity of the piston rod is V:

Consider annulus end

$$q = \left(\dfrac{\pi}{4}\right)(D^2 - d^2)V \tag{4.1}$$

For full bore end

$$Q + q = \left(\dfrac{\pi}{4}\right)D^2 V \tag{4.2}$$

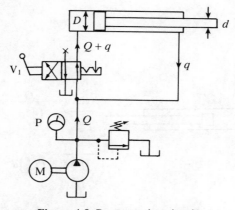

Figure 4.8 Regenerative circuit.

From equations (4.1) and (4.2)

$$Q = \pi \frac{d^2}{4} V$$

So

$$V = \frac{4Q}{\pi d^2}$$

$$= \frac{Q}{a}$$

where a = area of piston rod.

When the cylinder is being extended regeneratively, the pump delivery is effectively filling the volume vacated by the rod. The cylinder acts in a manner similar to a displacement cylinder (Figure 4.9). The nett forward thrust is the difference of the forces on the full bore and annulus sides of the piston.

$$\text{Forward thrust} = P\left(\frac{\pi D^2}{4}\right) - (P\pi/4)(D^2 - d^2)$$

$$= P(\pi/4)d^2$$

Thus the forward thrust is the product of the pressure and the piston rod area. This is similar to the forward thrust of a displacement cylinder.

The flow through the directional control valve on extend is $(Q + q)$ and the valve must be of sufficient size to carry this flow without malfunction.

When the directional control valve V_1 in Figure 4.8 is operated to the 'crossover' state, the cylinder will retract conventionally as the annulus end receives flow from the pump output and the full bore end is connected to tank.

Piston rod retract velocity is

$$\frac{\text{Pump delivery}}{\text{Annulus area}}$$

$$= \frac{Q}{(\pi/4)(D^2 - d^2)}$$

Retract thrust is

$$P(\pi/4)(D^2 - d^2)$$

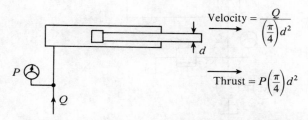

Figure 4.9 Equivalent displacement cylinder circuit.

> **EXAMPLE 4.5**
>
> The hydraulic cylinder used in Examples 4.3 and 4.4 which has a bore of 200 mm diameter and a rod of 140 mm diameter, is connected regeneratively as shown in Figure 4.8. (i) If the same flow rate of 157 l/min is used, calculate the extend speed. (ii) If the maximum system pressure is 100 bar, calculate the dynamic extend thrust.
>
> (i)
> $$\text{Piston rod area} = \frac{\pi \times 0.14^2}{4} = 0.0154 \text{ m}^2$$
>
> $$\text{Extend speed} = \frac{\text{Flow rate}}{\text{Piston rod area}}$$
>
> $$= \frac{157 \times 10^{-3}}{0.0154} \left(\frac{\text{liters}}{\text{min}} \times \frac{\text{m}^3}{\text{liter m}^2} \right)$$
>
> $$= 10.2 \text{ m/min}$$
>
> This compares with 5 m/min when connected conventionally (Example 4.3).
>
> (ii) For a regenerative system
> Dynamic extend thrust = 0.9 pressure × rod area
>
> $$= 0.9 \times 100 \times 10^5 \times 0.0154 \left(\frac{\text{N}}{\text{m}^2} \times \text{m}^2 \right)$$
>
> $$= 138.6 \text{ kN}$$
>
> This compares with a dynamic forward thrust of 283 kN as calculated in Example 4.4. As the area of the annulus is almost equal to that of the rod, the regenerative extend and conventional retract thrusts and speeds are almost the same.

Through rod cylinders

These are similar in construction to the standard double-acting cylinder, but have a cylinder rod extending through both cylinder end caps. It is possible to have a different diameter of piston rod at each end of the cylinder, but more common to have the same diameter. The main applications of through rod cylinders are when the same speed is required in both directions, when both ends of the rod can be utilized to do work, or where the non-working end is used to indicate or signal the position of the load. In some applications the rod is fixed at both ends and the cylinder body carrying the load moves on the rod.

A major problem in the manufacture of through rod cylinders is achieving correct alignment and concentricity of cylinder bore, piston, end caps and rods. Any misalignment will result in excessive seal wear and premature cylinder failure.

Standard metric cylinders

BS:5785 1980 gives tables of preferred sizes for the cylinder bore and rod diameter of metric cylinders. Most cylinder manufacturers have based their standard range of metric

Table 4.1 Recommended cylinder bore and rod sizes.

Piston Diameter (mm)		40	50	63	80	100	125	140	160	180	200	220	250	280	320
Piston rod Diameter (mm)	Small		20	28	36	45	56	70	90	100	110	125	140	160	180	200
	Large		28	36	45	56	70	90	100	110	125	140	160	180	200	220

cylinders on these recommendations, offering two rod sizes for each cylinder bore (Table 4.1).

A number of combinations have a piston rod to piston diameter ratio in the region of 0.7, which gives an annulus area of approximately one-half the full bore area. This area ratio is of use in regenerative circuits to give similar values of speed and thrust on both the extend and retract strokes.

4.1.4 Acceleration and deceleration of cylinder loads

Acceleration

To calculate the acceleration of cylinder loads, the equations of motion must be understood.

Let

u = initial velocity
v = velocity after a time t
s = distance moved during time t
a = acceleration during time t.

The standard equations of motion are:

$v = u + at$
$v^2 = u^2 + 2as$
$s = ut + \frac{1}{2}at^2$

and

$s = \frac{1}{2}(u + v)t$

The force F to accelerate a weight W horizontally with an acceleration a is given by

Force = Mass × Acceleration

or

$F = (W/g)a$

where g is the acceleration due to gravity and is 9.81 m/s^2. The force P required to overcome friction is given by $P = \mu W$, where μ is the coefficient of friction.

EXAMPLE 4.6

A mass of 2000 kg is to be accelerated horizontally up to a velocity of 1 m/s from rest over a distance of 50 mm. The coefficient of friction between the load and the guides is 0.15. Calculate the bore of the cylinder required to accelerate this load if the maximum allowable pressure at the full bore end is 100 bar. (Take seal friction to be equivalent to a pressure drop of 5 bar. Assume the back pressure at the annulus end of the cylinder is zero.)

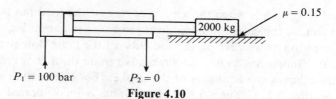

$P_1 = 100$ bar $P_2 = 0$

Figure 4.10

In this case $u = 0$, $v = 1$ m/s, $s = 0.05$ m and a is unknown.
Using equation $v^2 = u^2 + 2as$
$$1^2 = 0^2 + 2a \times 0.05$$
$$1 = 0.1a$$
$$a = 10 \text{ m/s}^2$$

Force to accelerate load is given by

$$F = (W/g)a$$

where $W = 2000 \times 9.81$ N.
Therefore,

$$F = \frac{2000}{9.81} \times 9.81 \times 10 = 20\,000 \text{ N}.$$

Force P to overcome load friction is given by

$$P = \mu W$$
$$= 0.15 \times 2000 \times 9.81 = 2943 \text{ N}.$$

Total force to accelerate load and overcome friction is $(F + P) = (20\,000 + 2943) = 22\,943$ N.

Cylinder area required for a given thrust is calculated from

Thrust = Area × Pressure

The pressure available is pressure at full bore end of the cylinder less the equivalent seal break out pressure.

Pressure available = $(100 - 5) = 95$ bar = 95×10^5 N/m²

$$\text{Area} = \frac{22\,943}{95 \times 10^5} \left(\frac{\text{N}}{\text{N/m}^2} \right)$$
$$= 0.002415 \text{ (m}^2\text{)}$$
$$= 2415 \text{ mm}^2$$
$$= \frac{\pi D^2}{4}$$

where D is the cylinder bore diameter.

So

$$D = [(4/\pi) \times 2415]^{1/2} = 55.4 \text{ mm}$$

The cylinder diameter is thus 55.4 mm. This neglects the affect of any back-pressure. The nearest standard cylinder above has a 63-mm diameter bore.

Load deceleration-cylinder cushioning

Cushions are fitted to cylinders when the kinetic energy of the load has to be absorbed within the component. A typical cushioning arrangement is shown in Figure 4.11.

A spike on the piston or a sleeve on the rod cuts off the main flow path of the fluid leaving the cylinder. Thus it has to find an alternative route through a restrictor, which 'meters out' the flow during the final part of the stroke. (For explanation of 'meter-out' flow control see Section 3.2, Chapter 3.) A check valve is incorporated to bypass the cushion restrictor for the return stroke.

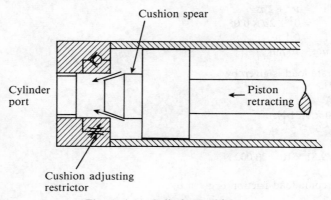

Figure 4.11 Cylinder cushions.

CUSHIONING PRESSURE

During deceleration extremely high pressure may develop within a cylinder cushion.

The action of the cushioning device is to set up a back-pressure to decelerate the load.

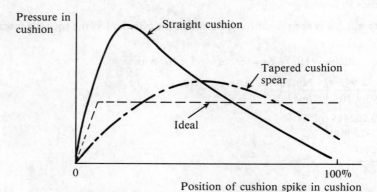

Figure 4.12 Pressure distribution in cushion.

Section 4.1 *Hydraulic cylinders* 143

Ideally the back-pressure will be constant over the entire cushioning length to give a progressive load deceleration. In practice the cushion pressure is highest when the piston rod has just entered the cushion (Figure 4.12).

Some manufacturers have improved the performance of their cushioning devices by using a tapered or stepped cushion spear.

EXAMPLE 4.7

A cylinder has a bore of 125-mm diameter and a rod of 70-mm diameter. It drives a load of 2000 kg vertically up and down at a maximum velocity of 3 m/s. The lift speed is set by adjusting the pump displacement and the retract speed by a flow-control valve. The load is slowed down to rest in the cushion length of 50 mm. If the relief valve is set at 140 bar, determine the average pressure in the cushions on extend and retract. (Neglect pressure drops in pipework and valves.)

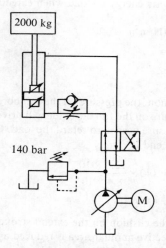

Figure 4.13

Kinetic energy = $\frac{1}{2}$ Mass × Velocity
Kinetic energy of load = $\frac{1}{2} MV^2$
$= \frac{1}{2}(2000) \times 3^2 = 9000$ Nm

Average force to retard load over 50 mm is

$\dfrac{\text{Kinetic energy}}{\text{Distance}}$

$= \dfrac{9000 \times 10^3}{50} = 180$ kN

The forces acting on the load will be as shown in Figure 4.14.

Load = 2000 kg = 2000 × 9.81 = 19.6 kN.

Annulus area = $\left(\dfrac{\pi}{4}\right)(0.125^2 - 0.07^2) = 0.0084$ m^2 = 8.4×10^{-3} m^2

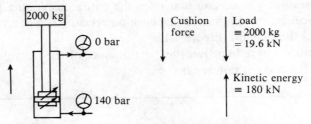

Figure 4.14

Full bore area $= \left(\dfrac{\pi}{4}\right)(0.125^2) = 0.0123 \text{ m}^2 = 12.3 \times 10^{-3} \text{ m}^2$

The kinetic energy of the load is opposed by the cushion force and the action of gravity on the load (see Figure 4.14).

Cushion pressure to absorb kinetic energy of load when extending is

$$\dfrac{(180 \times 10^3) - (19.6 \times 10^3)}{(8.4 \times 10^{-3})} \text{ (N/m}^2\text{)}$$

$= 19.1 \times 10^6 \text{(N/m}^2\text{)}$

$= 191$ bar

When the piston enters the cushion, the pressure on the full bore side of the piston will rise to relief valve pressure. This pressure on the full bore side will drive the piston into the cushions, and so increase the cushion pressure needed to retard the load. Cushion pressure to overcome hydraulic pressure on full bore end is

$$\text{Pressure} \times \dfrac{\text{Full bore area}}{\text{Annulus area}} = 140 \times \dfrac{12.3 \times 10^{-3}}{8.4 \times 10^{-3}}$$

$= 205$ bar

Thus, the average pressure in the cushion on the extend stroke is $(190 + 205) = 395$ bar.

During cushioning, the effective annular area is reduced as the cushion sleeve enters the cushion. This has been neglected in the calculation and in practice the cushion pressure will be even greater.

When the load is retracted, forces acting on the load will be as shown in Figure 4.15. The back-pressure owing to the flow control valve in the circuit will be minimal once the piston enters the cushion and will be neglected in this calculation.

The force in the cushion has to overcome the kinetic energy of the load, the weight of the

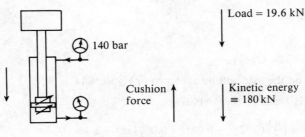

Figure 4.15

Section 4.1 Hydraulic cylinders

load and the force due to the hydraulic pressure:

Force owing to hydraulic pressure is Pressure × Annulus area

$$= (140 \times 10^5) \times (8.4 \times 10^{-3}) \text{ (N)}$$
$$= 117.6 \text{ kN}$$

Cushion force $= (180 + 19.6 + 117.6)$
$= 317.2$ kN

Cushion pressure is Force/Area

$$= \frac{317.2}{0.0123} \text{ (kN/m}^2\text{)}$$

$$= 25\,800 \text{ (kN/m}^2\text{)}$$

$$= 258 \text{ bar}$$

Average pressure in the cushion retracting will be 258 bar. Again this value will be somewhat higher as the cushion spike reduces the effective cushion area below that used.

Where high inertia loads are encountered the cylinder internal cushions may be inadequate but it is possible for the load to be retarded by switching in external flow controls (see Section 3.2). Deceleration can then take place over a greater part of the actuator stroke.

Cylinder maximum speeds

The maximum speed of a piston rod is limited by the rate of fluid flow into and out of the cylinder and the ability of the cylinder to withstand the impact forces which occur when the piston movement is arrested by the cylinder end plate.

In an uncushioned cylinder it is normal to limit maximum piston velocity to 8 m/min. This value is increased to 12 m/min for a cushioned cylinder, and 30 m/min is permissible with high-speed or externally-cushioned cylinders. Oversize ports are necessary in cylinders used in high-speed applications.

In all cases the maximum speed depends upon the size and type of load and it is prudent to consult the manufacturer if speeds above 12 m/min are contemplated.

When only a part of the cylinder stroke is utilized, the cushions cannot be used to decelerate the load. In such cases it may be necessary to introduce some form of external cushioning especially where high loads or precise positioning is involved.

Operating temperature

The maximum operating temperature should not exceed 80°C otherwise elastomer seals will rapidly deteriorate. In some applications it may be necessary to use heat shields to protect the cylinder from external heat sources such as furnaces, etc. At temperatures above 50°C rapid deterioration of mineral oil occurs. Problems may also be encountered in low-temperature applications.

The operating temperature range can be extended by substituting metal piston rings for elastomer seals.

4.1.5 Cylinder mountings and strength calculations

Cylinder mountings

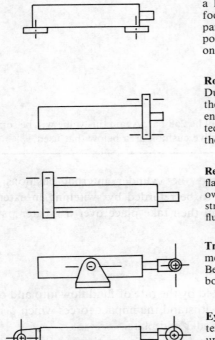

Foot mounting should be designed to give a limited amount of movement on one foot only to allow for thermal or load expansion, i.e. the cylinder should be positively located or dowelled at one end only.

Rod end flange or front flange mounting. During the extend stroke the pressure in the hydraulic fluid acts on the cylinder end cap, the force set up being transmitted to the front mounting flange through the cylinder body.

Rear flange, back flange or head end flange mounting. No stress in cylinder owing to load on extend stroke, only hoop stress present. The load acts through the fluid onto the rear flange.

Trunnion mounting allows angular movement. Designed to take shear loads only. Bearings should be as close to cylinder body as possible.

Eye or clevis mountings. There is a tendency for the cylinder to jack-knife under load. Side loading of bearings must be carefully considered.

Figure 4.16 Cylinder mountings.

Piston rod ends

The piston rod ends can be supplied with a male or female thread according to the manufacturer's specification. Rod end eyes with spherical bearings are available from some suppliers.

Protective covers

These are fitted to protect the piston rod when the piston is working in an abrasive environment or when the cylinder is not used for long periods and a heavy deposit of dust could accumulate on the rod. The protective covers are of telescopic or bellows form and completely enclose the rod at all times in the cylinder movement.

Bellows may either be moulded or fabricated. Moulded bellows are manufactured from rubber or plastic and owing to their construction are limited to a contraction ratio of about 4:1. An extended piston rod is required to accommodate the closed length of the

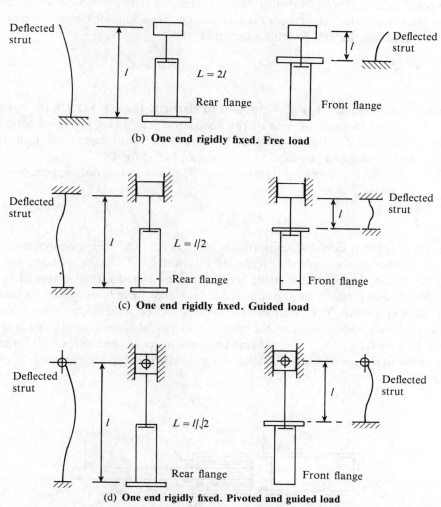

Figure 4.17 Relationship between piston rod, free buckling length (*L*) and method of fixing.

bellows. This increases the overall cylinder length and tends to restrict their use to relatively short stroke cylinders.

Fabric covers made of plastic, leather, impregnated cloth or canvas can have a contraction ratio of greater than 15:1. When a fabricated cover is used on a horizontal cylinder it must be supported externally to prevent the cylinder rod rubbing the cover.

Telescopic covers are made of a rigid material normally metal and are used in conditions where fabric covers are inadequate.

Piston rod buckling

The piston rod in a hydraulic cylinder will act as a strut when it is subjected to a compressive load or it exerts a thrust. Therefore the rod must be of sufficient diameter to prevent buckling. Euler's strut theory is used to calculate a suitable piston rod diameter to withstand buckling. Euler's formula states that

$$K = \frac{\pi^2 E J}{L^2}$$

where K = buckling load (kg), E = modulus of elasticity (kg/cm^2) (2.1×10^6 kg/cm^2 for steel), J = second moment of area of the piston rod (cm^4) ($\pi d^4/64$ for a solid rod of diameter d cm), and L = free (equivalent) buckling length (cm) depending on the method of fixing the cylinder and piston rod and is shown in Figure 4.17.

The maximum safe working thrust or load F on the piston rod is given by

$$F = \frac{K}{S}$$

where S is a factor of safety which is usually taken as 3.5. The free or equivalent buckling length L depends on the method of fixing the piston rod end and the cylinder, and on the maximum distance between the fixing points, i.e. the cylinder fully extended. In cases where the cylinder is rigidly fixed or pivoted at both ends there is a possibility of excessive side loading occurring. The effect of side loading can be reduced by using a stop tube inside the cylinder body to increase the minimum distance between the nose and the piston bearings (Figure 4.18). The longer the stop tube, the lower will be the reaction force on the piston owing to the given value of side load. Obviously the stop tube reduces the effective cylinder stroke.

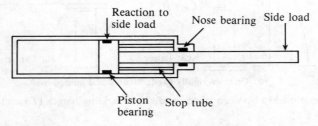

Figure 4.18 Use of stop tube to reduce side loading.

EXAMPLE 4.8

A regenerative circuit for an upstroking press is shown in Figure 4.19 together with the required load speed characteristics and mounting details. The press cylinder has to exert a force of 7 tonnes to lift the crosshead and tooling. When it closes, system pressure will increase and operate a pressure switch to change the circuit from regenerative to conventional operation. The pressure switch is set to operate at a pressure of 20% greater than that needed to move the crosshead and tooling. Maximum thrust required from the press is 20 tonnes with a stroke of 1.7 m. Determine a suitable standard cylinder (selecting one from Table 4.1), the pump delivery and the setting of the pressure switch. The system working pressure should not exceed 250 bar.

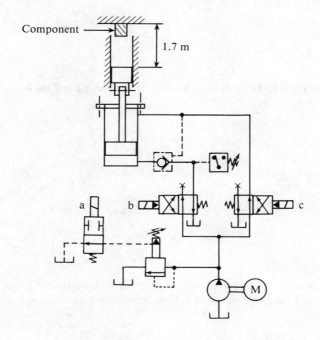

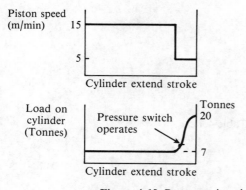

Operation	Solenoid a	b	c	Pressure switch
Pump unloaded	0	0	0	0
Regenerative extend	1	1	0	0
Conventional extend	1	1	1	1
Retract	1	0	0	0

1 = Energized
0 = De-energized

Figure 4.19 Regenerative circuit for upstroking press.

Solution

Piston rod diameter

The first step in the solution is to determine the minimum piston rod diameter for buckling strength.

$$\text{Buckling load, } K = \frac{\pi^2 E J}{L^2} \quad (4.3)$$

where $K = 20$ tonnes, (20 000 kg), $E = 2.1 \times 10^6$ kg/cm^2 and $J = (\pi d^4/64)$; d = rod diameter (cm) and L = free buckling length.

The cylinder is rigidly fixed by a front flange and the load pivoted and fully guided.

$$L = \frac{\text{Cylinder stroke}}{\sqrt{2}} \text{ (see Figure 4.17)}$$
$$= \frac{1.7}{\sqrt{2}}$$
$$= 1.2 \text{ m}$$
$$= 120 \text{ cm}$$

The value of E is given in kg/cm^2, so it is easier to work in kg and cm units than convert to newtons and metres.

From equation (4.3)

$$d^4 = \frac{64 \times L^2 \times K}{\pi^3 \times E}$$
$$= \frac{64 \times 120^2 \times 20\,000}{\pi^3 \times 2.1 \times 10^6}$$
$$= 283 \text{ cm}^4$$

Therefore

$$d = 4.1 \text{ cm}$$
$$= 41 \text{ mm}$$

This does not include any factor of safety which is usually taken as between 3 and 4. Recalculating using a factor of safety of 3.5, the maximum safe working load is $K/3.5 = 20$ tonnes. Therefore,

$$K = 3.5 \times 20 \text{ tonnes}$$
$$d^4 = \frac{64 \times 120^2 \times 20\,000 \times 3.5}{\pi^3 \times 2.1 \times 10^6}$$
$$= 283 \times 3.5$$
$$= 991 \text{ cm}^4$$

Therefore

$$d = 56 \text{ mm}$$

This is a standard size to BS5785: 1980 (see Table 4.1).

Cylinder bore

Maximum thrust required = 20 tonnes
Maximum allowable pressure = 250 bar
Assume dynamic thrust = 0.9 pressure × area

Section 4.1 Hydraulic cylinders

Piston area (A) is

$$\frac{20 \times 10^3 \times 9.81}{250 \times 10^5 \times 0.9} \left(\frac{Nm^2}{N}\right)$$

$$= 0.00872 \text{ m}^2$$

$$A = \frac{\pi d^2}{4} = 0.00872 \text{ m}^2$$

Therefore,

$$d = [0.00872 \times (4/\pi)]^{1/2}$$

The minimum piston diameter = 0.105 m = 105 mm.

From Table 4.1, the smallest suitable piston diameter is 125 mm with corresponding 70-mm diameter piston rod. The system pressure to give a dynamic thrust of 20 tonnes is given by

$$\text{Pressure} = \frac{\text{Thrust}}{\text{Area}} \times \left(\frac{1}{0.9}\right)$$

$$= \frac{20\,000 \times 9.81}{(\pi/4) \times 0.125^2} \times \left(\frac{1}{0.9}\right) \text{ (N/m}^2\text{)}$$

$$= 177.7 \text{ bar}$$

Extend force of 7 tonnes under regenerative conditions is equal to the maximum allowable pressure × piston rod area. Then, taking the dynamic thrust = (0.9 × static thrust), the required pressure is

(Load/Piston rod area) × 1/0.9

$$= \frac{7000 \times 9.81}{(\pi/4) \times 0.07^2} \times 1/0.9$$

$$= 198.4 \text{ bar}$$

The pressure switch is set to operate when the system pressure is 20% greater than that required to give a thrust of 7 tonnes. The setting of pressure switch = 198.4 × 1.2 = 238 bar. The fact that this pressure is higher than that used during pressing must be recognized and catered for by the electrical control circuit.

Flow rate required during regenerative extend is the area of the piston rod times the velocity, i.e.

$$\text{Flow for regenerative extend} = \frac{\pi \times 0.07^2}{4} \times \frac{15}{60} \text{ (m}^3\text{/s)}$$

$$= \frac{\pi \times 0.07^2}{4} \times \frac{15}{60} \times 60 \times 1000 \text{ (l/min)}$$

$$= 57.7 \text{ l/min}$$

Flow required during normal extend is the product piston area × piston velocity

$$\text{Flow} = \pi \times \frac{0.125^2}{4} \times \frac{5}{60} \text{ (m}^3\text{/s)}$$

$$= \left(\frac{\pi}{4}\right) \times 0.125^2 \times \left(\frac{5}{60}\right) \times 60 \times 1000 \text{ (l/min)}$$

$$= 61.3 \text{ l/min}$$

A pump with a delivery in excess of 61.3 l/min will give the required piston speeds.

4.2 SEMI-ROTARY ACTUATORS

These are devices used to convert fluid pressure energy into a torque which turns through an angle limited by the design of the actuator. With the majority of designs the angle of rotation is limited to 360° although it is possible to considerably exceed this when using piston-operated actuators.

4.2.1 Vane-type actuators

A vane-type semi-rotary actuator consists of one or two vanes connected to an output shaft which rotates when hydraulic pressure is applied to one side of the vanes. A single-vane unit is limited to approximately 320° rotation and a double-vane unit to approximately 150°.

There will always be some internal fluid leakage across the vanes and this increases with operating pressure and as the viscosity of the working fluid decreases. Internal leakage can cause problems where smooth speed control of the rotary motion is required. For all applications of vane-type actuators the manufacturer's recommendations regarding operating pressure and type of fluid must be followed. The maximum torque obtainable from currently available single-vane units is approximately 40×10^3 Nm and for double-vane units 80×10^3 Nm.

The principle of single- and double-vane semi-rotary actuators is shown diagrammatically in Figure 4.20. In some designs of double-vane actuators, cross-drillings

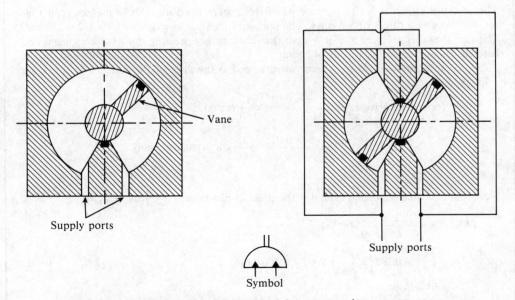

Figure 4.20 Single (left) and double (right) vane semi-rotary actuators.

Section 4.2 Semi-rotary actuators

through the shaft connect opposite sides of the vanes thus eliminating the need for the external pipework shown in the figure.

4.2.2 Piston-type actuators

A hydraulic cylinder is used to provide linear motion which is converted by a mechanism into angular motion. In several commercial designs the cylinder drives a rack and pinion gear and frequently the rack is an integral part of the piston rod (Figure 4.21). The angle of rotation depends upon the stroke of the cylinder and rack and the pitch circle diameter of the pinion. Although several complete revolutions are possible the majority of commercial units have a 360° angular movement. Some small amount of precise adjustment of the start and finish of the stroke by means of internal stops is normally available. Cushions may be incorporated for controlled deceleration at the end of the movement.

Where several revolutions are required a double-acting cylinder connected to an external rack-and-pinion mechanism often proves a simple solution. For fast speeds and high inertia loads rotated through precise angles, external adjustable stops incorporating cushions or some form of cushion circuitry are advizable to reduce shock loading.

Hydraulic fluid should not leak past the piston in this design, so the unit can be hydraulically locked in any position and may be operated at very slow speeds. The output torque available from rack and pinion type actuators is in excess of 800×10^3 Nm at a pressure of 210 bar.

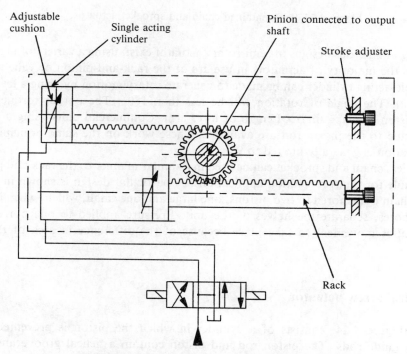

Figure 4.21 Rack and pinion semi-rotary actuator.

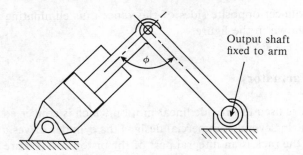

Figure 4.22 Lever arm.

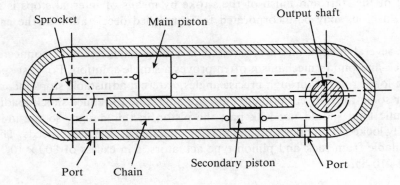

Figure 4.23 Self-contained chain and sprocket actuator.

Although numerous designs of semi-rotary actuator exist, using a variety of ingenious mechanisms the majority of actuators in use are of the rack-and-pinion or vane type.

A double-acting cylinder can be made to generate rotary motion by using a lever arm (Figure 4.22). The angle of rotation will be less than 180°. The output torque is the product: Piston thrust × sin ϕ × Length of lever arm. Commercial units using a slotted link connected to the piston rod are available. These work on the same principle, and angular rotation is generally limited to 90°.

An endless chain and sprocket can be used in a similar manner to the rack and pinion, and is suitable for multi-revolution applications. A particular design is shown in Figure 4.23. The chain is anchored to two pistons, one large and one small, which, when in their respective bores, separate the halves of the unit. Pressure applied to one port of the actuator causes movement to take place because of the difference in area of the two pistons.

4.2.3 Helical screw actuator

This type (Figure 4.24) consists of a cylinder in which the piston is prevented from rotating by guide rods. The piston rod and piston contain a helical groove and mate together in a manner analogous to a screw and nut. As the piston is driven along the barrel

Section 4.3 Hydraulic motors

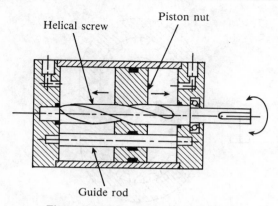

Figure 4.24 Helical screw actuator.

it causes the rod to rotate. Because of the difficulty in providing a hydraulic seal between the piston and rod, the design is limited to low-pressure applications. A particular feature is the ability to hold a position resisting rotation by external torques when both sides of the piston are exhausted. Angular rotation in excess of 360° is possible.

4.2.4 Control of semi-rotary actuators

Semi-rotary actuators are used to move objects through a controlled angle such as opening large butterfly valves in pipe lines and to bend and form tubes and bars. They are controlled by valves in a way similar to that in which linear actuators may be controlled, so far as torque, speed and direction of rotation are concerned. The arc of travel can be limited by the use of external or (in some designs) internal mechanical stops, with cushioning if required.

4.3 HYDRAULIC MOTORS

These are similar in design to hydraulic pumps and can be considered to fall into two basic categories:

(a) Those using a generated form for the driving element, such as gear, vane, gerotor, etc.
(b) Those using a piston or series of pistons as the driving element, such as axial piston, and radial piston.

4.3.1 Generated form types

These tend to be two-dimensional generated forms which can be cut to various widths to give different swept displacements. In the sectioned drawing of a hydraulic motor shown

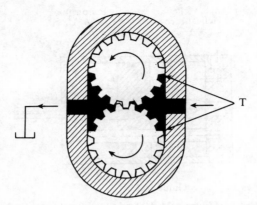

Figure 4.25 Generated form hydraulic motor – external gear type. Torque (T) is a function of pressure acting on one tooth on one gear (alternating between gears).

in Figure 4.25 the working element is a pair of meshing external gears. Once a particular size of working element has been designed, various capacities and hence various powers of motor can be simply produced using common end plates with the appropriate width of body and working element. With this kind of motor it is difficult to effect satisfactory sealing between the working element and the contacting surfaces on the end plates and housing. Leakage problems are always present in generated-form motors but are effectively reduced by operating the units at relatively low pressures – usually below 200 bar. To achieve sealing at even these pressures involves high interface forces and its attendant high friction. Thus generated-form motors have generally relatively low volumetric and mechanical efficiencies but with modern machining techniques volumetric efficiencies of precision units may exceed 95%.

Gear motors

Large motors are based on the gear-pump design with pressure-loaded side plates to give relatively good volumetric efficiency but high internal friction. Smaller units rely upon much closer clearances between the gears and side plates. The gears in some designs are centralized by the build up of hydrodynamic pressures between the gears and the side plates.

The torque developed in the gear motor is a result of the fluid pressure acting on the meshing teeth. The torque will vary depending upon instantaneous tooth-meshing position. The larger the number of teeth on the gear, the lower will be the torque variation, but there will be a lower volumetric displacement and thus less output power for a given size. The variation in torque can be as much as 20% depending upon the number of teeth and, together with the low inertia of the rotating element, limits the minimum speed for smooth running.

Minimum recommended gear motor speeds for smooth running vary between 400 and 1000 rev/min according to design and size of the unit. Slow speeds can only be attained from a gear motor by coupling it to a reduction gear but even this will not give a smooth starting torque.

Section 4.3 Hydraulic motors

External gear motors may be unidirectional or reversible. If a unidirectional motor is reversed there is likely to be damage to the shaft seal and the internal seals on the end plates.

Gear motors are used when relatively high speeds – typically up to 4000 rev/min – and low starting torques are needed, and when overall efficiency is not a critical factor. Their application is usually limited to an output power of about 10 kW but more powerful units are available.

Internal gear motors of the Gerotor type have a lower variation in output torque and thus are more suitable for low speed applications. A gerotor motor consists of an internal and external gear which intermesh.

In one type known as the 'orbit motor' the outer ring is fixed and the rotor orbits within the fixed ring. This is shown in Figure 4.26 which also shows the rotating valve distributing the fluid so that the pressurized section rotates with the orbiting of the rotor. The motion of the rotor is transmitted to the output shaft by a crowned spline coupling.

As pressurized fluid is fed in turn to the displacement chambers, first one and then another tooth goes fully into engagement. All the teeth will engage and disengage once while one tooth on the rotor moves from one tooth on the stator to the next. Meanwhile the rotor centre will have completed $\frac{6}{7}$ of an orbit and the output shaft $\frac{1}{7}$ of a revolution. So for one complete shaft revolution there will have been 42 tooth engagements and disengagements or power strokes. This gives a higher torque output and a lower smooth speed than a conventional gear motor. The smooth speed range claimed is from 10 to 2000 rev/min and the torque up to 300 Nm, dependent upon size. Gerotor motors are available with integral reduction gear boxes. These geared units can be operated at speeds below 1 rev/min with torques greater than 4000 Nm.

The distributor valve has two groups of ports situated alternately. One set feeds

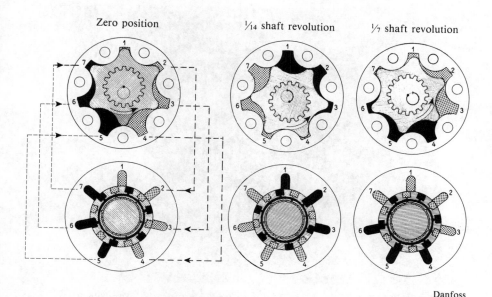

Danfoss

Figure 4.26 'Orbit Motor'.

pressure fluid to the appropriate displacement chambers while every alternate port provides a flow path for fluid being returned from the chambers not in use.

Note All gear motors, due to their design, are essentially fixed-capacity units. Their output speed can only be varied by controlling the fluid flow rate to the motor.

Vane motors

Conventional vane motors are similar in construction to the balanced vane pumps described in Chapter 2 but differ in that contact between the vanes and the eliptical cam ring is maintained by a coil or leaf spring. A vane pump cannot be supplied with pressure oil and made to act as a motor, although some vane motors can be driven and will perform as pumps. Torque is developed by the applied fluid pressure acting on the exposed surface of the vane. They are generally considered to be low-to-medium torque units (1600 Nm maximum) which perform best at speeds above 100 rev/min.

One special design (Figure 4.27) which gives a high torque (13 000 Nm maximum) has a smooth speed range of 0–150 rev/min and lower torque models operate up to 300 rev/min. In this the cam ring has four rises and the vanes reciprocate four times in each revolution. They differ from conventional vane motors in that pressure is only applied when the vanes are fully extended. The shaft is hydrostatistically balanced. Double rotor units are available as well as multiple displacement models, in which in the low-speed mode, fluid is ported into all the rises giving maximum torque capability. By using fewer of the displacement chambers, rotational speed can be increased for the same flow rate, but at the expense of torque.

Because of the high torque for small physical size this motor has been selected for a number of tunnelling and mining applications.

Dynex/Rivett

Figure 4.27 High torque vane motor.

Cam rotor motors

The principle, as can be seen in Figure 4.28 is similar to that of a conventional vane motor except that the vanes are in the stator and reciprocate against an elliptical rotor. Two rotating elliptical cams are staggered at 90° to each other, and rotate in a cylindrical housing. Each housing contains two vanes working on each cam, resulting in four separate pumping chambers with low flow/pressure pulsation. The shaft is hydrostatically balanced. The vanes are spring-loaded to facilitate start up and pressure oil on the back of the vanes keeps them in contact with the cam. The advantages are that it is quiet, with low pulsation and good starting torque. Maximum pressure is 175 bar continuous, 210 bar intermittent. The speed range is 50–3000 rev/min.

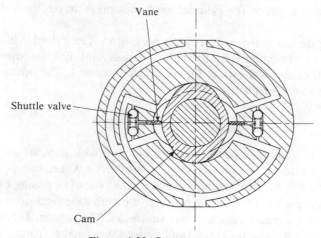

Figure 4.28 Cam rotor motor.

4.3.2 Piston-type motors

A circular piston running in a circular cylinder is capable of being simply manufactured to a very high degree of accuracy. Thus very fine clearances between the piston and the cylinder are achieved, giving low leakage rates and high volumetric efficiencies. The sealing efficiency of the piston and cylinder depends on the radial clearance, R_c, the length of the leakage flow path L (piston length), the working pressure P, and the viscosity of the fluid ν.

$$\text{Leakage} \propto \frac{R_c P}{L \nu}.$$

So piston motors can be designed to work at relatively high pressures, with leakage being reduced by increasing the length of the piston or increasing the viscosity of the oil.

In theory, all piston motors can be made to have a variable displacement by controlling the stroke of the pistons or varying the number of strokes per revolution. Controlling the piston stroke can give a stepless speed variation; varying the number of strokes per revolution will give stepped speed change and is less commonly used.

Axial piston motors

These are very similar to axial piston pumps and many manufacturers offer units with port plates, which can act as either pumps or motors. Both in-line and bent axis types are available. It must be noted that units with seated valves cannot have their functions reversed without altering the valving.

The in-line axial piston pump shown diagrammatically in Figure 2.6 in Chapter 2 will act as a motor if pressurized fluid is supplied to one of the ports and channelled to the appropriate pistons by the valve plate (kidney port plate). The fluid forces the pistons outwards and the reaction force set up against the fixed swash plate results in a tangential force causing the piston block and output shaft to rotate. When the piston reaches its maximum outward stroke, the flow path is switched by the kidney port plate to the outlet port and fluid is forced out of the cylinder as the piston is driven inwards by the swash plate reaction.

An alternative design is the bent axis piston motor. The cylinder block and output shaft are maintained in alignment by either a universal link or meshing gears. A pump/motor of this design is shown in Figure 2.7 in Chapter 2. The operating principle is similar to that of the in-line piston motor.

Ball motors

These can be considered as axial piston cam motors. A double-cam arrangement is shown in Figure 4.29 in the form of a wheel motor. Inside the wheel hub is a piston block assembly containing a number of cylinders each with a pair of opposing ball pistons. The pistons bear on cam plates. Distributor ports in the stub axle feed pressure fluid to the cylinders forcing out the pistons against the cam tracks. The tangential component of the reaction force causes the cylinder block (and hence the wheel) to rotate. Another set of distributor ports comes into play when the pistons are fully extended and used fluid is returned through these as the pistons are pushed back by the cam plates.

In a particular design which has nine cylinders, the cam plates have three lobes giving

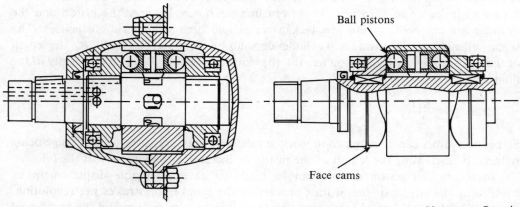

Mannesmann Rexroth

Figure 4.29 Ball motor.

27 working strokes per revolution resulting in a smooth torque characteristic. These motors have capacities of 160–764 cm^3/rev and a maximum speed of 500 rev/min.

Radial piston motors

Radial piston motors are capable of very high torque and smooth operation at low speed. There are several designs; Figure 4.30 shows a type which has a number of large bore pistons situated radially in a housing round the shaft which has an equally large eccentric cam. The piston assembly is forced under fluid pressure onto the eccentric causing rotation of the shaft. A bearing pad or slipper attached to the piston has to swing through a small angle to maintain correct contact with the eccentric. This is achieved by using a ball jointed con-rod inside the oscillating piston assembly. Pressurized fluid is fed through each piston to form a hydrostatic bearing between the bearing pad and the eccentric shaft. A valve block attached to the motor shaft connects the pistons to the pressure and tank ports in sequence, giving an almost constant torque characteristic. Piston motors of this type usually have five or seven cylinders but there are models with 10 cylinders. Output torques of up to 21 700 Nm and speeds up to 450 rev/min are obtainable.

Having an unbalanced configuration, the bearing loads and life expectancy are dependent upon the operating pressure and this may be a limiting factor in some applications.

The radial piston multi-lobe cam motor shown in Figure 4.31 is a balanced configuration which comprises a center rotating piston assembly and a fixed outer cam ring with eight lobes. A central valve distributes pressurized fluid to the cylinders which are forced outwards onto the cam and as they follow the cam profile, the piston block rotates driving the output shaft. Variations on this design allow groups of pistons to be

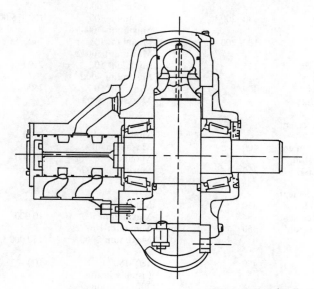

Vickers Systems/Staffa

Figure 4.30 Radial piston motor.

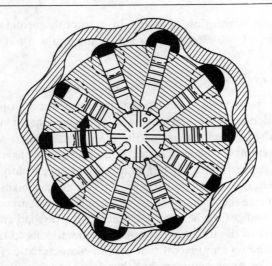

Figure 4.31 Radial piston multi-lobe cam motor.

Table 4.2 Summary of motor characteristics.

Motor type	Typical maximum operating pressure (bar)	Operating speeds (rev/min)	Typical maximum torques (Nm)
Gear	200 to a maximum of 300	Minimum smooth speed 400 Maximum 6000	500 Nm
Vane	140–200	Minimum 100 Maximum 4000	100–16 000
Gerotor	100–200	Minimum 10 Maximum 5000	2400 Nm
Cam rotor	175	Minimum 50 Maximum 4000	2500
Axial piston (swashplate) Fixed displacement	400	Minimum 50 Maximum 4000	2500
Variable displacement	400	Minimum 50 Maximum 4000	2500
Axial piston (bent axis) Fixed displacement	350	Minimum 50 Maximum 8500	10 000
Variable displacement	350	Minimum 50 Maximum 8500	10 000
Radial piston	450	Minimum 1 or less Maximum 2000	150 000
Wheel motors	450	180–1500 (usually in the 200–400 range)	1000–32 000

Note Variable-displacement radial piston motors are now being manufactured

Section 4.4 Hydraulic motor circuits

switched out of service giving a stepped speed increase but with a consequential reduction in torque.

By slightly pressurizing the casing and driving the pistons back into the rotor, the motor can free wheel — a feature which is particularly useful on mobile 'off-the-road' applications, where by switching the wheel drive motors to the 'free wheel' condition much greater towing speeds are possible during intersite transport.

A summary of the characteristics of various types of hydraulic motor is given in Table 4.2; the figures cover a range of sizes and makes. Maximum values of torque, speed and operating pressure will not be obtainable from any one motor of a particular type.

4.4 HYDRAULIC MOTOR CIRCUITS

Hydraulic motor circuits or hydrostatic transmissions fall into two categories — open or closed loop. In an open-circuit transmission, all the fluid discharged by the motor returns to the oil reservoir, whereas with a closed-circuit transmission, most of the fluid discharged from the motor is returned to the pump inlet. Some fluid may be taken out of the loop for conditioning, i.e. cooling and filtering. This is replenished by a make-up circuit.

4.4.1 Open-circuit transmissions

Figure 4.32 shows a fixed-speed, non-reversible open-circuit transmission. An anti-cavitation check valve is included to supply the motor with fluid if it runs on when the pump is stopped and hence prevents the motor cavitating. (For the definition of cavitation see Section 5.2.2 in Chapter 5.)

If:

n_p is the pump speed (rev/min)
D_p is the swept volume per revolution of the pump
n_m is the motor speed (rev/min)
D_m is the swept volume of the motor per revolution

then, theoretical volume of fluid delivered by pump is

$n_p D_p$

and the theoretical volume of fluid required by motor is

$n_m D_m$

Assuming there is no leakage, then

$n_p D_p = n_m D_m$

So motor speed is

$n_m = n_p D_p / D_m$ (4.4)

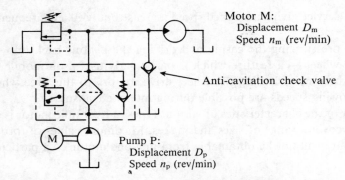

Figure 4.32 Fixed-speed, non-reversible, open-circuit transmission.

If a variable-displacement pump is used to replace the fixed displacement unit shown in Figure 4.32, then a variable speed drive results.

Assuming the pump drive speed is fixed, then the hydraulic motor speed from equation (4.4) becomes:

$$n_m = n_p D_m / D_p = \text{Constant} \times D_p$$

Consider the work done per revolution of the hydraulic motor neglecting all losses.

Work done = Torque × Angle turned through
= Fluid displaced × Pressure

then work done per revolution is

$$T_m \times 2\pi = D_m \times P_m$$

where T_m is the motor torque and P_m is the pressure drop across the motor. Then

$$T_m = D_m P_m / 2\pi$$

Thus the theoretical output power developed by the motor is

$$T_m n_m = \frac{D_m P_m n_m}{2\pi}$$

The characteristic curves for this transmission are shown in Figure 4.33. This particular drive configuration using a variable-displacement pump and a fixed-displacement motor is sometimes known as a 'constant torque transmission' and may be either an open- or closed-loop circuit.

Consider now, replacing the fixed-displacement motor in Figure 4.32 by a variable-displacement unit but retaining the fixed-displacement pump. Then, neglecting all losses, the motor speed is given by

$$n_m = \frac{n_p D_p}{D_m}$$

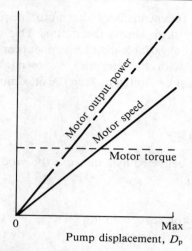

Figure 4.33 Characteristic curves for transmission with variable displacement pump and fixed displacement motor (constant torque transmission).

and the motor torque by

$$T_m = D_m P_m / 2\pi$$

Thus the motor output power is

$$T_m n_m = \left(\frac{D_m P_m}{2\pi}\right) \times \left(\frac{n_p D_p}{D_m}\right)$$

$$= \frac{n_p D_p P_m}{2\pi}$$

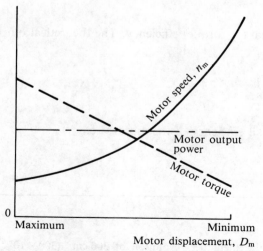

Figure 4.34 Characteristic curves for transmission with fixed displacement pump and variable displacement motor (constant power transmission).

The motor output power is independent of the motor displacement and is therefore constant for a given pressure drop across the motor. This drive configuration with a fixed-displacement pump and a variable-displacement motor is sometimes known as a 'constant power transmission' with characteristics as shown in Figure 4.34. The minimum displacement of the motor must be limited to keep the maximum motor speed down to its permissible maximum value.

Hydraulic motor efficiencies

The theoretical output power delivered by a motor is the product of the quantity of fluid flowing and the pressure drop:
Let

D_m = displacement of the motor per revolution
P_m = pressure difference across the motor
n_m = motor speed (rev/min)
Q_m = quantity flowing into the motor.

Theoretical quantity flowing is $D_m n_m$. The actual quantity flowing will be greater owing to leakage, i.e.

$$Q_m = \frac{D_m n_m}{{}_m\eta_v}$$

where ${}_m\eta_v$ = volumetric efficiency. The theoretical output torque is

$$\frac{D_m P_m}{2\pi}$$

Actual output torque is

$$\frac{{}_m\eta_t D_m P_m}{2\pi}$$

where ${}_m\eta_t$ = mechanical (or torque) efficiency. The theoretical output power is

$$Q_m P_m$$

Actual power output is

$$Q_m P_m {}_m\eta_o$$

where ${}_m\eta_o$ is the overall efficiency of the motor, i.e.

$${}_m\eta_o = {}_m\eta_t \times {}_m\eta_v$$

EXAMPLE 4.9

A motor has a displacement per revolution of 300 cm³ (300×10^{-6} m³) and a speed of 200 rev/min with a pressure drop of 200 bar (200×10^5 N/m²). The volumetric efficiency is 90% and the mechanical efficiency is 95%.

The $_m\eta_o = 0.9 \times 0.95 = 0.855$. Theoretical volume flowing per minute is

$$\frac{300}{1000} \times 200 \text{ (l/min)}$$

$$= 60 \text{ l/min}$$

Actual volume flowing into motor Q_m is

$$60/_m\eta_v$$

$$= 60/0.9$$

$$= 66.7 \text{ l/min}$$

Theoretical torque is

$$\frac{D_m P_m}{2\pi}$$

$$= \frac{300 \times 10^{-6} \times 200 \times 10^5}{2\pi} \text{ (m}^3 \times \text{N/m}^2\text{)}$$

$$= 955 \text{ Nm}$$

Actual torque $T = {}_m\eta_t \times 955 = 0.95 \times 955 = 907$ Nm

Actual output power is

$$2\pi n_m T$$

$$= 2\pi \left(\frac{200}{60}\right) \times 907 \left(\frac{\text{Nm}}{\text{s}}\right)$$

$$= 18\,996 \text{ Nm/s}$$

$$= 19 \text{ kW}$$

Alternatively, this may be calculated from

$$\text{Theoretical output power} = \frac{Q \text{ (l/min)} \times P \text{ (bar)}}{600} \text{ (kW)}$$

$$= \frac{66.7 \times 200}{600}$$

$$= 22.23 \text{ kW}$$

Actual output power is

Theoretical output power × overall efficiency
$= 22.23 \times 0.855$

$= 19$ kW (as before)

Reversible open-loop transmission

The direction of rotation of the hydraulic motor in Figure 4.35 is determined by the directional control valve. A cross line relief valve network is built into this circuit to deal with the pressure surge which will occur if the directional control valve is operated rapidly to reverse the motor drive.

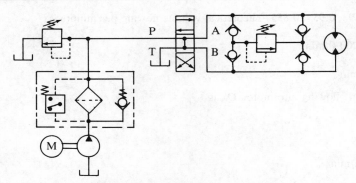

Figure 4.35 Reversible open-loop transmission.

In any fixed-displacement pump motor drive a flow control valve can be used to give 'meter-in' or 'meter-out' speed control, but this may result in a hot system, as the excess fluid will have to discharge across the relief valve.

The choice of directional control-valve center condition is important in these circuits. If the service ports A and B are blocked in the center or transition condition, very high pressure surges can occur on changeover. This can be particularly damaging if the motor does not have an external drain connection. With the open-center directional-control valve shown in Figure 4.35, the motor will freewheel when the valve is centered. This may be an undesirable feature and can be remedied by a brake or remote piloted back-pressure valve installed in the motor circuit as illustrated in Figure 3.15 in Chapter 3.

In the case of a reversible drive, a brake valve would be installed at each side of the motor. For simplicity only one such valve is shown in the circuit in Figure 3.15. When pressure is applied to the top line to the motor, the brake valve is piloted open, thus giving a free return flow from the motor. When the directional control valve is centralized, the brake valve closes, as the pilot pressure falls. This sets up a back-pressure equivalent to the brake valve spring setting and so slows the motor down. The brake valve should be set to slow the motor down as quickly as possible without hydraulic shock occurring. The check valve enables the motor to be driven in the reverse direction.

Note Brake valve is one of the many names for the pressure control valve described in detail in Section 3.1.2 of Chapter 3 under the heading 'over-center valve'.

4.4.2 Closed-loop transmissions

There is always some designed leakage in the pump and motor units used in hydrostatic transmissions. When these are built into a closed-loop configuration (Figure 4.36), a separate fluid supply has to be provided to make up the leakage. This is usually achieved by using a 'make-up' pump to feed the low pressure side of the loop.

The make-up pump supplies filtered fluid at a low pressure, set by the relief valve, to the check valve network. This network allows make-up fluid to flow to the return side of the main pump whilst isolating the high-pressure side. The pressure setting of the make-up circuit relief valve has to meet the requirements of the main pump. It may be very low

Section 4.4 *Hydraulic motor circuits*

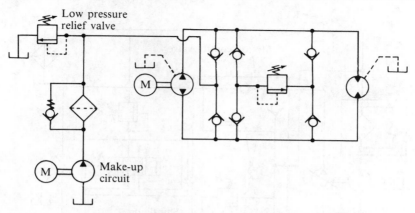

Figure 4.36 Closed-loop hydrostatic transmission with make-up pump.

(approximately 1 bar) but some designs call for back-pressures in the order of 20 bar. In non-reversing, closed-loop transmissions it may be possible to use an elevated reservoir to supply make-up oil so eliminating the necessity for a make-up pump and associated circuitry.

In the circuit shown in Figure 4.37 a standard pressure line filter can be used as there is no reversal of flow. A check valve situated after the filter prevents any possibility of the motor causing reverse flow through the filter. A brake valve is shown to give controlled retardation of the driven load and an anti-cavitation check valve is placed across the motor.

When a transmission is constantly running or used for hydrostatic braking there may be excessive heat generation and it is desirable to bleed off some of the fluid for conditioning. This bleed off must be from the low-pressure side of the transmission otherwise more heat energy will be generated. The cooling and filtering of the fluid can be by a separate conditioning loop or as part of the make-up circuit.

In the circuit shown in Figure 4.38 RV_1 is set slightly higher (about 1.5 bar) than RV_2. When the hydraulic motor is being driven the shuttle valve connects the low-pressure side of the circuit to RV_2 which sets the boost pressure at the pump and bleeds off excess oil from the circuit through the cooler. (This is sometimes referred to as 'scavenging' or

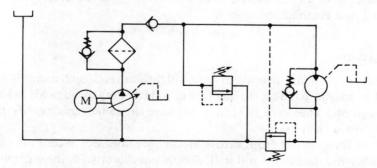

Figure 4.37 Closed-loop hydrostatic transmission with gravity feed make-up.

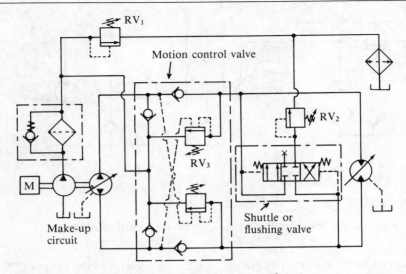

Figure 4.38 Closed-loop reversible transmission with scavenging circuit.

flushing). When the pump is set to 'neutral' the make-up circuit relieves across RV_1. In some circuits the scavenging circuit passes the fluid through the pump and motor casings to help in cooling those components. When designing systems, the sizing of the make-up circuit is extremely important. RV_3 is a motion control valve which functions both as a cross-line relief valve and a brake valve.

Closed-loop transmission characteristics

In the case of non-reversible drives these are similar to the open-loop characteristics. The characteristics of reversible drives are mirror images of the non-reversible drives; typical examples are shown in Figure 4.39.

4.4.3 Multi-motor circuits

More than one motor can be driven from one pump, with motors connected in either series, parallel or a combination of both.

Series connection

Neglecting leakage the same quantity of fluid will flow through each motor in Figure 4.40. The effect of leakage is to reduce the quantity of fluid flowing through M_2 to less than that flowing through M_1. Thus even if M_1 and M_2 have the same capacity, M_2 will run at a slightly lower speed than M_1.

The sum of the pressure drops across M_1 and M_2 must not exceed the setting of the relief valve otherwise the motors will stall. Should one motor stall, the other will also stop because there is no flow.

Basic outline circuit   **Characteristics**

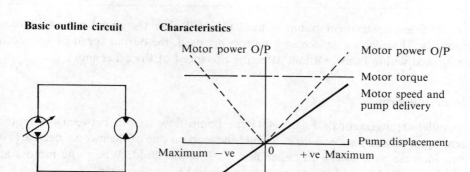

(a) *Variable pump, fixed motor*

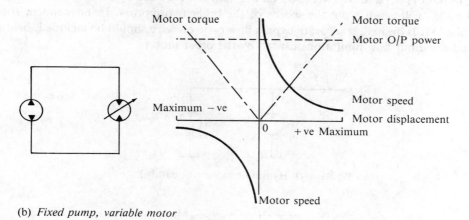

(b) *Fixed pump, variable motor*

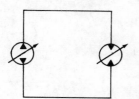

The characteristic is a combination of the previous two, as both pump and motor displacement can be varied. Any desired characteristic can be obtained to suit the specific application.

(c) *Variable pump, variable motor*

Figure 4.39 Characteristics of reversible closed-loop transmissions.

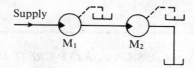

Figure 4.40 Hydraulic motors in series.

If a variable-displacement pump is used this will alter the speed of both motors proportionally. If variable-displacement motors are used, the output speed of one motor can be adjusted within limits without affecting the speed of the other motor.

Parallel connection

In the parallel arrangement of Figure 4.41 the pump flow is split between the motors dependent upon the resistance to flow. Should the load on one of the motors decrease, its speed will increase as it offers an easier flow path to the fluid. Where the motors are mechanically coupled to each other, their speeds will be synchronized.

Flow-control valves or flow dividers may be used to split the flow and provide a degree of synchronization. The circuit of Figure 4.42 uses three-port preferential flow valves to regulate the speeds. Motor M_1 has first call on the pump delivery with the excess flow going to motors M_2 and M_3. M_3 takes all the fluid not required by M_1 and M_2 and so the speed of M_3 will result from the speeds of the other two motors. If independent speed control of M_3 is desired, a separate bypass flow control valve should be included, bearing in mind the input flow limitations caused by the other motors.

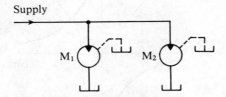

Figure 4.41 Hydraulic motors in parallel.

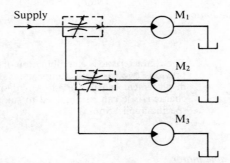

Figure 4.42 Hydraulic motors in parallel with preferential flow control.

4.5 *MOTOR CIRCUIT DESIGN EXAMPLES*

EXAMPLE 4.10: DRUM DRIVE DESIGN FOR CONCRETE MIXER TRUCK

The inside of the mixer drum contains a series of spiral vanes. Counterclockwise rotation drives the material into the drum giving a mixing action. High-speed clockwise rotation is

Section 4.5 Motor circuit design examples

used to discharge the mixed concrete. The pumps are to be driven by a power take off from the truck engine and are driven at a speed which can vary between 600 and 2000 rev/min. During normal running an approximate drum speed of 5 rev/min is needed for mixing; this has to be independent of the vehicle engine speed. For discharge, a drum speed of 20 rev/min is required. This can be at the engine's maximum speed as the vehicle will be stationary and it will only be required for a short period. A 20:1 reduction gear is fitted between the hydraulic motor and the drum. (Assume a gear efficiency of 90%.) The torque required to drive the drum is 12 000 Nm and the maximum permissible circuit pressure 207 bar (3000 psi).

With the 20:1 reduction gear box, the torque output T_m of the hydraulic motor will be

$$T_m = \frac{12\,000}{20 \times 0.9} = 667 \text{ Nm}$$

The speed of the hydraulic motor n_m required during discharge will be

$$n_m = 20 \times 20 = 400 \text{ rev/min}$$

The performance figures and curves for a range of radial piston motors are given in Figure 4.43. From the details for the M3 unit at 207 bar and 400 rev/min, torque = 850 Nm, and at 207 bar and 100 rev/min, torque = 900 Nm. The M3 has a displacement of 280 cm^3/rev. At 400 rev/min and 170 bar the volumetric efficiency is 98.5%. At 100 rev/min and 170 bar the volumetric efficiency is 98%.

Flow required by motor at 400 rev/min is

$$\frac{280 \times 400 \times 10^{-3}}{0.985}$$

$$= 144 \text{ l/min}$$

Flow required by motor at 100 rev/min is

$$\frac{280 \times 100 \times 10^{-3}}{0.98}$$

$$= 29 \text{ l/min}$$

A variable pump with constant-volume control can be used to provide an almost constant mixing speed whilst the vehicle is being driven around (i.e. independently of engine speed). Additional flow necessary to discharge the load can be supplied when necessary by a fixed-displacement pump.

The Volvo V30B-35 axial piston pump is suitable for the working pressure and has a maximum displacement of 35.4 cm^3/rev. With a constant volume control, 29 l/min will be available at engine speeds in excess of 29/35 × 1000 rev/min (i.e. 828 rev/min).

A specification for the Volvo V30B series is given in Table 4.3.

For the fixed-displacement pump a simple gear type can be used. The flow required is 114 − 29 = 85 l/min.

The pump drive operates up to a maximum speed of 2000 rev/min. The details of Dowty gear pumps are given in Chapter 2 (Section 2.3, Table 2.4). A suitable pump would be the 2PL 146 having a maximum operating pressure of 210 bar and a nominal delivery of 67.3 l/min at 1500 rev/min. This is equivalent to a delivery of 89 liters/min at 2000 rev/min which will be satisfactory.

The pressure drop over the hydraulic motor to deliver the required torque is 170 bar. Assume a total pressure drop in the circuit components pipe work, etc. of 30 bar. The maximum pressure the pumps will have to operate against will therefore be 200 bar.

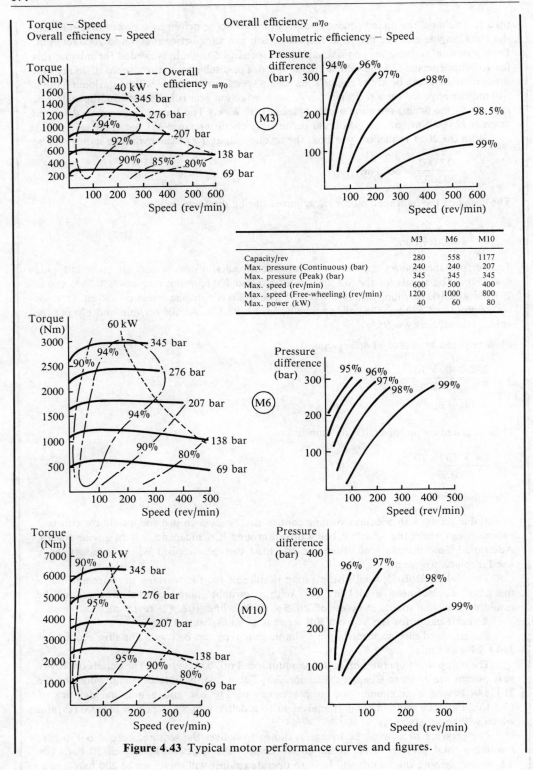

Figure 4.43 Typical motor performance curves and figures.

Section 4.5 *Motor circuit design examples*   **175**

Table 4.3 Specification for Volvo V30B series axial piston pump.

		Size of unit		
Specification parameter		35	66	128
Displacement	(cm^3/rev)	35.4	66.0	128.0
Theoretical torque at 100 bar	(Nm)	57.4	107	207
Maximum speed boosted	(rev/min)	3000	2500	2100
Maximum speed unboosted	(rev/min)	2000	1800	1500
Flow at 1450 rev/min	(liters/min)	50	93	180
Input power at 1450 rev/min and 250 bar	(kW)	22	43	84
Minimum continuous speed	(rev/min)	150	150	150
Maximum continuous pressure	(bar)	250	250	250
Maximum intermittent pressure	(bar)	420	420	420

Available as a fixed- or variable-displacement unit. Controls available are:
Handwheel
Servo control
Pressure compensated
Constant flow
Constant power
A displacement limiter is fitted on the variable-displacement units.

Volvo Hydraulics Ltd

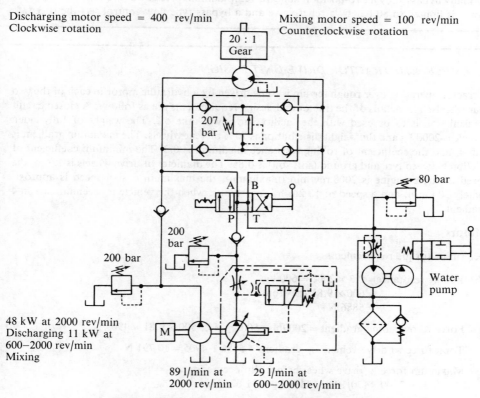

Figure 4.44 Drum drive for concrete mixer truck.

By calculation the input power requirements from the diesel engine drive are:

(a) for gear pump (assuming overall efficiency is 80%)

$$\text{Power} = \frac{200 \text{ (bar)} \times 89 \text{ (l/min)}}{600 \times 0.8} = 37 \text{ kW}$$

(b) For piston pump (assuming overall efficiency is 90%)

$$\text{Power} = \frac{200 \times 29}{600 \times 0.9} = 10.75 \text{ kW}$$

Thus the total power required is (a) + (b) = 47.75 kW.

A suggested circuit is shown in the Figure 4.44. A four-port, three-position, hand-operated valve with detents is used as the main control valve. In the center position both pumps are dumped to tank. For counterclockwise rotation (mixing), the valve connects P to B and A to T. Flow from the piston pump drives the motor whilst the gear pump is dumped to tank. With the valve in the 'tramline' condition P to A, B to T, both pumps feed the motor giving high-speed clockwise rotation to discharge the mix. Each pump is fitted with its own relief valve and a check valve to prevent the pump motoring under reverse flow. A hydrocapsule containing a relief valve and check valve network is used for motor protection. Return-line filtration has been chosen and the flow rate through this will be reasonably constant.

The circuit incorporates a further refinement which is a water pump driven by a hydraulic motor (a small high-speed low-torque unit). As this is only occasionally required, it is normally bypassed by a two-position hand valve. Maximum pressure at the water-pump drive motor is limited to 80 bar by a relief valve and a bypass-type flow control sets the speed.

EXAMPLE 4.11: TRACTOR DRIVE UNIT DESIGN

A tractor operating over rough terrain is to be driven by a hydraulic motor in each of the two rear wheels. The details of the tractor and design requirements are as follows: A closed circuit transmission is to be used with the facility for power take off. The weight of fully-laden vehicle is 2000 kg and the weight distribution is 70% on rear wheels. The maximum gradient is 1 in 4, and the coefficient of rolling resistance maximum is 0.3. The minimum coefficient of friction between tyre and ground (soft soil) is 0.85. The diameter of drive wheels is 1.2 m, the speed of drive engine is 2000 rev/min (maximum), the maximum design speed is approximately 20 km/h, and a speed of 10 km/h is acceptable when the vehicle is ascending a 1 in 4 gradient.

Motors

Determine torque requirements.

(i) Rolling resistance = 0.3 × 2000 = 600 kg
 = 600 × 9.81 (newtons)
 = 5886 N

Force to overcome gradient = 2000/4 = 500 kg, i.e. 500 × 9.81 = 4905 N.

Total force to drive vehicle up gradient = 5886 + 4905 = 10 791 N

Maximum force at drive wheel before slip occurs is
 0.85 × 2000 × (70/100) = 1190 kg = 11 674 N.

Thus the vehicle will just climb a 1 in 4 slope without slipping under design conditions.

Total torque required at drive wheels is

force × radius
= 10 791 × 1.2/2
= 6475 Nm

As there are two drive wheels the torque per wheel is 3237 Nm. Select a motor from the characteristic curves (Figure 4.43). A size M10 unit is suitable.

(ii) Required vehicle speed is 20 km/h. The wheel diameter is 1.2 m, so the wheel speed is

$$\left(\frac{20 \times 10^3}{60}\right) \times \left(\frac{1}{\pi \times 1.2}\right)$$

= 88.4 rev/min

(iii) For flow required per motor, volumetric efficiency (from chart for M10 at 88 rev/min and a pressure differential of 100 bar) is approximately 99%. The flow required is

$$\frac{\text{Motor capacity} \times \text{Speed}}{\text{Volumetric efficiency}}$$

$$= \frac{1.177 \times 88.4}{0.99} = 105 \ l/min$$

Total flow for both motors is 210 l/min for maximum design speed.

(iv) For vehicle operating on level terrain, only rolling resistance need be considered:

Total force required = 5886 N
Torque at drive wheels = 5886 × (1.2/2) = 3532 Nm
Torque per wheel = 1766 Nm

With the M10 motor this requires a pressure differential of 103 bar.

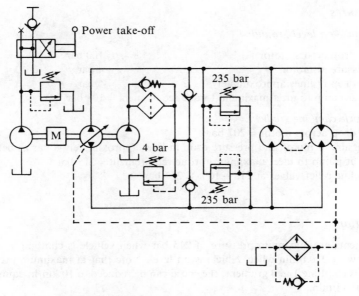

Figure 4.45 Hydraulic circuit for tractor drive.

(v) For vehicle descending a 1 in 4 gradient. The gradient now aids the drive:

Force required = (rolling resistance) − (gradient force)
= (5886 − 4905)
= 981 N

Torque at drive wheels = 981 × (1.2/2) = 588 Nm
Torque per wheel = 294 Nm

If the rolling resistance reduces below the gradient force the vehicle will tend to run away but a closed-loop transmission enables engine braking to be used.

(vi) For vehicle starting on gradient, starting torque must be greater than torque required to drive the vehicle up the gradient. The maximum torque which can be applied at the wheels is that which will cause the wheels to slip.

Force at which slip occurs = 11 674 N. This is equivalent to a motor torque of

$$11\ 674 \times \left(\frac{1.2}{2}\right) \times \left(\frac{1}{2}\right) = 3502 \text{ Nm}$$

The force which is available for acceleration is the maximum force at which drive wheel slip occurs less the total force needed to drive the vehicle up the gradient. Therefore the available force for acceleration is (11 674 − 10 791) = 883 N.

$$\text{Acceleration} = \frac{\text{Acceleration force}}{\text{Mass}} = \frac{883}{2000} = 0.44 \text{ m/s}^2$$

For the motor to develop a torque of 3502 Nm, it will require a pressure drop of 207 bar. The suggested hydraulic circuit for the tractor drive is shown in Figure 4.45.

Pump

Pump pressures

(1) *With vehicle on level ground*:

Pressure drop across motor	= 100 bar
Back-pressure at motor	= 4 bar
Pressure drop in lines approximately	= 4 bar
Pressure at pump to meet maximum torque requirement	= 108 bar.

(2) *With vehicle climbing gradient*:
Pressure drop across motor = 207 bar
Back-pressure at motor and pressure drop in lines approximately 8 bar as before
Pressure at pump to meet maximum torque requirement = 215 bar
Set cross line relief valves at 215 + 10% = 237 bar.

Pump selection

The requirements are maximum pressure of 215 bar when vehicle is climbing gradient and maximum flow of 210 l/min when vehicle is on level. Note that at maximum pressure, i.e. when vehicle is climbing 1 in 4 gradient, the speed can be reduced to 10 km/h, equivalent to a flow rate of 105 l/min.

At maximum flow rate, the vehicle is travelling on level. The pressure required is 100 bar.

Section 4.5 *Motor circuit design examples*   **179**

Pump requirements

Flow is 210 l/min at 108 bar and 105 l/min at 215 bar. Thus approximately the same hydraulic power is required under both conditions. Select a constant power pump to operate at a maximum drive speed of 2000 rev/min.

From Table 4.3 Volvo V30B 128 pump has a maximum flow rate of 180 l/min at 1450 rev/min. Thus at 2000 rev/min it has a maximum rated delivery of approximately 240 l/min. The maximum system pressure required of 215 bar is below the operating maximum for the V30B pump unit. Select a manual servo-control unit and set the stroke limiter to give a maximum flow of 210 l/min. The main pump must be boosted to operate at 2000 rev/min and this is accomplished by the make-up pump circuit.

Input power to pump

When the vehicle is operating on level ground, the pump is at full delivery and operating at 110 bar. The theoretical power required is

$$\frac{\text{Flow (l/min)} \times \text{Pressure (bar)}}{600} \text{ (kW)}$$

$$= \frac{210 \times 108}{600} \text{ (kW)}$$

$$= 37.8 \text{ kW}$$

Assuming an overall efficiency of 0.9, the input power will be approximately $37.8/0.9 = 42$ kW.

When the vehicle is climbing a gradient the pump will be set to half delivery using the manual servo but the system pressure will be approximately double, hence requiring the same input power.

Considering now the make-up circuit, this pump has to supply sufficient fluid to make up the case leakage for the pump and motors. The motor volumetric efficiency at 88 rev/min and 205 bar is approximately 97.5% from Figure 4.43. Thus the case leakage for the two motors will be given by

Leakage = (capacity per revolution) × (Revolutions/min)
  × (1 − Volumetric efficiency) × 2(motors)
  = 1.177 × 88 × (1 − 0.975) × 2
  = 5.2 l/min

The pump leakage is not given so assume it is equal to the total motor leakage. Select a make-up pump with a delivery of, say, 12 l/min to operate at a pressure of up to 4 bar. The theoretical drive power required by the make-up pump is

$$\frac{12 \text{ (l/min)} \times 4 \text{ (bar)}}{600}$$

$$= 0.08 \text{ kW}$$

CHAPTER FIVE
FLUIDS FOR HYDRAULIC SYSTEMS

5.1 HYDRAULIC FLUIDS

5.1.1 An outline of the development of hydraulic fluids

Water was the first fluid used for the transmission of fluid power. High-pressure water mains which existed in many cities were used to power lifts and machinery. The poor lubricating properties, the limited range of working temperature and the rust-promoting tendency of water have limited its modern usage to very large systems where its easy availability, low cost or fire-resistant properties outweigh its disadvantages.

Although mineral oils were readily available at the beginning of the twentieth century, it was not until the 1920s that any great use was made of them in hydraulic power systems. In the 1940s, additives were first used to improve the physical and chemical properties of hydraulic mineral oils. The first additives were developed to counter rust and oxidation, i.e. a corrosion inhibitor and an anti-oxidant.

Mineral oils are highly flammable and when used in a high temperature, constitute a severe fire risk. This has led to the development of fire-resistant fluids which are mainly water-based with resultant limitations on operating conditions.

The need for extremes of operating temperatures and pressures has led to the development of synthetic and water-based fluids.

Classification of fluid

In order to identify basic types of hydraulic fluid, standards organizations have classified them according to composition and properties and allocated a series of letter symbols (Table 5.1). It is hoped that manufacturers will incorporate these codes into their reference numbers or fluid descriptions.

5.1.2 Properties of hydraulic fluids

Relative density or specific gravity

This is the weight of a given volume of fluid at a given temperature relative to the weight of the same volume of water. The relative density of mineral oil is in the region of 0.9 (the

Section 5.1 Hydraulic fluids

Table 5.1 Classification of hydraulic fluid to ISO 6743 and BS 6413 Part 4; 1983.

Symbol (ISO)	Composition and properties
HH	Non-inhibited refined mineral oil
HL	Refined mineral oil with improved anti-rust and anti-oxidation properties
HM	Oils of type HL with improved anti-wear properties
HR	Oils of type HL with improved viscosity–temperature properties
HV	Oils of type HM with improved viscosity–temperature properties
HS	Synthetic fluids with no specific fire-resistant properties
HFAE	High water-based fluid maximum of 20% combustible materials (greater than 80% water content)
HFAS	Chemical solutions in water, greater than 80% water content
HFB	Water-in-oil – water droplets in a continuous oil phase (60% oil/40% water)
HFC	Water polymer – water glycol (35% water minimum, 80% maximum)
HFD	Water-free pure chemical fluids
HFDR	Phosphate esters
HFDS	Chlorinated hydrocarbons
HFDT	Mixture of HFDR and HFDS

exact value is dependent upon the origin of the base oil and the additives used). Synthetic fluids can have a relative density of greater than 1. The relative density of a hydraulic fluid is important when designing the layout of the pump and reservoir and must be considered when calculating the static pressure at the pump inlet.

Viscosity

This is the resistance of a fluid to flow or the internal friction of the fluid. Viscosity is one of the most important properties of a hydraulic fluid and it is usually the viscosity requirement of a system that determines which grade is selected. It has to be viscous enough to enable the system to be sealed and prevent external leakage, but have sufficiently low a viscosity to flow easily. If a fluid of too high a viscosity is used, it will require greater energy to overcome the internal friction resulting in heat generation. Should too low a viscosity fluid be used there will be increased leakage and a fall in volumetric efficiency.

The viscosity may be either *dynamic* or *kinematic*. In power hydraulics the kinematic viscosity is used and is expressed in mm^2/s in SI units. This corresponds to the older but still predominantly used centistoke (cSt) unit. The viscosity may be measured by noting the time taken for a given volume of fluid to flow through a capillary tube under set conditions.

The viscosity of a fluid used to be given in Redwood or Saybolt seconds, or in degree Engler. These units are no longer acceptable, but conversion tables are available.

The viscosity of some fluids varies considerably with temperature and for this reason the viscosity is stated at a standard temperature (40°C for the ISO specification). Thus a hydraulic fluid referred to as HH32 is a non-inhibited mineral oil hydraulic fluid with a viscosity of 32 mm^2/s at 40°C.

VISCOSITY INDEX

The viscosity of all oils reduces with an increase in temperature and increases with a

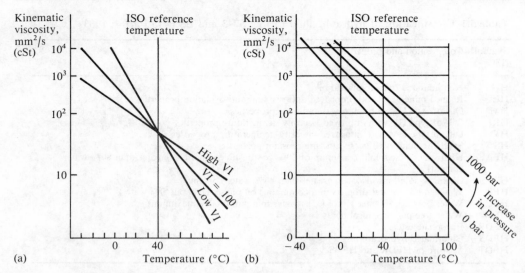

Figure 5.1 (a) Temperature–viscosity and viscosity index. (b) Temperature–viscosity and pressure.

reduction in temperature. The severity of this change is defined by the value of the viscosity index given to the particular oil. An oil which shows relatively little change in viscosity with temperature is said to have a high viscosity index. An oil with a viscosity index of 80 is said to have a high viscosity index value. Medium viscosity indices are between 40 and 80 and low viscosity index below 40. Most hydraulic oils have a viscosity index greater than 80.

The viscosity index of a mineral oil depends very much upon the origin of the base oil. Some hydraulic systems require a higher viscosity index than can be obtained even by careful selection of the base oil and viscosity index improvers have to be added. Viscosity–temperature curves for typical hydraulic oils of differing viscosity index are shown in Figure 5.1(a).

VISCOSITY–PRESSURE CHARACTERISTICS

As the pressure to which a hydraulic oil is subjected increases, the viscosity of the oil also increases. With the ever-increasing operating pressures of hydraulic systems (presses operating at 1000 bar and over are not unusual) great care must be taken in the selection of the hydraulic fluid. At very high pressure, the characteristics of an oil can be drastically altered and it may become virtually solid. Temperature–viscosity curves for a typical hydraulic oil at various pressures are shown in Figure 5.1(b).

Shear stability

In practically all hydraulic systems there are sections where the fluid is subjected to very high rates of shear. This may be at the tips of the vanes in a vane pump or the fine clearances of control orifices in valves. High shear rate affects the oil and the viscosity index additives in two ways: the oil in the areas of high shear rate suffers a drop in viscosity but recovers, as soon as it leaves the high shear area. Viscosity index additives suffer some permanent breakdown which results in a gradual fall in viscosity of the oil during the first few hours of service. The viscosity reaches a constant level after the initial breakdown and this loss of viscosity is never recovered.

Foaming characteristic

All fluids contain dissolved air, the amount depending upon temperature and pressure. Typically, a mineral-based oil can contain up to 10% by volume of dissolved air. As the temperature of the fluid increases or the pressure reduces, air is driven out of solution. If this occurs at the pump inlet, cavitation will result.

Entrained air exists in a fluid as discrete bubbles and is not to be confused with *dissolved air*. Entrained air will make a fluid more compressible which leads to sponginess; it will also cause a pump to cavitate (see Section 5.2.2). This effect is serious enough on the efficiency of the system and the increased wear owing to cavitation hammer, but if these bubbles of air or foam pass through the pump to the high-pressure side, the relatively high incompressibility characteristic of the fluid is lost because of the presence of a highly-compressible gas. Air can be entrained into the fluid at any point where fluid is moving in contact with air. Badly designed fluid reservoirs, porous pump suction lines and leaking cylinder seals are some of the points where air can infiltrate.

When fluid containing entrained air returns to the reservoir, the bubbles of air rise to the surface and cause foam. If the foam builds up, it can cause severe problems in a hydraulic system; consequently most hydraulic fluids contain foam-depressant additives which cause the rapid breakdown of the foam. These additives unfortunately reduce the rate at which the bubbles rise to the surface. The exact composition and amount of this additive has to be very carefully controlled.

Fluid contamination by air and methods of combating its harmful effects are discussed more fully in Section 5.2.

Pour point

The pour point of a fluid is the temperature 3°C above that at which the fluid ceases to flow. As a general rule the minimum temperature at which a fluid operates should be at least 10°C above the pour point.

Compressibility

All fluids are compressible and for a mineral oil there is an approximate reduction in volume of 0.7% for every 100 bar increase in pressure. The compressibility of any fluid is dependent upon the temperature and pressure. Air bubbles in the fluid will considerably increase the compressibility.

The bulk modulus B of the fluid is the ratio of the volumetric stress to volumetric strain of the fluid:

$$B = \frac{\text{Volumetric stress}}{\text{Volumetric strain}}$$

$$B = \left(\frac{\Delta P}{\Delta V / V}\right)$$

where ΔV is the volume change in the fluid volume V when subject to a change in pressure ΔP. The value of B varies, dependent on the fluid and conditions. Typically for water, $B = 20\,000$ bar and for mineral oil, $B = 15\,000$ bar. When the pressure of a large volume of fluid in a container is rapidly released, severe decompression shock will result. This can be avoided by the use of decompression valves as described in Section 3.3 of Chapter 3. Any entrained air exacerbates the effect of compressibility.

Thermal expansion

When the mass temperature of a fluid changes there is a corresponding change in the fluid volume. The increase in volume of a hydraulic fluid is typically 0.7% per $10°C$ rise in temperature. This is of importance where large volumes of fluids are being used. For example if an oil reservoir contains 400 litres at $15°C$, when the oil warms to its operating temperature of $55°C$, there will be an increase in volume of approximately 11 liters.

Lubrication

This must prevent excessive wear and provide a film of lubricant between moving parts. In simple hydraulic systems, hand-operated pumps, cylinders etc., a fluid providing thick film lubrication is adequate. In high speed, highly-loaded components such as vane pumps, additives have to be incorporated in the fluid to increase the load-carrying properties and provide boundary-layer lubrication.

When two metal surfaces rub at high loads and speeds, metal-to-metal contact takes place between minute high spots giving rise to very high local temperatures which can result in point welds. This continual local welding and tearing of the metal soon gives rise to heavy wear and seizure.

Additives used in hydraulic fluids

These can be classed as follows.

POUR POINT DEPRESSANTS
These are used to inhibit formation of wax crystals in mineral oils.

VISCOSITY INDEX IMPROVERS
When cold these long-chain polymers adopt a coiled form and so have little effect on the fluid viscosity. When the fluid temperature increases the polymers uncoil and intermesh causing a thickening effect in the oil. The improvement in viscosity index may be at the expense of shear stability.

DEFOAMERS
The formation of foam is reduced by the addition of a few parts per million of a suitable additive typically a silicone polymer.

Section 5.1 Hydraulic fluids

OXIDATION INHIBITORS
Oxidation of the oil causes an increase in both viscosity and organic acids. As the temperature increases, the rate of oxidation increases exponentially. Any dirt or water in the oil may act as a catalyst. An additive with a greater affinity for oxygen than the oil possesses is frequently employed. This additive will be used up with time.

RUST/CORROSION INHIBITORS
These are additives which form a continuous film on metal surfaces and prevent water or organic acids reacting with the surfaces.

ANTI-WEAR AGENTS
These are used to reduce wear and scuffing under boundary lubrication conditions. The additives are either long-chain polymers which are adsorbed on the metal surface or extreme pressure (EP) additives which are deposited on surfaces where a high local temperature occurs causing the surface to become polished.

5.1.3 Hydraulic fluids survey

In this section actual hydraulic fluids will be described in relation to the requirements considered in the previous section and their advantages and disadvantages which make them more or less suitable for particular applications.

Hydraulic fluids

MINERAL OILS

Mineral oils are probably the most widely used hydraulic fluids. In their favour it can be said that they are relatively inexpensive, widely available, and can be offered in suitable viscosity grades. They are also usually adequate from the point of view of lubrication ability and are not corrosive (though mineral oils themselves do not prevent corrosion from occurring owing to the presence of water and air).

Provided that the operating temperatures are reasonable ($40°C$ for optimum fluid life) mineral oils are chemically stable. At higher temperatures they suffer chemical breakdown, forming acids, varnishes, resins and sludges with a loss of lubrication properties. Premium grade mineral hydraulic oils contain a package of additives to combat the effects of wear, oxidation, foam formation, and to improve viscosity index and lubricity.

There are however certain disadvantages of mineral oils which at present cannot be remedied by incorporating additives. The two most important are the flammability characteristic and the increase in viscosity at high pressures. Fire risk precludes the use of mineral oils in hazardous areas such as injection and plastic moulding machines, coal mines and in the vicinity of furnaces. Viscosity–pressure characteristics limit their use to pressures below 1000 bar.

Mineral oils are compatible with most sealing materials with the exception of butyl rubber.

OIL-IN-WATER EMULSION HFAE

Tiny droplets of oil are dispersed in a continuous water phase. The dilution is normally between 2 and 5% oil in water and the characteristics of the fluid are more similar to water than oil. Because of this, it is extremely fire resistant, highly incompressible with good cooling properties. The main disadvantages are poor lubrication properties (which may be slightly improved by additives) and a low viscosity resulting in high leakage losses. This fluid is most commonly used in very large systems such as mining applications with high-displacement slow-running pumps and components which are suitable for water. It cannot normally be used as a direct replacement for mineral oils.

WATER-IN-OIL EMULSION HFB

Water-in-oil (invert) emulsions are the most popular fire-resistant fluids. They have a continuous oil phase in which tiny droplets of water are dispersed. The characteristics more nearly approach oil than water but the lubrication properties are very much reduced. This can again be partially overcome by running pumps at reduced speeds but larger displacement pumps are necessary to obtain the required flow. The usual dilution is 60% oil/40% water. For optimum life, operating temperatures should not exceed 25°C but intermittent operation up to 50°C is permissible. At the higher temperature, water content is affected owing to evaporation with a resultant decrease in the emulsion's fire-resistance properties.

The internal vapor pressure of water-in-oil emulsions approximates to that of water and consequently these fluids are extremely susceptible to cavitation. The suction pressure drop at the pump inlet should not exceed 0.05 bar, and positive head conditions are preferable.

When the system has been idle for long periods there is a tendency for the oil and water to separate. However, during running, the 'churning' action of the pump will re-emulsify the fluid.

The material used in the system must be checked for compatibility with the fluid; there are problems with magnesium and cadmium. Cork must not be used as it is softened by water-in-oil emulsion and butyl seals should be avoided. Most hydraulic components designed for use with mineral oils are suitable for HFB fluids but the system should be de-rated.

WATER–GLYCOL FLUIDS HFC

These fluids were developed primarily for use in aircraft because of their very low flammability characteristics. However, their application is limited, since they cannot be used at high temperatures (because of their water content). Their lubricating ability is inferior to mineral oils. Care needs to be exercised from the point of view of corrosion inhibition because of their aqueous nature and the inevitable presence of air. Corrosion and oxidation inhibitors for water–glycol systems need to be selected carefully with regard to the metals present. Zinc, magnesium, cadmium and aluminum components cannot be used and this fluid also attacks most paints.

Water–glycol fluids are very stable with respect to shear because of the low molecular weight of their constituents. Good anti-freeze properties make them particularly suitable for low-temperature applications.

Section 5.1   Hydraulic fluids

PHOSPHATE ESTERS HFDR

These fluids have remarkably good fire-resistant properties and are used in industries such as plastic moulding and die-casting where unusually great fire risks occur. Their lubricating ability is similar to that of mineral oil, and there is evidence that in certain circumstances they may be superior.

Elastomers used in conjunction with phosphate esters have to be chosen carefully. Some silicone polymers and butyl rubber are suitable. Certain metals, particularly aluminum, and most paints are vulnerable to attack.

The excellent performance characteristics of phosphate ester synthetic fluids are outweighed in many applications by the environmental and health hazards which they present.

Selection of fluid

Environmental conditions determine the need for a fire-resistant fluid. The working temperature range must be considered, taking into account minimum and maximum ambient temperatures. Local availability of spare fluid is another factor.

The viscosity of fluid used has been traditionally determined by the pump requirements but in more sophisticated systems with hydraulic motors and servo-valves the viscosity requirements of these components must also be satisfied. If doubt exists on the best fluid for an application, the component and fluid manufacturer should be consulted.

Particular care must be taken when handling synthetic fluids as many are toxic or give off toxic fumes. Recent legislation has restricted the manufacturer of certain types but the banned ones may be encountered in old systems.

Changing type of fluid used in a system

When a change from a mineral oil to a fire-resistant fluid becomes necessary, 'water-in-oil' emulsions, water–glycols or phosphate esters can be used provided the following precautions are taken:

- Check suitability of seals, metals and paints.
- Derate the system by running the pumps at lower speeds and pressures. (See Table 2.6 in Chapter 2.)
- Finally on changing over a system it must be thoroughly flushed to remove all traces of the original fluid.

Note Fluids containing water are only suitable for a restricted temperature range.

A summary of properties and features of various hydraulic fluids is given in Table 5.2.

5.1.4   Future developments

With the spread of hydraulic power to a very wide range of industrial applications, the need for a relatively cheap and possibly expendable fire-resistant hydraulic fluid has become apparent.

Considerable development is being concentrated on chemical solutions in water

Table 5.2 Hydraulic fluids.

Properties	Agent				
	Mineral oil	Water–glycol	Invert emulsion	HWBF	Phosphate ester
Density	0.85 to 0.9	1.05	0.95	1.00	1.2
Viscosity	Very low to very high			1 cSt	Very low to very high
Typical viscosity at 40°C (cSt)	30–70	45	65	1	60
Viscosity index range	50–200	140–275	150–210		<0–200
Average viscosity index	100	150	160		15
Vapour pressure	Low	High	High	High	Low
Corrosion resistance					
liquid phase	Good	Good	Good	Good	Good
vapour phase	Fair	Fair–Poor	Fair–Poor	Fair–Poor	Good (unless contaminated by water)
Stability	Excellent	Good	Poor	Poor	Good
Lubricity	Excellent	Limited	Limited	Limited	Excellent
Bulk fluid temperature (°C)	−30 to 80	−30 to 65	−10 to 65	−10 to 65	20 to 150
Optimum operation temperature (°C)	40	25	25	25	65
Fire resistance	None	Good	Good	Excellent	Excellent
Compatibility with:					
seals	Neoprene, Nitrile (Buna N), Viton, *not* butyl	As for mineral oil, *not* leather	As for mineral oil, *not* leather	As for mineral oil, *not* butyl, ethyl-propylene, silicon	Butyl, Viton, Teflon
metals		*not* zinc, cadmium, magnesium, aluminum	*not* magnesium, cadmium	*not* copper, aluminum	Can be problems with aluminum
paint		Attacks most paints			Attacks most paints
Other materials		*not* cork, leather, paper	*not* leather, cork, asbestos, paper		
Special precautions and problems		Down rate systems to 40%	Down rate systems to 50%	70 bar maximum, 1500 rev/min maximum. Good filtration required. High leakage. Pump inlet conditions are critical.	Down rate to 80–85%. Toxic. Too high viscosity at low temperatures.

Section 5.2 Fluid contamination control

(HFAS fluids) and the oil-in-water microemulsions (HFAE fluids where the oil is dispersed as microscopic globules in the water phase). In order to use high water-base fluids, special corrosion-resistant materials are being developed for hydraulic components and bearings.

5.2 FLUID CONTAMINATION CONTROL

All substances suffer from contamination or impurities. Hydraulic fluids are no exception, and even in the cleanest, finest filtered system available, a degree of contamination is always present. Contamination is very difficult to define; the easiest way is to consider that it is an alien substance or effect. Contamination can be by:

(a) Substances such as solids, liquids or gases.
(b) Energy: noise, radiation, heat, light.

Hydraulic systems are more usually concerned with a substance affecting the fluid, but excessive heat energy can cause severe degradation of the fluid with subsequent system failure. Even electromagnetic radiation can contaminate a hydraulic system by causing a malfunction of electronic controls. The noise generated by a hydraulic system contaminates the environment and has to be considered in system design and component selection.

This section will deal mainly with the substances which contaminate hydraulic systems and in particular solid and liquid contaminants, but the other forms of contamination and their effects must be understood.

5.2.1 Energy contamination

Heat energy

Too high or too low a fluid temperature can cause a system to malfunction. If the fluid over-heats it may:

(a) Give off vapor and cause cavitation of the pump.
(b) Increase the rate of oxidation causing rapid deterioration of the fluid by producing sludges, varnishes, etc. thus shortening its useful life.
(c) Reduce the viscosity of the fluid resulting in increased leakage both internal and external.
(d) Cause thermal distortion in components.
(e) Damage seals and packings owing to embrittlement.

As the fluid temperature decreases its viscosity increases. This in turn causes an increased pressure drop in a pipe or components for a given flow rate. Should this occur in the pump suction line, cavitation may result in extreme cases. Extremes of temperature are almost certain to shorten the service life of system components.

When more energy than is required is put into a system, the excess is converted into heat energy due to pressure drops across components.

- Major heat generation is inherent in pressure and flow control valves where the excess power consumed is dissipated as heat into the fluid or to a lesser extent by localized heating of the component.

Additional but usually less critical sources are:

- Resistive pressure drops in pipes, orifices and passageways through components.
- Leakage flows in pumps, motors and valves.
- Friction between moving mechanical parts, seals etc.
- Rapid cycling of gas-charged accumulators which can increase the gas temperature above that of the fluid causing a heat transfer into the fluid.
- Compression heat created in pumps by entrained air.

Heat may be absorbed into the system from outside sources – prime movers, adjacent furnaces etc.

Hydraulic systems should be designed so that a heat balance occurs at a satisfactory operating temperature. This may be achieved by dissipating excess heat to the environment or rejecting it into the coolant in a heat exchanger. In a cold environment, it may be necessary to install heaters to maintain the fluid at a suitable minimum temperature.

Magnetism

Magnetic fields cannot be avoided in hydraulic systems; they will occur near electric motors and solenoid valves and will attract magnetic particles. The effects can be minimized by correct component design but for the user the only answer is to filter out the particles.

Electrostatics

Again the system designer has little control over this problem which is fundamental to some components. An electrostatic charge is generated when there is friction between dissimilar materials followed by their separation. The material need not be solid as it is possible for static electricity to be generated as a gas or liquid moves across another gas, liquid or solid surface. The sudden discharge of an electrostatically-charged component can produce a spark which in the presence of flammable vapours may cause an explosion. However, the major problems associated with hydraulic systems are the blocking of orifices and clearances and alterations to filter efficiency owing to the cohesion of charged particles. Pitting of highly-stressed areas results from the weak electrolytic solution formed and one effect particularly associated with phosphate ester fire-resistant fluids is corrosion and erosion of orifices and seals.

5.2.2 Gaseous contamination

Most hydraulic systems work in an atmospheric environment and therefore air is the commonest gaseous contaminant. Air can exist in a hydraulic fluid either in a dissolved state or as a suspension in the form of small bubbles or air pockets. Dissolved air will not cause any difficulties in hydraulic systems provided it remains dissolved. Free air in suspension creates many problems:

(a) Cavitation – generally associated with pumps and motors.
(b) Loss of system stiffness – owing to compressibility of the air.
(c) Reduction in life of fluid – increased oxidation rate.
(d) Noisy operation – air-bubble collapse.
(e) In extreme cases it will cause over-heating, loss of power and reduced lubricity.

Cavitation can be defined as the formation of cavities within a substance. In fluids these cavities will be air or vapour bubbles and it is the collapse of these bubbles on entering a high pressure zone which causes cavitation damage. The damage can be very severe as the metal surface is 'chipped' away and a component may become unserviceable within seconds. Hydraulic pumps and motors are particularly at risk. Cavitation is also the formation of a partial vacuum between a fluid and a solid surface moving relatively to each other. Ships' propellers are susceptible to damage from this source.

All fluids contain dissolved gas, the amount of gas depending upon the type of fluid, the particular gas, the temperature and pressure in the system. If a fluid is fully saturated with air, it cannot dissolve any more air at that pressure and temperature. The amount of air which can be dissolved increases with pressure and decreases with temperature. Hydraulic oil at atmospheric pressure and at $20°C$ will dissolve 10% by volume of air. If the pressure is doubled the equivalent free volume of air which can be dissolved is also doubled.

Aeration control

As stated before, the ever-present dissolved air does not present a problem as long as it remains dissolved. It will, however, come out of solution if there is an increase in fluid temperature or a decrease in pressure. The level of air saturation is set in the fluid reservoir which is normally at atmospheric pressure; any reduction below this pressure will generate free air. The main point where lower pressures are found is in the pump suction line. Care must be taken in the design to avoid low suction pressure.

Any increase in temperature may cause air to come out of solution. Local increases in fluid temperature occur in the circuit at flow and pressure control valves; however the system pressure is higher than atmospheric and the air remains dissolved. Air will come out of solution when the pressure is reduced, i.e. in the return lines to the tank. If this occurs, foaming of the hydraulic fluid results.

Air may be entrained in the fluid in the reservoir or by leaks in the pump suction lines; it can also enter past the cylinder rod seals. Tanks with return lines terminating below the fluid level and the use of a wire mesh air separator will reduce air entrainment. A large fluid surface area in contact with air and excessive agitation encourages aeration but a rubber diaphragm can be used to separate the fluid from the air. Fluids also have built-in additives to reduce foaming. If return lines have anti-syphon holes, the jet of fluid emerging from these can cause unnecessary aeration. A strategically placed deflector plate will significantly minimize this problem.

The reservoirs in Figure 5.2 include some of the features used to control aeration.

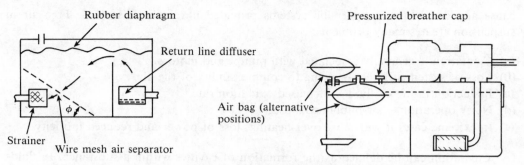

Figure 5.2 Reservoir features which reduce aeration.

Effects of air bubbles entrained in the fluid in the reservoir can be reduced by:

(a) Anti-foaming additives in the fluid.
(b) Terminating return lines with diffusers below the fluid surface.
(c) The use of a suction strainer.
(d) The use of a wire mesh separator.

For (d) the wire mesh should be between a 60 and 200 mesh size (apertures between 250 and 75 μm square). The angle of inclination ϕ of the mesh should be between 20° and 40° and the entire separator should at all times be below the surface of the oil. Its purpose is to encourage the bubbles to combine together and hence disperse more easily. The flow velocity perpendicular to the mesh should not be greater than 1 m/min. A suction strainer helps prevent air bubbles entering the pump inlet.

A sealed reservoir using a diaphragm, bladder or piston to separate the fluid from air, can eliminate absorption. The bladder etc. is used to take up any change in fluid volume which occurs within the reservoir. An additional advantage of a sealed reservoir is elimination of the ingress of dirt. The main disadvantage is a limit to volume variations because any fluid leaks which occur must be immediately made up in the reservoir.

Air can enter the system past piston rod seals and collects in the cylinder resulting in a spongy movement. The air then has to be bled out of the cylinder.

When a filter element is changed it is possible for a large volume of air to enter the system and this may prove difficult to clear.

5.2.3 Liquid contamination

Contamination by a liquid may be by the residue of pickling or cleaning agents left in the system, but water is the most usual liquid contaminant. Water can enter a system either by leakage from coolers, by condensation in the oil reservoir, or even directly. (It has been known for hydraulic systems to be cleaned down using high-pressure water jets, and for reservoirs located outside to be left open without lids.)

Minute quantities of water dissolve in mineral oils as discrete molecules. The amount depends upon the type of oil, viscosity and temperature but is typically between 100 and 1000 parts per million. When the saturation point is exceeded, any additional water will appear as free droplets. Sometimes these droplets combine with the oil and form an emulsion when the mixture is agitated in a hydraulic circuit. The emulsion formed will tend to be 'permanent' if the interfacial tension is low (below 25 mN/m). Additives, oxidation products and dirt tend to stabilize the emulsion. There will be some demulsification if the interfacial tension is greater than 35 mN/m. Sometimes it is possible to drain free water off in the reservoir after the mixture has been standing.

The ability of the mixture to separate out the water is effected by additives in the oil. Additives may exist in the oil or be added to it, to either maintain the emulsion or disperse it.

Water-in-oil is extremely detrimental to the operation of the system in the following ways:

(a) It causes corrosion and rust.
(b) It increases wear rate owing to reduced lubricating film thickness and accelerated fatigue of metal surfaces caused by corrosive etching.
(c) It reacts with the oxidation inhibitor additives reducing their protective properties, and generally causes fluid breakdown.
(d) It reacts with some additives, in particular with anti-wear additives to form a sludge or slime which rapidly blocks filters and valves having fine clearances.
(e) In low-temperature systems, the water will become minute ice particles which act as dirt contamination causing system malfunction.
(f) It promotes the growth of bacteria in the fluid.

Water removal

This is extremely difficult because the oil dissolves the water and also forms emulsions which may be very stable. It is far better and cheaper to guard against water contamination than try to remove water from a contaminated oil.

If the fluid is left standing there may be some separation of the water in the oil reservoir; it can then be drained off. Settling will not remove the dissolved water. A centrifuge will increase the separation rate and is often used to remove free water; it is not very effective on emulsions.

Natural evaporation of the water from the fluid in the system depends upon the fluid temperature and the humidity of the air. As most of the evaporation takes place within a closed reservoir, the air above the fluid will soon become saturated. Natural evaporation has almost no effect on the water content of the fluid.

A method of drying oils is to pass a dry gas through the oil. If the gas used is air, this oxidizes the oil, so it is best to use nitrogen.

Adsorption and absorption can be used: *adsorption* relies on the intermolecular attraction of the water to certain materials such as Fuller's earth, silica gel etc. *Absorption* is the mechanical removal of water by trapping water in a media such as dry paper or a polypropylene medium containing a water-absorbing polymer.

Many methods of drying will remove some, or all, the additives from the oil. Frequently, if the oil is contaminated with water it is cheaper to refill with clean oil and, if the quantity warrants it, send the contaminated oil to a company specializing in oil reclamation. Some oil companies operate a mobile oil-reclamation service which involves bringing equipment to the site and cleaning the oil *in situ*. Even this will not return the oil to the original condition as there will be a loss of additives. This method is only suitable for very large systems (2000 liters or above).

A filter containing water-absorbing polymers does not remove additives and will retain free and emulsified water but not fine particulate contamination. Methods of dealing with the latter are discussed at length in this chapter. The two should be used together as part of a contamination-control system.

5.2.4 Microbiological contamination

Microorganisms exist practically everywhere, but they multiply and grow where they can find food, moisture and warmth. In hydraulic systems at least two of these prerequisites for growth exist: warmth and food in the form of hydrocarbons. If the fluid is water-based, the third requirement is fulfilled. If the fluid is a 'dry' mineral oil, there will be no bacteriological growth, but only a few parts per million of water in the oil will be sufficient to start growth.

Bacteriological growth has proved to be a problem in water-based fluids particularly the oil-in-water emulsions. Bacteria range in size from below 1 μm to 100 μm clumps. A bacterium can reproduce itself every 20 minutes. If this growth rate were sustained for 48 hours, it would produce 4000 times the mass of the earth. Fortunately it is not!

Fungi are found in many forms – yeasts, moulds, mildews, mushrooms – and are from a few micrometres to over a meter. Protozoa and algae are found in water and range in size from 1 μm to several meters.

Microbial contamination occurs as slimes and sediments. A fluid so contaminated may have a brown or green slime-like appearance. The effect on a system is to reduce fluid and filter life, increase corrosion, and produce obnoxious smells. Viscosity and lubrication properties are also affected. It is extremely difficult to control, but oil companies are carrying out intensive programs into developing suitable germicide additives.

With the increasing use of water-based and synthetic fire-resistant fluids, the contaminants already discussed are becoming of greater consequence. Nevertheless, 'dirt' or particulate contamination is the best known and most usual problem in hydraulic systems. More effort and research has been devoted to this aspect than any of the others. There are, however, occasions when the other forms are more serious. Generally it is far easier to remove particulate contaminants than it is to clean the other types of contamination from a system. An installation should be looked at with a view to protection from contamination generally and not just dirt filtration.

5.2.5 Particulate contamination (dirt)

Every hydraulic system is contaminated by dirt. It is present even before the components are assembled together. Solid contaminants or 'dirt' may originate as metallic particles from wearing parts, weld scale, and oxides, silicates from fabrications and castings, and elastomers from seals and hoses. Environmental dust is ingested through breathers and cylinders and rod seals. More dirt enters whenever the system is broken into for maintenance purposes. New oil from a manufacturer is not hydraulically clean. Particles of dirt are always present; the important thing is to filter out the harmful sizes.

Particle characteristics

PARTICLE SIZES

Particles are measured in 'micrometers', sometimes abbreviated to 'microns'. The symbol is μm:

$$1 \text{ micrometer} = \frac{1 \text{ meter}}{1\,000\,000} = \frac{1 \text{ millimeter}}{1\,000} = \frac{1 \text{ inch}}{25\,400} = 0.000\,039 \text{ inch}$$

$$0.001 \text{ inch} = 25.4 \; \mu\text{m}$$

The smallest spherical particle which a person with good eyesight can resolve is 40 μm diameter. An average human hair, which is about 75 μm (0.003 inch) diameter is visible mainly owing to its length. Other examples are: red corpuscles in human blood, which are 7.5 μm diameter, white corpuscles in human blood (25 μm diameter), and a 100 mesh has holes which are 149 μm diameter. The full stops on this page are each approximately 400 μm diameter.

FLUID SAMPLING AND PARTICLE COUNTING

Before analysis can take place it is important that the sample obtained is truly representative of the system fluid. There are various sampling procedures which are described in Chapter 7.

Counting may be either by using a microscope or an automatic particle counter. For the microscope method, a sample of the fluid is filtered through a membrane and the particles captured by the membrane can be sized, counted statistically and the type of contaminant determined. When carried out manually this is a slow process and requires an experienced interpreter. There are machines which will automatically scan the slide and provide more accurate results.

More sophisticated machines use a light interception method to count particles as they pass a window. The counting head is designed to withstand high pressures and therefore can be connected directly into a hydraulic system pipework enabling contamination levels to be monitored whilst the system is working. The operator is able to select various size bands and determine the number of particles in each, thus obtaining a profile of the size distribution.

CLEANLINESS STANDARDS

There are a number of contamination codes and particulate count classifications in use most of which have defense or aerospace origins: Def. Std 05/42, NAS 1638 and SAE 749 are commonly used. Table 5.3 is based on NAS 1638 where a classification number is given according to the number of particles of given sizes in a 100-ml fluid sample. It assumes a fixed particle-size distribution and problems arise when the pattern differs from that of the standard. Changes in the distribution profile are accommodated when using what is rapidly becoming the Industry Standard − ISO/DIS 4406 (BS 5540 part 4 and CETOP RP 70H are equivalent). In this standard, numbers are allocated to cover various

Table 5.3 Contamination classification to specification NAS 1638.

Class	\multicolumn{5}{c}{Size range (μm) for number of particles per 100-ml sample}
	5–15	15–25	25–50	50–100	100+
00	125	22	4	1	0
0	250	44	8	2	0
1	500	89	16	3	1
2	1000	178	32	6	1
3	2000	356	63	11	2
4	4000	712	126	22	4
5	8000	1425	253	45	8
6	16 000	2850	506	90	16
7	32 000	5700	1012	180	32
8	64 000	11 400	2025	360	64
9	128 000	22 800	4050	720	128
10	256 000	45 600	8100	1440	256
11	512 000	91 200	16 200	2880	512
12	1 024 000	182 400	32 400	5760	1024

bands or ranges of particle numbers. The bands have been devized using a step factor of two which keeps the total number of ranges to a manageable but meaningful level (Table 5.4).

The contamination code consists of two numbers. The first represents the band of particles greater than 5 μm and the second a band of particles greater than 15 μm, e.g. a 100-ml sample which has 200×10^3 particles greater than 5 μm and 7.5×10^3 particles greater than 15 μm will have the code number 18/13.

An alternative and more usual method of presentation is shown graphically on a standard form based on log/log^2 graph paper (see Figure 5.3). Results of particle counts on a fluid sample are plotted and where the line crosses the 5 and 15 μm lines gives the code, e.g. 18/13. Although the standard produces a profile which is intended for fluid specifications, it does not recommend or specify acceptance levels. It basically classifies levels for silt and chip size particles (these are defined later in this section) and the customer can use the code to specify target levels for his particular equipment.

New hydraulic oil contains large numbers of dirt particles of a size which again will

Table 5.4 ISO Solid contaminant code (ISO 4406).

Band No.	1	2	3	4	5	6	7	8	9	10	11	12
No. particles in 100 ml fluid sample From	1	2	4	8	16	32	64	130	250	500	1×10^3	2×10^3
To	2	4	8	16	32	64	130	250	500	1×10^3	2×10^3	4×10^3

Band No.	13	14	15	16	17	18	19	20	21	22	23	24
No. particles in 100 ml fluid sample From	4×10^3	8×10^3	16×10^3	32×10^3	64×10^3	130×10^3	250×10^3	500×10^3	1×10^6	2×10^6	4×10^6	8×10^6
To	8×10^3	16×10^3	32×10^3	64×10^3	130×10^3	250×10^3	500×10^3	1×10^6	2×10^6	4×10^6	8×10^6	16×10^6

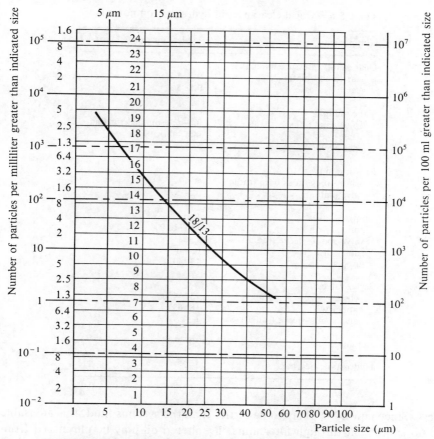

Figure 5.3 ISO solid contaminant coding graphical representation.

damage systems. A typical dirt size analysis of new 'clean' hydraulic oil is given in Table 5.5. This corresponds to NAS 1638 Class 9 (Table 5.3). Samples vary considerably from batch-to-batch, company-to-company and with size of container. The size bands chosen for this particular analysis do not coincide with those used in ISO/DIS 4406 but Class 9 is approximately equivalent to 18/15.

In general, most new oil as supplied by the manufacturer will have a contamination level of 16/11. Usually large containers have less contaminants per unit volume than small

Table 5.5 Particle analysis from a 100 ml sample of new hydraulic oil.

Range (μm)	Number of particles
5–10	128 000
10–25	42 000
25–50	6 500
50–100	1 000
>100	92

Table 5.6 Typical clearances in hydraulic components.

Component	Typical clearance (μm)
Gear pump (pressure-loaded)	
gear to end plate	0.5–5
gear tip to case	0.5–5
Vane pump	
tip of vane to case	0.5–1 (estimated for thin film lubrication)
sides of vane	5–13
Piston pump	
piston to cylinder bore (radial)	5–40
valve plate to cylinder	0.5–5
Control valves	
control orifices	130–10 000
valve spool–sleeve (radial)	1–23
rotary disc type	0.5–1
poppet type	13–40
Servo valves	
orifice	130–450
flapper wall	18–63
valve spool–sleeve (radial)	1–4
Actuators	50–250
Hydrostatic bearings	1–25

containers. Since new oil is likely to be dirtier than the systems fluid, it is advisable to fill and top up equipment through filter units. Prefiltered oil may be purchased from some manufacturers but the cost is usually prohibitive and it can easily become contaminated during handling before it reaches the system.

From examination of Tables 5.5 and 5.6 it is obvious that any hydraulic system must contain a large number of particles of various sizes. The more nearly the particle sizes approach the component critical clearances, the more likely they are to cause a *catastrophic* failure such as the jamming of a valve spool or a pump rotor. The smaller particles tend to cause *degradation* failure, that is a reduction in efficiency or increased internal leakage in components over a period owing to wear. Obviously component degradation can eventually lead to a catastrophic failure.

Abrasive wear has been found to be predominantly caused by particles in the size range 1 to 5 μm. These sizes are within the category sometimes referred to as 'silt'. Particles above 10 μm are referred to as 'chips' or settling size.

Origin of dirt in hydraulic systems

In the manufacture of hydraulic components, valves, pumps, actuators etc., moulding sand, machine swarf, and grinding paste will be left in the component. Machine turnings left inside the housing of a solenoid valve can cause the spool to jam, burning out the solenoid. More dirt in the form of rust, paint, rubber particles, lint etc., is introduced into

Section 5.2 Fluid contamination control

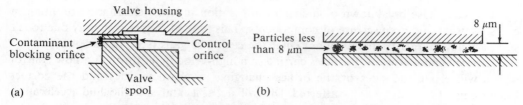

Figure 5.4 Dirt build-up. (a) Particles blocking orifice. (b) Silt in clearance.

the system during assembly. Whenever a threaded connection is made, some metal particles will be shed particularly from plated components. It must be remembered that cleaning components with cloths or air blast will not remove the fine dirt which is so damaging to hydraulic systems.

Dirt is generated within the hydraulic system wherever metal surfaces in contact are in relative motion. If there is dirt within the system, the wear process accelerates and becomes a chain reaction. A high flow velocity through a control orifice can erode the surface of the orifice and cause a system to malfunction.

In a system using cylinders, the surface level of fluid in the oil reservoir is constantly changing as the cylinders extend and retract. To ensure that the oil reservoir remains at atmospheric pressure, a combined air breather/filler is fitted into the tank top. This allows air to flow in and out of the tank. The breather allows some dirt to pass through, which adds to the contamination within the system. An alternative to the tank top breather is a rubber diaphragm or balloon which expands or contracts as the oil level varies as shown in Figure 5.2. The diaphragm completely seals the tank and eliminates the ingress of air carrying dirt particles. Pressurized oil reservoirs are used in applications where a positive head is essential at the inlet port of the pump. The reservoir is pressurized by compressed air which must be dried and filtered. Dirt will also enter the system through cylinder rod seals, badly fitting oil tank joints, open filler caps, etc. Preventative maintenance can reduce the ingress of dirt quite considerably.

Maintenance work such as replacing components, topping up fluid etc., introduces large quantities of contaminants.

Dirt falls into three size categories:

1. Settling size which is greater than 10 μm.
2. Silt size which is in the range 0.5–10 μm.
3. Smoke size which is less than 1 μm.

Silt-sized particles can be just as damaging as larger particles because of the minute clearances in the hydraulic components. It is estimated that in a typical hydraulic system there are more than 10 million particles greater than 1 μm in each liter of hydraulic oil.

Effect of dirt on a system

A hydraulic fluid has to transmit power, lubricate surfaces, transfer heat energy from the system to the reservoir and to seal clearances between close-fitting components. Dirt will affect those functions. A build-up of dirt across small control orifices alters their characteristics and eventually may cause a malfunction (see Figure 5.4(a)). This type of failure is caused mainly by chip size particles greater than 10 μm.

Dirt can cause breakdown of fluid film lubrication in small clearances, resulting in excessive wear and even seizure of components. This type of failure is mainly due to silt size particles. Larger particles cannot enter the clearance gaps (see Figure 5.4(b)).

Dirt can cause erosion. Particles carried in a high-velocity jet through a flow control orifice will erode the sides of the orifice changing its characteristics and the correct functioning of the valve may be affected. Dirt will act as a catalyst in the fluid accelerating oxidation, sludge formation and fluid degradation.

To summarize system failures; these fall into two distinct categories:

(a) *Catastrophic Failure* These occur without warning and are sudden system breakdowns usually caused by large particles. A spool valve jamming, a pump or motor seizing, a control orifice blocking would be catastrophic failures.

(b) *Degradation or Gradual Failures* These are caused by a large number of small particles over a long period of time. Failure occurs when the performance of a component falls below an acceptable level and has to be replaced. An example is the gradual reduction in pump output as a result of increased internal leakage. Degradation failure can bring about catastrophic failure; a valve spool can start sticking, resulting in a solenoid burning out and then complete machine breakdown.

Thus to fully protect a system, both large and silt-size particles must be removed. Extraction of particles of 10 μm and above should prevent catastrophic failure. This is considered adequate for many systems, but where solenoid valves, precision-flow control valves, or servo valves are involved, filtration down to 3 μm or finer is recommended.

5.3 FILTER CONSTRUCTION AND FILTRATION TECHNOLOGY

5.3.1 Filter construction

Filter housing

Particle removal is by means of a filter such as is shown in Figure 5.5. This consists of a housing which is permanently installed in the system. It has a detachable bowl containing a replaceable filter element to capture the dirt particles. The filter unit depicted has a

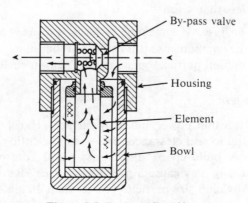

Figure 5.5 Pressure line filter.

low-pressure relief valve which will open and bypass fluid flow when the element becomes clogged. Most filter units incorporate an indicator to show when the element needs replacing.

The type of housing depends upon the position it is going to occupy in the circuit. If it is going to be subjected to high pressure it will be made from a steel forging or casting. For return-line and suction-line applications low-cost pressed-steel casings are used.

Filter elements

Elements are broadly classified into two types: surface and depth.

SURFACE-TYPE ELEMENTS

In this type all the pores are approximately the same size and of regular distribution. They have a sharp cut-off and tend to be effective against all particles greater than the rated size. Minimum particle retention depends upon the form of construction. Scalloped or etched discs (5 μm), wire wound tubes (5 μm), sintered metal (3 μm), porous ceramic (1 μm), woven fabric (10 μm), perforated metal (10 μm), woven metal screen (5 μm), and cellulose membrane (submicrometer).

Some surface-type elements can be cleaned in ultrasonic baths and by back-flushing.

In a woven wire edge-type filter, particles larger than the aperture between wires are trapped on the surface of the element, building into a layer. Initially it will block only particles larger than filter rating, building up a layer on the mesh. This produces a caking effect and the filter stops progressively smaller particles as the layer forms.

DEPTH-TYPE ELEMENT

Most modern elements are manufactured from inert fibers impregnated and bonded with epoxy resin. The fluid has to flow through an appreciable depth of element and dirt is trapped in the tortuous flow pattern within the material. The holes must not be a uniform size so that particles are captured in the various zones inside the material. Therefore all the zones contain some holes greater than the filter rating and it is possible for particles to migrate through the media particularly when subjected to flow and pressure surges. Consequently its rating which might be submicrometer is essentially statistical and it will stop a considerable number of particles less than the rated size but pass some which are larger. The main advantage of this type of element is its large dirt-holding capacity. The material is usually supported on both sides by a woven wire mesh to prevent collapse and it is pleated to present as large a surface area as possible. This and some other types of depth filter absorb contaminants and retain particles by mechanical means, particles being trapped in pockets and at pore openings and constrictions in the media. Small particles in stagnant areas attach themselves to the fibers by electrostatic forces, or are held in place by fluid pressure. Particles which are larger than the openings become wedged and are held by friction forces.

Minimum particle retention of this and similar types of depth filter again depend upon element media and construction: glass fiber and epoxy (0.5 μm), paper (5 μm), felt (10 μm), and yarn wound spools with graded fibers (2 μm). The figures are very much affected by flow conditions.

Another type of depth filter contains powder material such as diatomaceous earth, Fuller's earth, charcoal, active alumina or clay. These have sub-micrometer ratings and

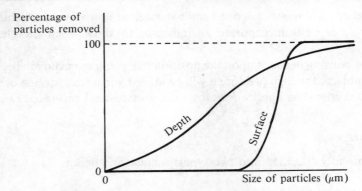

Figure 5.6 Characteristics of surface and depth filters.

absorb contaminants retaining them by molecular adhesion. The differing characteristics of surface and depth filters are shown in Figure 5.6. Surface types have a definite 'cut-off' at a specific particle size whilst the performance of depth-type elements is of a statistical nature.

Generally surface-type filter elements can be cleaned by back-flushing whilst most depth-type elements cannot and are thrown away when dirty. Certain filter elements exhibit characteristics of both types, e.g. sintered and ceramic elements are sometimes classified as depth filters.

Magnetic separators

Magnetic separators are sometimes used to catch large ferrous particles. These may consist of magnetic plates fitted into the oil reservoir or magnetic plugs built into filters or components. It must be remembered magnetic separation can only remove large ferro-metallic particles well above silt size. A range of magnetic filters is commercially available but must only be used in hydraulic systems together with conventional filtration methods.

Electrostatic filtration

Static electricity is a fluid contaminant but paradoxically the principle is used for oil purification. An electrical current is passed between two electrodes immersed in the fluid. Particles in the fluid become electrically charged and attach themselves to a collection plate situated between the electrodes. Manufacturers claim that particles as small as 0.05 μm can be removed by this method, in which case care must be taken as certain additives are likely to be removed either directly or 'on the back of' other particles.

5.3.2 Filtration technology

Filter capacity

The rated flow through a filter will depend upon the pressure drop over the element. Surface-type elements have a large percentage open area (30% is typical) which results in a relatively low pressure drop for given flow rate compared with a depth-type filter.

As the filter removes dirt from the fluid, the flow passages become partially blocked and the pressure drop over the element increases. A bypass valve is built into many filters; this valve opens when the pressure drop over the element reaches a predetermined value.

Many filters are fitted with an indicator (either visual or electrical) which shows the state of the element and indicates when the filter needs changing. The indicator must operate before the bypass opens otherwise dirty fluid will pass into the system. The 'contamination capacity' of a filter is an indication of the weight of contaminant the filter can hold for a given pressure drop under specified conditions. It does not give any indication of the life of a filter element in the circuit as this depends upon the cleanliness of the fluid and the environmental conditions.

NOMINAL RATING

Although derived from an obsolete US military specification, this is an arbitrary micrometer rating indicated by the filter manufacturer. Owing to the lack of standardization and reproducibility, this rating is of little value.

ABSOLUTE RATING

This is the diameter of the largest hard spherical particle which will pass through the filter element under specified test conditions. It is an indication of the largest opening in the filter element.

BUBBLE POINT TEST

This is a simple method for accurately determining the absolute rating of wire mesh filters. It will also give an approximate absolute rating for disposable filter elements. The test consists of immersing the element under test in a fluid of known surface tension, applying air pressure to the inside of the element whilst it is immersed in the fluid. The air pressure is gradually increased whilst the element is rotated slowly until a stream of bubbles appears. For a wire cloth the pressure is inversely proportional to the maximum pore opening.

MEAN PORE SIZE OR MEAN FILTRATION RATING

This is a measure of the average size of pores in a filter element.

All the above filter ratings apply to clean elements with Air Cleaner Fine Test Dust (ACFTD) (see below) or hard spherical dirt as the contaminant. Dirt particles are seldom spherical and may be long slivers of metal or strands of material. They could be 10 μm in diameter and 200 μm long and could pass through a 10 μm absolute element. The chance of this occurring is reduced as caking of the element progresses or by using a depth-type filter which has tortuous paths.

BETA RATINGS

The *beta ratio* indicates the efficiency of contaminant removal from the fluid flowing through filter elements.

$$\text{Filtration ratio } \beta_x = \frac{\text{Number of upstream particles larger than } x \text{ } \mu\text{m}}{\text{Number of downstream particles larger than } x \text{ } \mu\text{m}}$$

i.e. a filter having a β_{10} ratio equal to 2 will only stop 50% of particles of 10 μm and greater; a filter having a β_{10} ratio equal to 10 will stop 90% of particles 10 μm and greater. The higher the ratio the more efficient the filter element.

The β ratio is obtained using the ISO multi-pass test which uses an artificial contaminant (ACFTD). This has a known consistent particle shape and size distribution. Tests are carried out at a steady flow rate with artificially-high contamination levels. In practice the filter performance can be significantly affected by cyclic flow, differential pressure, cold starts, bypass leakage, filter-housing construction and fire-resistant fluids.

The micron size at which $\beta_x = 75$ is generally considered to be equivalent to the absolute rating of the filter.

Filter-element collapse pressure

The majority of filter elements are designed to withstand a maximum pressure difference across the element of 10 bar. This does not mean that they are limited to low-pressure applications. These elements may be used in filters where the circuit pressure could be, say, 400 bar. Provided the pressure drop across the element does not exceed 10 bar then the element will remain intact. They are used in pressure, return or suction-line applications and must always be protected by a bypass valve which usually opens at a pressure differential of 3–4 bar (0.1 bar in suction filters).

Specially-reinforced disposable elements which can withstand a differential pressure of 210 bar are available for use in housings without a bypass valve. As the element gets dirty, the pressure drop across it increases blocking the fluid flow; consequently they are referred to as a 'dirt fuse'. The main disadvantage, apart from the cost (usually three to four times the cost of a low-pressure element), is that if the element gets extremely dirty, a severe reduction in output power and excessive heat generation will occur. A differential pressure of 7 bar is considered to be a suitable figure at which the element should be changed. Recleanable sintered metal elements are available with very high burst and collapse pressure ratings. They are often used as 'last-chance' filters in dirt-sensitive components such as in the pilot supply to a servo valve.

Bypass valves

The maximum permissible pressure drop across a filter element under contaminated conditions is sometimes referred to as the 'terminal pressure'. For an element which has a collapse pressure of 10 bar, the terminal pressure would be in the order of 3 bar. Under ideal conditions an indicator would give warning at 2.5 bar and bypass valve crack open at 3.5 bar.

Bypass valves are in three basic forms:

(a) A check valve built into the filter housing head (see Figure 5.5). As the filter blocks, the pressure differential across the element approaches the value of the check-valve spring and the valve opens allowing some of the flow to bypass the element.
(b) A check valve built into the filter element. With some versions of this type, a new bypass valve is automatically fitted with each element change.

(c) The element is spring-loaded and as the differential pressure increases the element moves until it opens allowing part of the flow to be bypassed.

Whenever a filter is bypassing, unfiltered fluid will be fed into the parts of the circuit which the filter is intended to protect. Unfortunately, in even the best filters the bypass valves tend to leak, especially when subjected to pressure and flow surges. This significantly reduces the filter efficiency.

Condition indicators

Some means of monitoring the condition of a filter element is essential and indicators are available for most makes of filter. Three basic types are used:

(a) A pressure gauge is in the filter inlet port. A reading has to be taken when the filter is new and working under normal conditions. The gauge is then marked to show at what pressure the element should be changed. This is a very unsatisfactory method and only suitable for low-pressure applications.
(b) A mechanical device works off the bypass valve. As the element clogs and the bypass gradually opens, a mechanical linkage operates an external pointer indicating the filter condition. The pointer may be used to operate an electrical micro-switch. In some designs the linkage has to be adjusted each time a new element is fitted and if this is overlooked false indication occurs.
(c) A differential pressure switch monitors the pressure drop across the filter element. This may be a mechanical visual indicator or an electrical pressure switch. These tend to be the most accurate types of indicator. Generally the device incorporates a spring-loaded piston which has the pressure upstream of the element connected to the end opposite the spring whilst the pressure downstream of the element assists the spring. As the filter gradually blocks, the upstream pressure increases relative to the downstream pressure and the piston moves compressing the spring. In some designs a magnet is built into the piston to actuate the visual indicator button or electrical switch. This not only prevents leakage, as the magnetic field operates through a closed end, but also provides a 'snap action'. A further feature available is a bimetal thermal 'lock out' which prevents the switch operating under cold start conditions. (At low temperatures the viscosity of the fluid increases causing a higher pressure drop across the element and without a thermal lock-out the indicator will interpret this as a 'dirty' element.)

On some installations where a bypassing filter is likely to cause damage to sensitive components, dual indicators may be fitted. The first gives a warning that the element needs changing and if this is unheeded, the second which operates at a slightly higher pressure closes the system down before the bypass valve opens.

Where filters without indicators are used, the elements must be changed at regular service intervals.

Filter elements which are subject to cyclic pressure changes may fail by fatigue. If an element has ruptured the indicators will show a 'clean' element. For this reason elements should be replaced at regular service intervals (say annually) even if a change has not been shown by the indicator.

A filter with bypass valve and electrical indicator can be shown symbolically as

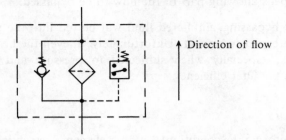

Filter contamination–time curves

Where a filter is fitted in a circuit, it will reduce the fluid contaminant to a particular level and maintain that level. If the filter is removed or starts to bypass the element, the concentration of dirt rapidly increases. Figure 5.7 shows a typical contamination–time curve (plotted over a number of weeks) on a piece of equipment originally fitted with a 25-μm nominal filter element. When it is replaced with a 3-μm absolute element a rapid initial decrease in the dirt count is achieved followed by further gradual improvement. Changing back to a 25-μm nominal element soon causes the cleanliness level to deteriorate back to its original level.

It is therefore absolutely essential that as soon as a filter element blocks or bypasses, a new element of the correct rating be installed.

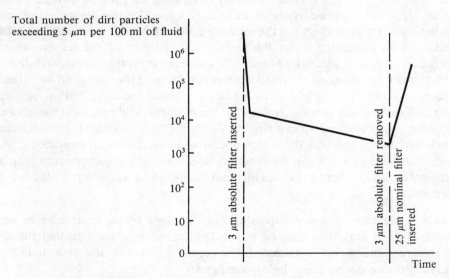

Figure 5.7 Typical contamination–time curve.

5.3.3 Filter location

As dirt is generated at all points in a hydraulic circuit, the only way to ensure that a component has 100% protection is to place a filter immediately upstream of the component. This is done in the case of a component which is particularly susceptible to contamination damage, for example, a servovalve usually has its own filter. Generally filtration is a compromise and attempts to give adequate protection to all components.

Figure 5.8 shows various positions in which filters can be inserted in a circuit. It is not intended that they should be used in all these positions – just one or two are most usual.

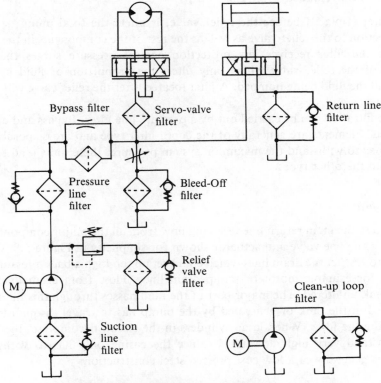

Figure 5.8 Some possible filter locations.

Suction-line filter

In the suction line between the reservoir and the pump, the pressure drop across a suction filter has to be very low to avoid pump cavitation, and a low-pressure (0.1 bar) bypass valve is essential. Elements have absolute ratings, in the range 25–125 μm depending upon the permissible pressure drop. Although the typical suction filter of 74 μm seems rather coarse, the effect of caking on the element removes a large number of particles much smaller than the absolute rating. The element is often woven wire and can be cleaned. A suction filter is to be recommended for applications where it is essential to protect the pump; it may not offer adequate protection to components in the system.

Because of the low pressures involved, the housing is frequently a low-cost light metal pressing. Some types of filter are mounted within the reservoir.

A *suction-line filter* must not be confused with a *suction strainer* which is a very coarse mesh about 125–250 μm fitted in the oil reservoir in the suction line. The strainer stops large debris and helps prevent air bubbles from entering the pump inlet. It does not usually have a bypass valve.

In some large systems a boost pump is used to feed the main hydraulic pump. In this case a fine rated filter may be used in the line between the boost pump and main pump.

Pressure-line filter (placed after the pump)

When the filter is placed before the relief valve, it filters the total pump delivery and affords protection to the relief valve as well as the rest of the components in the circuit. In this position the filter receives less protection against pressure surges than a filter downstream of the relief valve which only filters the proportion of fluid used by the system, not all the fluid being pumped. A filter located after the relief valve will be subject to flow variations and surges.

Very fine filtration can be carried out by a pressure line filter. Bypass and non-bypass types are used. Elements are generally of the depth-filter type and are disposable. As the filter casing has to withstand the maximum system pressure, these units tend to be more expensive than the other types.

Return-line filter

This is located in the main return line receiving flow from all the major components in the circuit not just the one valve and actuator shown for simplicity in Figure 5.8. Care must be taken not to restrict the drain lines which could set up too high a back-pressure in some components, for example motor or pump drain lines. Flow from drain lines is often returned to tank unfiltered. The major part of the fluid passes through this filter, but any dirt introduced in the tank or generated by the pump has to travel through the system before reaching the filter. When large cylinders in the circuit decompress, the filter may have to withstand very high flow rates. Since the casing only has to withstand low pressures, it is generally of a low-cost pressed-steel construction.

Relief-valve filter

This is a return-line filter and operates at lower pressures than a pressure-line filter and can therefore be of cheaper construction. As the pressure drop across the filter is not usually a prime consideration, a fine degree of filtration, down to 3 μm or less can be used. The flow over the relief valve is spasmodic so the filter is subject to flow surges which can cause migration of contaminant through the element and even degradation of the filter media. The filter only cleans a part of the fluid flowing and this is not a constant fraction of the total flow. It is difficult to envisage any advantage for this method when compared with other possible locations.

Bypass and bleed-off filters

On exceptionally large systems, where the cost of a full flow filter is considered to be prohibitive, partial-flow filtration is used. The theory is that passing some of the flow through a small filter will eventually clean up the system. This must not be used where sensitive components are involved, as quite large particles will miss the filter.

Section 5.3 Filter construction and filtration technology

The *bypass type* is a pressure-line filter with a built-in venturi device which puts part of the flow through the filter and part directly to the system. As the filter element blocks, more and more flow is bypassed.

The *bleed-off method* uses a return-line filter. A flow-control valve bleeds-off part of the flow through the filter to tank. The major disadvantage is power wastage because the filtered flow drops from system pressure to atmospheric pressure across the flow control and filter.

Servo valve filter or component filter

These are pressure-line filters used to give protection to special components having extremely fine clearances. A main-line filter is also used and the component filter removes any contaminant generated between the main filter and the component, or contaminant which has passed through the main filter. The component filter should have a 3 μm absolute rating or less ($\beta_3 \geqslant 75$) to give adequate protection and a non-bypass type 'dirt fuse' is preferable.

Miniature last-chance chip filters

Very small sintered discs and miniature filters are often fitted into the control ports of sensitive components as a final line of resistance. These tend to require frequent cleaning but in some instances may be self-cleaning. For example, when fitted into the service ports of a sensitive valve to prevent contamination from the actuator reaching the valve, they are back-flushed every time the actuator is reversed.

Reverse-flow filters

If conventional filters are fitted into the main lines of a hydrostatic transmission, when the transmission is reversed all the accumulated dirt will be washed out of the filter into the pump suction – not a very satisfactory state of affairs. However, filters are available to overcome this problem and are shown symbolically in Figure 5.9(a). Reverse-flow filters have a special valve arrangement which allows fluid to pass through the element in one

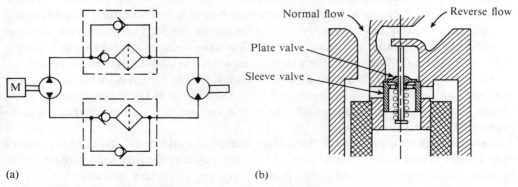

Figure 5.9 Reverse flow filters. (a) In hydrostatic transmission circuit. (b) Section.

direction but to by-pass the element when flow is reversed, preventing flow through the element.

Figure 5.9(b) illustrates a section through the head of a reverse flow filter. The head contains two check valves: a plate type, gravity return; and a sleeve type, spring return. As shown it is under *no-flow* conditions. The sleeve is held by the spring in the upper position blocking a set of cross-ports and the plate is resting on top of it blocking the ports through the sleeve.

Under *normal-flow* conditions, fluid passes through the filter element and up through the center, the plate-type check valve being lifted by flow forces.

When subjected to *reverse-flow* conditions, the plate and sleeve are both forced down against the spring and fluid passes through cross-holes in the head, bypassing the filter element.

A single filter can be used to condition flow in both directions by using a series of check valves in a bridge network (Figure 5.10). Flow through the filter is always in the same direction.

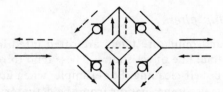

Figure 5.10 Filter and check valves in bridge network.

Clean-up loop filter

A secondary pump filter circuit is often used in large systems where conventional filter locations are unacceptable from size and location cost considerations. To ensure an adequate clean-up rate, the flow through the filter per minute should be at least 10% of the system capacity. A satisfactory cleanliness level can be maintained if the clean-up loop has a minimum flow rate of 20% of the main system flow rate. Clean-up loop filters are also used for initial filling of systems and for cleaning a system during commissioning. There are pump filter units commercially obtainable which can be plugged into a reservoir as needed, or can be used as transfer units to pump and filter fluid from drums into the reservoir. These units can be hired for commissioning and reconditioning large systems.

Separate clean-up loop filters are gaining favour on smaller systems because the filter operates most efficiently, owing to a steady flow rate without pressure surges, pulses, shock or vibrations. They will continue filtering when variable-displacement pumps are off-load and may be left operating whilst the main pumping system is switched off. Filter elements can be changed while the main system is running. If the clean-up pump has a higher flow rate than the main pump, the output can be used to supercharge the main pump suction with clean fluid.

Another advantage is that the 'loop' may incorporate additional conditioning devices such as coolers and water-removal units. The latter may take the form of a standard filter housing containing a water-absorbing polymer in a polypropylene element.

Air breather filter

Air entering a reservoir as the fluid level changes should be cleaned to the same micrometer level as that of the working fluid. Standard hydraulic spin-on elements may be used as air breathers; the media functions more efficiently with air as the fluid than it does with a liquid. An element having a removal rating of 25 μm absolute for a liquid is rated at 5 μm absolute for air breather applications. A 3 μm absolute element has a submicronic rating for air and is impervious to fungus spores and bacteria provided the media is dry. However, care must be taken when sizing these components as an inadequately-sized or blocked air filter can cause pump cavitation and fatigue damage to the reservoir which should be protected by a pressure/vacuum relief valve. The element should be mounted on a length of tube above the reservoir clear of splashes. Flow capacity and filtration efficiency are both impaired if the media is wet. A good-quality air breather is particularly important on systems operating in hostile environments.

Conclusions regarding filter position may be stated as follows:

- Most of the various filter positions have specific advantages and disadvantages.
- A suction filter protects the pump which is often the most expensive component in the system, but fine filtration is difficult at this point because of the risk of pump cavitation. One solution is to supercharge the main pump suction by using a boost pump and fine filter.
- In the event of pump failure, a pressure-line filter protects the rest of the circuit. However, its efficiency is reduced by the flow and pressure transients frequently encountered in this position. If it is situated some distance from the pump the hoses can act as an accumulator and operation of a valve in the circuit may then subject it to instantaneous flow rates equivalent to many times the pump flow rate.
- Even higher flow rates may be presented to a return-line filter when large cylinders decompress. On the other hand, only a return-line filter will catch dirt ingested through a cylinder seal before it can reach the reservoir and be recirculated into the system.
- Large volumes of air may enter a system when filter elements are changed. In this respect return-line and clean-up loop filters are least likely to cause problems.
- The β ratio gives a good indication of filter efficiency and value. An above-average filter may be rated at 5 μm absolute and have efficiency ratios of $\beta_{10} = 500$ and $\beta_3 = 10$. This means that under steady-state test conditions it will filter out 99.8% particles above 10 μm and 90% of particles greater than 3 μm. These conditions can be fulfilled in a clean-up loop. The efficiencies will be degraded when subjected to the pressure and flow surges encountered in pressure and return lines.
- The ideal filter location is one in which there is a constant flow through the element at a steady pressure. It should be mounted on a structure which is isolated from mechanical vibration as this may release particles previously captured by the element.
- It would appear that the use of a separate clean-up loop would be the ideal solution, but this does not give specific protection to sensitive system components; it only provides an overall contamination control. Thus the best system may well be a composite of clean-up loop filter with a filter in the pump suction line to protect the pump and

a last-chance filter in front of any sensitive components, or a clean-up loop plus a pressure-line filter to protect the circuit from effects of a pump failure.
- Wherever a filter is located in a system it should be mounted in an accessible position with the condition indicator clearly visible. If elements can be changed with a minimum of inconvenience, good maintenance will be encouraged.

5.3.4 Filter sizing

Micrometer rating

It is impossible to state a precise micrometer rating needed for a system as it depends upon a number of parameters which will have varying degrees of importance.

1. The sensitivity of the components in the system to contamination and type of fluid.
2. The operating pressure and temperature.
3. The system duty cycle.
4. Life expectancy.
5. The degree of reliability required.
6. The environmental conditions.
7. Safety liabilities.

Target cleanliness levels

A series of field studies carried out under the UK Fluid Power Contamination Control Programme has produced some general target cleanliness levels for arious industries and applications (Table 5.7). These give guidance to average contamination levels which are acceptable from the point of view of reliability.

The system designer should attempt to provide filtration equipment which will achieve or better these figures but it is the responsibility of the user to maintain them by monitoring levels and changing elements at the proper times. A system cannot be too clean, but the finer the degree of filtration, the higher the initial cost and the higher the energy absorbed in the filter. Contrary to popular belief, the use of fine filters does not necessarily mean more frequent element changes once the system has been cleaned. With a lower contaminant content there will be less self-generated dirt in the system.

Table 5.7 Recommended acceptable contamination levels.

Application	Target contamination class to ISO/DIS 4406
Sensitive laboratory or aerospace equipment	13/10
Machine tools	15/9
Injection moulding machines	16/11
Mobile	18/11
Mechanical handling	18/13
Marine	17/12

Flow capacity

The flow capacity of a given filter will depend upon:

1. The permissible pressure drop through the housing.
2. The permissible pressure drop over the filter element.
3. The viscosity of the fluid and hence the operating pressure and temperature.
4. The specific gravity of the fluid.

Most manufacturers give recommended flow rates for their filters but these are for a fluid having a certain specific gravity and viscosity at a particular temperature. These may not be appropriate to the fluid being used, in which case they have to be adjusted *pro rata* bearing in mind the minimum operating temperature.

Proper account has to be taken of cold start conditions when the viscosity may be several times the normal operating viscosity, in which case the pressure drop through the element will increase in direct proportion.

A general rule of thumb is that the filter capacity per minute should be at least twice the pump rated flow (per minute) or one-third the reservoir capacity in liters, whichever is the greater.

A more accurate method is to determine a maximum pressure drop across a clean filter which will be acceptable for the system. Usual figures are:

- 1 bar for pressure filters having a bypass valve.
- 1.5 bar for non-bypassing pressure filters.
- 0.5 bar for return-line filters.
- 0.05 bar for suction filters.

For this purpose some manufacturers quote a range of differential pressures and flow rates for clean elements and housings. A size of filter is selected, which has a quoted higher flow capacity at the appropriate differential pressure than the peak flow rate which it will encounter in the system. This depends upon the position of the filter in the circuit. In general, the peak flow rate for a pressure-line filter will be the actual pump flow rate. It may be higher where accumulators are used or if it is situated a long way from the pump. Return-line filters often see flow rates many times higher than the pump flow, owing to cylinder differential piston areas and cylinder decompression, so allowances have to be made for these conditions. The larger the filter surface area, the better. A large filter will

- Give a lower pressure drop.
- Improve the system's overall efficiency.
- Improve contaminant removal.
- Increase the period between element replacement.
- Be less affected by cold start conditions.
- Improve system performance and reliability.

Although the capital outlay is greater, this will be offset by lower running costs. The importance of having the correct filter cannot be overstressed. It is one of the most important and unfortunately often most neglected components in any system.

5.4 LEAKAGE CONTROL

In most people's opinion hydraulic systems are messy things which *always* leak. With correct design, installation and maintenance, leakage can generally be controlled and often eliminated. If all circuits leaked we would not dare fly in an aeroplane.

Hydraulic-fluid leakage occurs both internally within components and externally. Excessive *internal leakage* reduces system efficiency and generates heat with subsequent deterioration of the fluid. A certain amount of internal leakage is designed into components for lubrication, control of compensators etc. *External leaks* are not only messy but often constitute a hazard. The fluid can damage equipment and possibly the company product. They are expensive because not only has the lost fluid to be replaced but machine down-time increases and production performance falls. There are no benefits from external leaks.

Leaks are caused by *vibration* and *shocks* which may be both mechanical and hydraulic. The effect is to loosen fittings and components and may lead to fatigue failures. *Wear* is another cause, taking place between mating parts and increased leakage results. Another is *seal deterioration* which is usually due to the effects of excessive temperature, contaminated fluid chemically attacking the seal, and seal wear as a result of dirt, excessive loads and lack of lubrication.

5.4.1 Hydraulic conduits

The fluid in hydraulic systems is generally conveyed through either rigid tubing or hoses. The correct specification of these conduits and in particular their connections is as important as that for the rest of the parts of the system and essential if adequate leakage control is to be achieved.

Hydraulic-tubing

Historically, in the UK thick-wall tube was used and classified by Imperial nominal bore sizes. The actual outside diameters corresponded with those of traditional water and steam pipes. In the USA, tubes were classified by outside diameter (Imperial). These, coupled with a range of Continental metric sizes, have caused a lot of confusion with fittings particularly because of the variety of threads in use. However, a range of metric tubes (Table 5.8) is rapidly becoming the industry standard. They are classified by outside diameter with a range of wall thicknesses for various working pressures.

The most commonly used material is carbon steel with composition and properties to DIN 2391/C (BS3602:1962, CDS23) and tolerances to BS 980:1950. Also used for certain applications is stainless steel, titanium, copper, tungnum and bundy, all having relative merits and disadvantages.

Connections are made using flared fittings, compression fittings, staple fittings, flanges and welded joints.

In selecting tube sizes, it can be seen from Table 5.8 that the standard tubes are

Section 5.4 Leakage control

Table 5.8 Cold-drawn seamless carbon steel tube for pressure purposes to DIN 2391/C.

Outer diameter × wall thickness (mm)	Approximate weight (kg/m)	Maximum working pressures (bar)		
		Safety factor 2.5:1	Safety factor 3.0:1	Safety factor 4.0:1
6 × 1.5	0.166	703	586	441
6 × 1.0	0.123	428	359	269
8 × 1.5	0.240	496	414	310
8 × 1.0	0.173	310	255	193
10 × 3.5	0.561	1089	903	676
10 × 2.0	0.395	531	441	331
10 × 1.5	0.314	386	317	241
10 × 1.0	0.222	241	200	152
12 × 2.5	0.586	552	462	345
12 × 1.5	0.388	310	262	193
14 × 2.5	0.709	324	282	217
14 × 1.5	0.462	262	221	166
15 × 2.5	0.771	428	352	262
15 × 1.5	0.499	241	207	152
16 × 3.0	0.962	490	407	303
16 × 2.5	0.832	393	331	248
16 × 2.0	0.691	310	255	193
16 × 1.5	0.536	228	186	145
18 × 1.5	0.610	200	166	124
20 × 4.0	1.58	531	441	331
20 × 3.0	1.26	379	317	234
20 × 2.5	1.08	303	255	193
20 × 2.0	0.888	241	200	152
20 × 1.5	0.684	179	152	110
22 × 3.0	1.41	338	283	214
22 × 2.0	1.07	221	179	138
22 × 1.5	0.758	159	138	103
25 × 4.0	2.072	394	263	210
25 × 3.0	1.63	297	248	186
25 × 2.0	1.13	193	159	117
28 × 4.0	2.37	359	297	221
28 × 3.5	2.11	310	255	193
28 × 2.5	1.57	214	179	131
28 × 2.0	1.28	166	138	103
28 × 1.0	0.666	83	69	52
30 × 4.0	2.56	332	276	207
30 × 3.0	2.00	241	200	152
35 × 3.0	2.367	242	161	121
38 × 5.0	4.07	324	269	228
38 × 4.0	3.35	255	214	159
38 × 3.0	2.59	186	159	117
40 × 6.0	5.03	379	317	234
40 × 5.0	4.32	310	255	193
42 × 3.0	2.885	201	133	101
48 × 6.0	6.21	310	255	193
65 × 8.0	11.24	303	255	186

classified by outside diameter. However, the hydraulics engineer is more interested in the bore which has to be suitable for the flow rate and the wall thickness which will be dependent upon the working pressure.

It is desirable for the fluid flow to be laminar and for pressure drops to be minimal.

This can be achieved by keeping fluid velocities low. The suitable velocity range for suction and return lines is 0.6 – 1.2 m/s and for delivery lines, 2.1 – 4.6 m/s.

EXAMPLE 5.1

A pump delivers 50 l/min, the system maximum working pressure is 200 bar, and the return-line maximum pressure is 60 bar.
Pump suction line (say allowable velocity is 0.8 m/s).

$$50 \text{ l/min} = \left(\frac{50}{60}\right) \times 10^{-3} \text{ m}^3/\text{s}$$

$$\text{Flow area} = \frac{50}{60} \times \frac{10^{-3}}{0.8} \text{ m}^2$$

$$= 1.042 \times 10^{-3} \times 10^6 \text{ mm}^2 = \pi d^2/4$$

$$\text{Bore diameter, } d = \left(1042 \times \frac{4}{\pi}\right)^{1/2} = 36 \text{ mm bore}$$

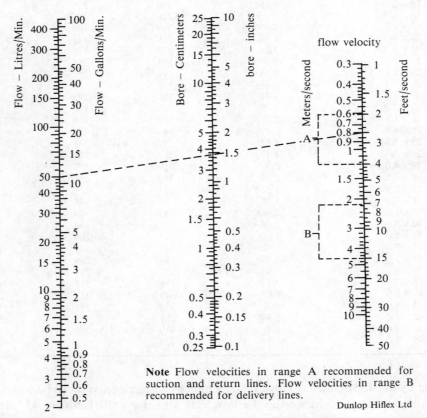

Note Flow velocities in range A recommended for suction and return lines. Flow velocities in range B recommended for delivery lines.

Dunlop Hiflex Ltd

Figure 5.11 Nomogram for flow rate, flow velocity and tube bore. The example is 50 l/min at flow velocity of 0.8 m/s = 3.7 cm diameter.

Section 5.4 Leakage control

From Table 5.8 a standard available tube is 42 mm O.D. × 3 mm wall thickness, giving a 36-mm bore tube. With a factor of safety of 4:1 this has a safe working pressure of 101 bar so it will also be suitable for the return lines.

For delivery lines, if we select a velocity of 3.2 m/s, which is four times the suction velocity 0.8 m/s, then the bore will be half that calculated for the suction tubes i.e.

Delivery tube bore = 36/2 = 18 mm diameter

Again from Table 5.8 a 25-mm O.D. tube with 3-mm wall has a bore of 19 mm and with a 3:1 factor of safety is suitable for 248-bar working.

Alternatively a 28 × 4.0 tube gives a higher factor of safety but the pipework installation costs will be increased. Most tube and hose manufacturers issue tables and charts which facilitate the selection of tube bores. Figure 5.11 reproduces a nomogram from Dunlop Hiflex Ltd.

5.4.2 High-pressure tube connections

Tubes may be connected permanently by welding or by a variety of disconnectable couplings and threaded joints.

Flared fittings

These are usually used on equipment of USA manufacture and are produced to JIC (Joint Industrial Council) and SAE (Society of Automotive Engineers) standards. The end of the tube is preformed by use of a 'dolly' and hammer or a hydraulic flaring press. The majority of fittings use a 37° (74° inclusive) flare. Because of the deformation of the tube a strong mechanical assembly is achieved, although fatigue failures do occur. The tube-nut thread type is UN (Unified).

In the Parker Hannifin Triple-lok tube fitting (Figure 5.12) the flared end of the tube is clamped between the end component and a sleeve by the tube nut. Mechanical support is provided by the sleeve and hydraulic sealing takes place on the contacting surface between the inside of the flare in the tube and the correspondingly tapered nose of the end fitting. With this design, by changing the sleeve, the remaining components of the fittings can be used for both inch and metric size tubing.

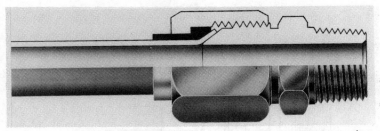

Parker Hannifin Corporation

Figure 5.12 Parker Hannifin Triple-lok tube-fitting.

Care must be taken in identifying flared fittings as the SAE 45° flared components commonly found on Automotive and Refrigeration Systems use the same thread range as the SAE 37° fittings but the flare angle is 90° inclusive. A Japanese industrial standard utilizes a 30° (60° inclusive) flare and BSPP threads.

Bite ring-type compression fitting

This is the most widely used type of connection (Figure 5.13) with interchangeable parts available from several manufacturers. It has been written into many European equipment standards and is usually referred to as the DIN 2353 coupling. (This is the DIN standard on which BS 4368 is based.) In fact the three individual pieces of the coupling are each covered by a separate DIN standard; DIN 2353 primarily covers the end fittings.

When the joint is made the bite ring (sometimes referred to as the cutting ring), is forced into a 24° inclusive taper and bites into the tube to produce both a mechanical coupling and a hydraulic seal.

Several attempts have been made to improve on this troublesome coupling usually by incorporating seals and/or serrated cutting rings. The 'Oclau' fitting used a serrated cutting ring which mated with preformed grooves on the tube. Although technically a very effective coupling its use tended to be limited to specialist applications, because of the extra work involved in preparing the end of the tube. The HF system now produced by the Ravitt Division of Aeroquip has a polymide sealing ring combined with the malleable steel cutting ring which is also serrated. Both systems connect directly with the 24° coned recess in DIN 2353 components.

Problems with most compression-type couplings often arise from incorrect assembly. Preparation of the tube end is important. This should be cut square and preferably sawn as tube cutters tend to deform the end. Fash should be removed from both the outside and inside edges of the cut end and a lubricant applied. Although tubes may be directly assembled with a fitting, under- or over-torquing is possible. Best results are achieved by preassembling the bite ring and tube nut into the tube using a presetting tool. This is a hardened steel dummy end fitting which is held in a vice. To make the joint, push the tube fully home into the recess in the presetting tool and tighten the nut by hand. Continue tightening using a spanner, allowing the tube to rotate with the nut. Once the tube stops rotating, mark the position of the nut and tighten by a predetermined amount as specified by the coupling manufacturer, usually ¾ to 1¼ turns. Unscrew the joint and inspect, visually checking that the ring has cut into the tube and cannot be rotated or moved. Final

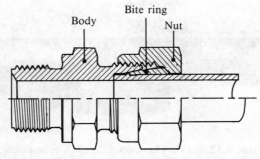

Figure 5.13 Bite ring-type compression fitting (DIN 2353).

Section 5.4 Leakage control

assembly of the tube into its proper coupling should then be possible by hand, with the final $\frac{1}{4}$ to $\frac{1}{2}$ turn using a spanner. Thin-wall tubes may necessitate a special presetting tool fitted with a mandrel.

Where a large number of joints are to be made, a presetting machine is advantageous. This is a pneumatically or hydraulically operated device which squeezes the bit ring onto the tube with considerable savings in speed and effort and to a consistent quality. Further recommendations on tube installations are included later in this chapter.

'O'-ring compression fitting

Some fittings incorporate a split collet to achieve the mechanical grip with a separate 'O'-ring to effect the hydraulic seal. The KR tube coupling (Figure 5.14) is an example of this type.

In the Aeroquip ORS-BT coupling a hardened steel serrated ferrule provides both the mechanical connection and hydraulic seal onto the tube. An 'O'-ring in a groove in the end of the adaptor body seals on the machined end face of the ferrule.

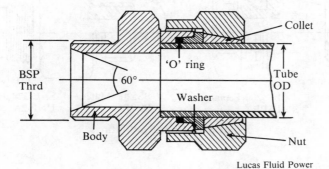

Lucas Fluid Power

Figure 5.14 KR tube coupling.

Conversion couplings

The Walterscheid Poziflare coupling in Figure 5.15 enables a SAE 37° flared tube to be mated with the ubiquitous DIN 2353 fittings which have a 24° connecting taper. The

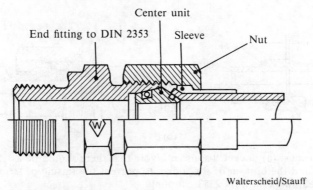

Walterscheid/Stauff

Figure 5.15 Walterscheid Poziflare conversion coupling.

center unit which makes the transition has 'O'-rings to seal on both tapers, whilst the flare makes a strong mechanical connection to the tube.

BS 5200 connector

Many fittings have coned ends and a variety are illustrated in Figure 5.16. DIN 2353 metric threaded components have a 24° inclusive cone which accepts the appropriate tube, bite ring and tube nut. BS 5200 connectors are threaded BSPP (British Standard Pipe Parallel). The male adaptor has a 60° inclusive internal cone which mates with a corresponding external cone within the female adaptor. The nut on the female adaptor is retained but can be rotated on its fitting to tighten the joint. The hydraulic seal is purely dependent upon the metal-to-metal contact between the cones. This type of coupling is predominantly used on hose ends and adaptors in the UK.

Care must be taken as there is unfortunately a similar metric 'Globeseal' fitting DIN 7631/7647 which uses a 60° cone.

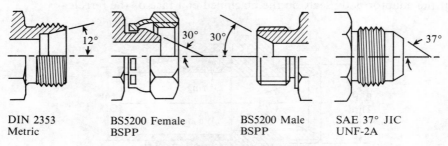

DIN 2353   BS5200 Female   BS5200 Male   SAE 37° JIC
Metric   BSPP   BSPP   UNF-2A

Figure 5.16 Coned end fittings.

Welded connections

Any coupling, no matter how good, is a potential leak point, so use should be limited to parts of the circuit where it is necessary to be able to effect a disconnection. The remaining joints are best permanently welded. Where pipework is welded, the welding must be carried out under controlled conditions, particularly when using butt-welded components.

An inert gas, such as argon or helium, should be used to flood the interior of fittings and tubing whilst welding is taking place. It is similar to the gas shield used in MIG and

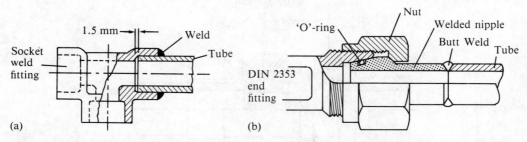

Figure 5.17 Weld fittings. (a) Socket 'tee-piece'. Note that there should be a 1.5 mm gap between the end of the tube and the bottom of the socket to prevent the raising of stresses when the weld contracts. (b) Welded nipple in DIN 2353 coupling.

Section 5.4 Leakage control

TIG welding processes and will prevent the formation of scale and combustion contaminants on the inside of the joint. It is important that the flow rates and pressure of the backing gas should be sufficient to purge the system and there should not be any openings through which air can be drawn into the system.

Joints should be properly prepared and the work carried out by certificated welders. If butt-welding is carried out without a backing gas, the component must be pickled afterwards.

Figure 5.17(a) shows a socket weld 'tee-piece'.

Welded nipples

One of the most effective disconnectable joints consists of a machined nipple which is butt-welded onto the end of the tube and which mates with the standard DIN 2353 end fitting (Figure 5.17b). The fitting reproduces in one piece the contour of a tube with a cutting ring in place. A machined groove takes an 'O'-ring which seals on the 24° taper of the end fitting. The 'O'-ring should only be fitted after the weld has cooled.

Note Remember to slip the tube nut onto the fitting before welding.

Problems can arise due to excessive penetration of the butt-weld restricting the tube bore. This may have to be machined away after welding.

Flange connections

Although some manufacturers may claim otherwise, it is very difficult to make a satisfactory compression join on tubes above 38 mm O.D. The problem is applying sufficient torque to tighten the joint.

For large-diameter pipes, flanged joints are recommended. Unfortunately they tend to be rather bulky and are more expensive than compression fittings. Many components (pumps, motors, filters, manifold blocks etc.) are machined to accept standard flanges to CETOP and SAE specifications (Figure 5.18). The collets can be welded to tubes or take the form of hose end fittings. Hose and pipe positioning is facilitated because the flange is

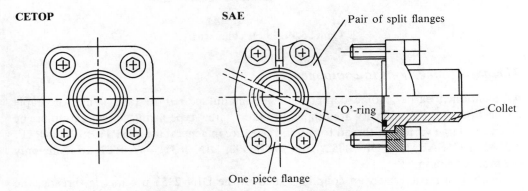

Figure 5.18 CETOP and SAE flange configurations.

separate from the collet. Sealing is by an 'O'-ring set in the flange face. Ranges of collets are available for butt or socket welding.

SAE flanges are the most commonly used. They are in two ranges: SAE 3000 lb/in^2 working pressure (207 bar) and SAE 6000 lb/in^2 working pressure (414 bar). Within each group both one-piece and split-flanges are used. Some sizes within the SAE 3000 range are suitable for higher pressures.

Increasing use is being made of flanges to CETOP recommendation RP63H. This covers a square flange suitable for 400 bar working with a range of collets for 100 bar, 250 bar and 400 bar. The 250-bar collets are interchangeable with SAE 3000 collets and the 400-bar collets with the SAE 6000.

One method of achieving a breakable joint on large-diameter tubes (above 50 mm) which combines the benefits of a flanged connector and a compression fitting is the KR flange coupling. In this the flange halves are attached to the tubes using a variation on the KR coupling (Figure 5.14). The act of tightening the flange bolts compresses the split collets onto the tubes and after initial fitting the joint can be separated and remade at the flange face without the need for special tools, long spanners or high tightening torques.

Staple connectors

A connector commonly used in the mining industry consists of a probe and socket locked together by a square section staple as shown in Figure 5.19. An 'O'-ring on the probe provides the hydraulic seal. Connection is quick and simple and the joint can swivel. A particular safety feature is that the joint cannot be disconnected when the line is pressurized. Some hydraulic components are available with the port cavities machined to accept the probes directly.

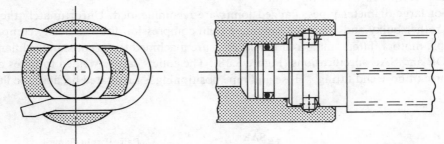

Figure 5.19 Staple connector.

System Stecko Ltd

Threaded connections to components

As mentioned earlier, some components are manufactured to take flange or staple connections but the majority are tapped to take a screw-type fitting. Great care must be taken as a large variety of thread forms and sizes are in general use: BSPT, BSPP, NPTF, NPSM, JIC, SAE, UN, DIN 3852, DIN 3901/3902, DIN 7631/7647, PTT, to name only the most common.

Although most European couplings such as the DIN 2353 use metric threads, the accepted standard for tapped ports in components, valve bases etc. is BSPP (British

Section 5.4 Leakage control

Standard Pipe Parallel). Japanese connections tend to be BSP or Metric and 'JIS Tapered PT' is identical to BSPT and 'JIS Parallel PF' is identical to BSPP.

Metric and United States standard threads have a 60° thread angle but BSP threads use a Whitworth form with a 55° thread angle. They were originally machined on pipe ends and the nominal size alludes to the pipe bore. Consequently the thread pitch diameter is approximately $\frac{1}{4}$ inch larger than the nominal size. It appears in two forms: taper (BSPT) and parallel (BSPP). An American taper pipe thread NPTF (National Pipe Thread Fuel) evolved in a similar manner. The seal is affected by deformation of the threads and its use in hydraulic applications is not recommended by the National Fluid Power Association. It is however widely used in fluid power systems throughout the world. NPSM (National Pipe Straight Mechanical) is the parallel version of NPTF.

It is difficult to distinguish between BSP and NP types as there are only minor differences in major diameter thread form and thread pitch. On some sizes the latter varies by only half a thread per inch.

Where fittings are designed specifically for screwing into tapped ports in components the ends may be plain, but where they are also intended for connection to other fittings the ends are coned. Figure 5.16 shows the 24° and 60° female coned ends for use with compression and BS 5200 fittings. The SAE male cone is also depicted.

In the United States, parallel-threaded SAE fittings having a unified thread form are widely used. A popular type for general hydraulic applications is the SAE 37° also known as JIC. This can be identified by its 74° inclusive angle male coned end, which will mate with a flared tube. A similar connector having a 45° (90° incl.) male cone is available but is usually found on automotive and refrigeration applications. Further confusion is bound to arise as the Japanese JIS 30 has a 60° inclusive male cone and BSPP threads.

In conjunction with the different thread types there are a variety of different ways of sealing the connection when it is screwed into a hydraulic port. With the taper thread type of fitting which may be BSPT, NPTF, DIN 3852 metric taper sealing is affected on the thread. Where taper fittings are used, sealing by PTFE tape or hydraulic air exclusion liquid thread sealant is acceptable. It should be ensured that thread sealants do not enter the component bores. NPTF is a taper thread which seals on the crest of the thread rather than the flank and does not need a sealing tape.

Most fittings use a parallel thread in conjunction with a sealing ring or washer. BSPP and DIN 3852 parallel fittings (Figure 5.20) may use:

- Metal-to-metal seal (cutting edge)
- Captive elastomer seal
- Bonded washer
- Copper washer
- 'O'-ring with retaining ring
- 'O'-ring with retaining ring and back-up washer.

They all seal onto a flat surface. Sometimes a spot face is used which also serves to locate the sealing element (i.e. bonded washer).

The SAE fitting uses an 'O'-ring seal located on the undercut beneath the fitting head. This seals within an accurately machined chamfer at the mouth of the tapped hole. The thread form is Unified.

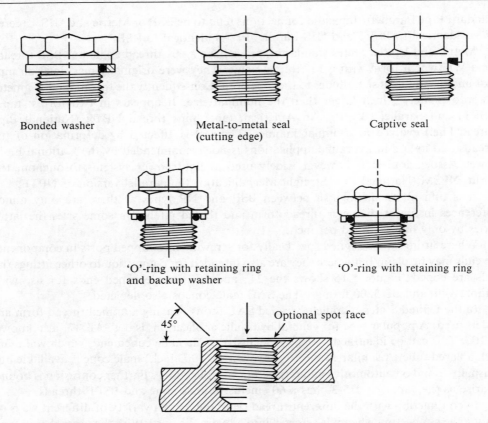

Figure 5.20 Methods of sealing BSPP and DIN 3852 fittings with parallel threads.

Installing tubing

There are a number of rules which should be adhered to when installing tubing:

- Maintain a high standard of cleanliness at all times.
- Keep tubes plugged until they are to be cut or manipulated.
- After cutting remove any fash and debris.
- Inspect threaded fittings and remove any fash before assembly.
- Continuous runs of tube should be used and bends incorporated wherever possible to allow for thermal expansion and contraction. The minimum bend radius for a tube is normally three times the tube outside diameter when bent hot, or five times the outside diameter when cold worked. Hot working should be avoided because scale formed on the inside of the tube will have to be removed by subsequent treatments such as pickling.
- When making a bend, allow a straight length of pipe of at least twice the nut length at the tube end to facilitate assembly of the fitting.
- Design systems with the minimum number of fittings to eliminate potential leakage points as well as for cost and hydraulic shock considerations. Avoid external stress by supporting the tube with clips and brackets designed for the purpose.

Section 5.4 Leakage control

- If a tube springs out of position when the fitting is disconnected it has been incorrectly aligned or bent and will cause undue stresses.
- Heavy components, valves, filters etc. should be bolted down, and not supported by the pipework. In applications subject to heavy shocks or vibration, hoses, accumulators and pulsation dampers can be used to alleviate these shocks.
- Check the tubing before final installation and clean by pulling through a lint-free cloth.

Pickling

Where welding has been carried out without an inert purging gas and scale has formed inside the tube, it will be necessary to pickle the pipework. This involves the use of dangerous chemicals and acids which must be stored and handled in accordance with appropriate Health & Safety regulations. Consequently this type of work is best carried out by a specialist contractor, either at his own premises or on site. It is not usually practical to pickle an installed system because it is almost impossible to remove all the acids which can be retained in fittings by capillary action.

The pickling process consists of:

1. Degreasing all parts.
2. De-rusting in a tank of a commercial de-rusting solution at a controlled temperature for the minimum period necessary to remove all the scale and rust. This is usually based upon the contractor's experience.
3. Rinsing in cold running water.
4. Neutralizing in a second tank which contains a neutralizing solution. This should be carried out at a controlled temperature and for a period specified by the solution manufacturer.
5. Rinsing in hot water.
6. Blow drying using dry filtered air.

Afterwards the parts should be plugged and sealed until ready for installation.

Pickling can sometimes cause more problems than it has prevented, so it is a process which should be avoided if possible.

Rigid tubing should only be used to connect components which are completely stationary with respect to each other. The suction line between the oil reservoir and a pump driven through a flexible coupling should be in hose as there will be relative movement. A rigid tube in this situation will set up stresses with subsequent problems. Alternatively the tube must pass through a flexible connector in the tank top.

5.4.3 Hydraulic hoses

Hydraulic hoses are constructed from layers of rubber or thermoplastic, reinforced by layers of steel wire or textile braiding, with a final protective layer of rubber or plastic. Figure 5.21 shows the construction of a SAE 100R2 hose which is probably the most widely used type. It has two layers of steel wire braiding and is frequently described as 2-wire hose. The SAE 100R1 (1-wire) hose is of similar construction but has only one layer of reinforcing and is consequently used for lower pressure work. For very high pressure

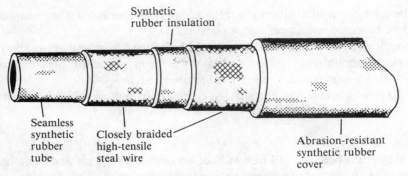

Figure 5.21 Hydraulic hose with two layers of braiding.

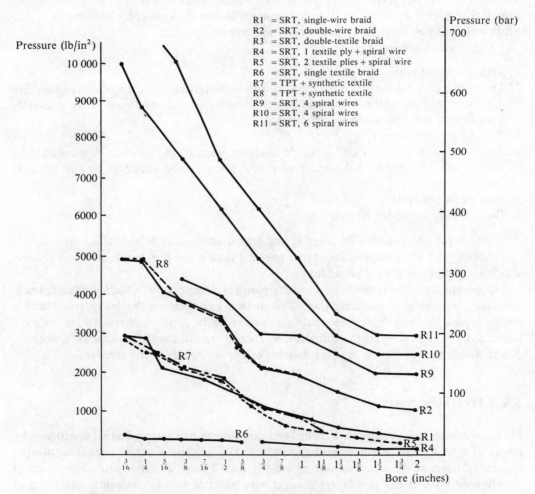

Figure 5.22 Hydraulic hose working pressures; SRT = synthetic rubber tube; TPT = thermoplastic tube.

applications, multi-spiral hoses are available. These are similar in construction to braided wire hoses but the reinforcement is 4, 6 or more layers of spirally wound high tensile wire separated by rubber insulation layers.

Some modern hoses, generally smaller bore types, use thermoplastic materials for the inner core and the covering, with textile braiding for the reinforcement. They are primarily used in the automotive industry.

Working pressure

The working pressure of a particular type of hydraulic hose is different for each bore size. Figure 5.22 shows this relationship for various types of hose manufactured to SAE J517C–100R.

Minimum burst pressure

Quite high factors of safety are used and the minimum design burst pressure for a hose is generally about five times the working pressure with the actual burst pressure much higher.

Hose ends attachments

A variety of connections are available for attaching hoses to components and rigid pipework. They fall into two families according to the manner in which they are fixed to the hose. *Swaged fittings* are pressed onto the hose using a hydraulic or screw-type swaging machine. *Reusable fittings* can be fixed using simple tools. The term 'reusable' is a misnomer because they are seldom used more than once.

In a swaged assembly the hose wall is sandwiched between the inner and outer sleeves of the end fitting and then literally squeezed together by a hydraulic swaging machine.

An adequate end fitting on R1, R2 and equivalent hoses can be achieved using reusable fittings. After suitable preparations an outer sleeve or ferrule is screwed onto the hose (left-hand thread) and then an inner sleeve or insert is screwed into the ferrule (right-hand thread) sandwiching the hose wall between the two sleeves. To facilitate assembly, the components should be lubricated with hydraulic fluid of the type which is to be used in the system.

Note With some types of hose it is necessary to strip off a section of the hose outer covering before fitting the ferrule. With non-skive hose the ferrule fits directly onto the outside cover. Assembly of reusable fittings to thermoplastic tube requires special pre-forming tools. These can also greatly facilitate assembly of conventional synthetic rubber hoses.

Manufacturers offer a huge variety of hose end fittings. Each class of hose may have over thirty different end fittings for each of a range of sizes and as swaged or reusable types. Some such as the SAE 6000 flange end are only suitable for swaged connections because of the high-pressure ratings. Similarly swaged fittings are predominantly used with multi-spiral hose. In all cases it is recommended for ease of fitting that at least one hose end is a swivel type.

Hose installation

Particular care should be taken with hose installation, although mostly it is a matter of using commonsense.

- Avoid sharp bends and twists — hose end fittings should be in the same plane.
- Do not allow hoses to rub against each other or against steel work.

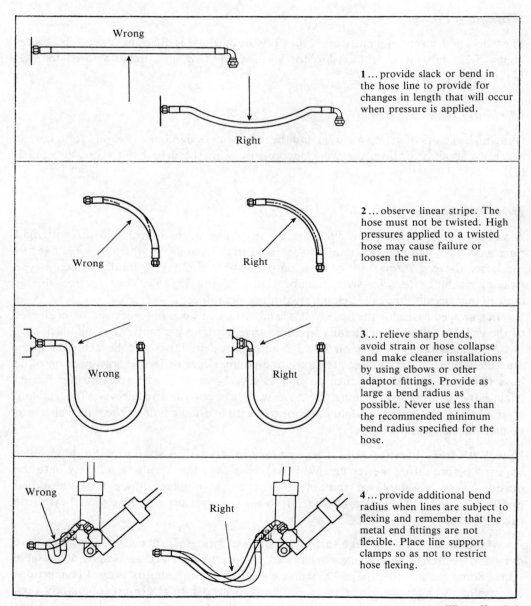

Figure 5.23 Hose layout.

- Select the correct type of end fitting.

Figure 5.23 shows some of the common pitfalls associated with hose layouts and how they should be overcome. Wherever there is a bend in a hose, even in static installations, it is essential to avoid tension at the connections by having a short length of straight hose adjacent to each end fitting.

Manufacturers issue details of minimum bend radii and recommendations for the minimum straight length adjacent to the fitting for the various types of hose both for static and flexing conditions. These should be strictly adhered to because, when a hose is subjected to bends smaller than recommended, operating pressure, limits and service life are reduced.

Coil wire spring guards are available to help control the bend radius. Similarly suction hoses can be supported by internal coils.

Hose failures

An occasional hose failure is to be expected – they do not last indefinitely. Hose is usually marked with a date or code which indicates when it was made. It is difficult to predict the actual life because this is dependent upon the application and environmental conditions.

Repeated failures of a particular hose can generally be attributed to incorrect selection, assembly or installation. Failure to determine the true cause may prove costly. A manufacturer can readily determine the most likely reason for failure from examination of the offending component but his diagnosis will be made much easier if full details of the application are furnished.

To summarize: a hose and its fittings must be:

- Compatible with the fluid being transmitted.
- Suitable for the working pressure bearing in mind surges and fluctuations.
- Able to withstand the variations in fluid and ambient temperatures pertinent to the application.
- Correctly sized to keep pressure losses to a minimum and to avoid damage due to excessive turbulence and heat generation.
- Care must be taken that the end fittings are suitable for the hose and correctly connected using assembly techniques and where necessary special tools, mandrels etc. as specified by the manufacturer.
- They should be installed properly orientated, avoiding sharp bends, kinks, flattening, abrasion, twists, vibration and stretching. Bear in mind that an approximate 4% reduction in length occurs under pressure. Any motion should be bending not twisting.
- Hose runs should be routed to avoid possible sources of mechanical damage, hot manifolds, furnaces etc. Protect where necessary with coil sleeves or fire-protective sheaths and support at appropriate intervals.
- Environmental conditions which can affect the service life of a hose include ultraviolet light, ozone, chemicals, salt water, solvents, corrosive liquids and air pollutants.
- Some applications require the hose to be partially conductive in order to discharge static electricity which may be created by the transmission of certain fluids. In other circumstances statutory regulations may demand that the hose be non-conductive.

5.4.4 Supports and clamps

For long and multiple tube and hose runs, clamps are recommended, spaced and positioned to give adequate support. They will help to absorb shocks and vibration, reduce noise and lessen the risk of leakage.

Commercial clamps are inexpensive and easily fixed. One type consists of a split-rubber bush retained in a pressed-steel clamping plate. A range of bushes accommodate the standard tube sizes.

Another variety utilizes thermoplastic moulded blocks, again covering the range of standard sizes. They can be mounted singly on weld plates or mounting rails or stacked and grouped.

Correct spacing of clamps is important and recommended minimum spacing intervals for various size tubes are given in Table 5.9.

Supports should be placed as close to bends as possible, preferably both sides of the bend. Hose clamps should be positioned to prevent the natural flexing of the hose when hydraulic pressure is applied. A stable back-up support has to be provided for the weld plates and rails.

Table 5.9 Clamp spacing intervals.

Pipe OD (mm)	Spacings (m)
6–12.7	0.9
15.9–22.2	1.2
25.4	1.5
31.8–38.1	2.1
48.3	2.7
60.3	3.0
73	3.4
88.9	3.7
101.6	4.0
114.3	4.3
141.3	4.9
168.3	5.2
219.1	5.8
273	6.7
323.9	7.0
355.6	7.6
406.4	8.2
457.2	8.5
508	9.1
558.8	9.8

Stauff Ltd

5.4.5 Circuit accessories

Quick-release couplings

Where it is necessary to make and break a hydraulic line, quick-release connectors are often used. These are in halves: a male nipple or probe and a female carrier or socket. Generally each half contains a check valve which closes and seals as the coupling is

Section 5.4 Leakage control

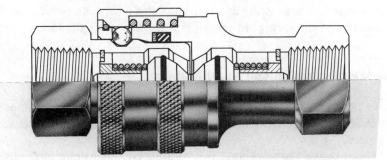

Parker Hannifin Corporation

Figure 5.24 Industrial Quick Coupling.

disconnected. The construction is such that when the halves are joined together the check valves are forced off their seats, creating a flow path through the coupling.

Figure 5.24 shows a Parker Hannifin Industrial Quick Coupling. Variations include designs where only one half is sealed. With most quick-release couplings high-pressure drops occur because of the restricted flow through the unit. This has to be taken into account when selecting the appropriate size of coupling.

TEST POINTS

These are small quick-release couplings which can be fitted into appropriate parts of a circuit for pressure and temperature measurement, fluid sampling and air bleeding. The female half of the fitting contains a check valve and is permanently installed in the system; the corresponding probe which is attached to the test apparatus is open ended.

Velocity fuses

Velocity fuses, anti-burst valves or hose break valves, as they are sometimes known, are incorporated into hydraulic circuits as a safety device (Figure 5.25). This fitting contains a valve and is usually screwed directly into a cylinder port. Typically, it consists of a poppet or plate valve which is held open by a light spring. Under normal flow velocities, fluid can pass in either direction. In the event of a sudden leak such as a burst hose, the flow forces induced by the rush of fluid through the valve in a direction opposing the spring cause it to operate. Consequently, according to the design, it will either shut off completely or close down to a controlled flow.

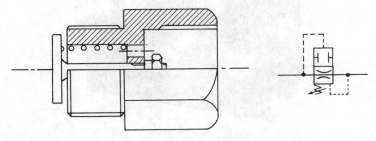

Figure 5.25 Velocity fuse.

In some designs, the flow rating at which the valve closes can be adjusted by altering the precompression of the spring. In other cases the flow rating is factory preset.

5.4.6 Pipework elimination and simplification

Use the correct fitting for the job and you will avoid 'Christmas trees' of adaptors. A bent tube has better flow characteristics than a 'bend' pipe fitting with the added advantage of no leakage points.

Manifolds can be used to advantage in hydraulic circuits to minimize pipework, and hence the possibility of leaks. They may take the form of integrated circuits which are machined blocks with cavities for cartridge valves, and faces machined to accept CETOP standard subplate valves. Much external pipework can be eliminated by cross-drillings, etc. Because of the machining costs they are usually considered only to be economical for batch production circuits. However, subsequent savings in down-time may justify them for one-off applications. Wherever manifolds are used, the designer has to be certain of his circuit beforehand as mistakes will be expensive and later modification may not be possible, or at the best difficult. There are proprietary manifold systems where a range of individual blocks with various valve interfaces can be assembled to form a complete system. These more readily permit modifications. The increased use of computer aided design (CAD) and manufacturing systems will make manifolds more practicable for one-off applications.

Integrated Hydraulics Ltd/Wandfluh

Figure 5.26 Valve stacking assembly.

Section 5.4 *Leakage control*

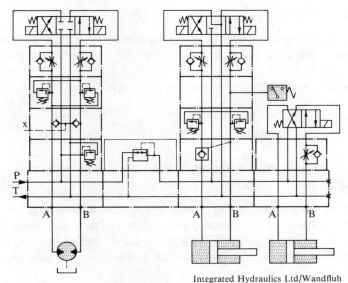

Integrated Hydraulics Ltd/Wandfluh

Figure 5.27 Hydraulic circuit for Figure 5.26.

Where manifolds are used test points should be incorporated to simplify fault finding at a later date. Test points are essential on cartridge valve manifolds as during commissioning individual cartridge valves may have to be tuned and the dynamic behavior optimized by experimentation with different control orifices and springs. This will involve taking pressure and flow readings.

Intelligent use of valve stacks (Figures 5.26 and 5.27) can eliminate much pipework. Most manufacturers have a range of sandwich valves with CETOP 3, 5 and 8 interfaces which can be stacked together both vertically and horizontally. Figure 5.26 shows a valve stacking assembly. All the interconnections are internal, the only external connections being to pump, tank and actuators. Figure 5.27 shows the hydraulic circuit for this valve assembly.

5.4.7 Seal protection

Wear of dynamic seals can be reduced by:

- Good filtration.
- Good surface finish on mating parts (quality of component initially selected).
- Prevention of side loads on cylinder rods, pump and motor shafts.
- Protection of cylinder rods, pump and motor shafts by means of scraper rings, gaiters, bellows.
- Running at correct temperature and speeds.

  Static seals can be protected by:

- Ensuring a good surface finish on housing and cavities.
- Using correct pre-tension on mounting bolts.

Seal deterioration can be prevented by:

- Operating at correct temperature; although seals will perform satisfactorily over a wide range of temperatures, best results are obtained within a quite narrow band (20 to 60°C for a mineral oil).
- Ensuring that seals and fluid are compatible with each other.

5.4.8 Rules for leakage control

In conclusion it is suggested that by the application of the following simple rules external leakage can be dramatically reduced.

- **Use as few fittings as possible** – Every joint is a potential leakage point
- **Use manifolds rather than individual pipes** – Reduce pipework between individual components with the circuit within the manifold. Also manifold mounted components are less susceptible to leakage than piped components because of the use of flanges and 'O'-rings.
- **Use welded joints where possible** – A comprehensive study carried out jointly by the British Steel Corporation and the National Coal Board (British Coal) has shown welded nipples and welded flanges to be superior to the majority of types of compression joint.
- **Use parallel threaded components with seals rather than taper threads** – Particularly effective when a connection has to be disturbed and then remade.
- **Tighten fittings and component mounting bolts to the correct torques** – It is possible to over-tighten compression fittings.
- **Allow for movement** – Always use a length of hose between components which may move relatively to each other or which are subject to vibration.
- **Support pipes to minimize vibration** – This applies to both rigid tubes and hoses. Fix pumps and motors on anti-vibration mounts.
- **Select components to minimize shock** – Use accumulators, cushioned cylinders and valves, pulse dampers and hoses. Sometimes where accumulators are used flow shocks occur as the system is loaded and unloaded. In such cases, the circuitry must include effective devices to limit the discharge rate of the accumulator and hence the rise and fall of pressure in the system.
- **Run at a constant temperature suitable for the components** – As temperature increases, oil viscosity reduces causing it to flow (and hence leak) more easily.
- **Use commonsense in layout of pipes and hoses** – Avoid sharp bends, hoses rubbing against each other, and the 'Christmas trees' of fittings

Note The cost of leakage is high. It has been calculated that a leak of one drop per second is equivalent to 1500 liters per year and a 5-mm diameter continuous stream would result in a mammoth 50 000 liters per year.

CHAPTER SIX

HYDRAULIC SYSTEM DESIGN

6.1 DESIGN CRITERIA

The main criteria in the design of any system are:

Simplicity
Reliability
Efficiency
Cost effectiveness
Maintainability.

The above are all interdependent, simplicity of design having probably the greatest effect of all the criteria. A simple design will have the least number of components and therefore fewer components to fail, so improving the reliability. As there are pressure drops through each component in a circuit, the design with the least components will have the least losses and so the highest efficiency. In general the lower the number of components, the lower the cost and so the higher the cost effectiveness. Simplicity of design can reduce maintenance costs but not necessarily make maintenance easier. Ease of maintenance has to be built into a circuit — by careful selection of components, layout of pipework, component positioning and inclusion of test facilities in the circuit.

The designer must have a thorough knowledge of the characteristics of all the hydraulic components available, keep up to date with new developments, and completely understand the function of the system to be designed, taking into consideration the environmental, ambient conditions and the type of labor available for operating the system.

The designer should obtain all the system requirements personally or from a very reliable source. Sometimes when a machine is made it works perfectly in the test bed but fails under actual working conditions. This may be owing to subtle differences in the work which the machine is performing. As an example, a machine to extrude grease worked perfectly under test conditions where the ambient temperature was 20°C but failed in the plant during the winter when the ambient temperature was much lower. The failure was a result of the considerable increase in viscosity of the grease and the corresponding increase needed in the extruding force.

Another problem the designer will meet is that the requirements are changed, or the client wants to use the machine for a different purpose. A plate folder may be designed to bend 10 mm plate, but the client will try to bend 15 mm plate.

It is advisable to over-design systems to give some allowance for overloads.

Safety is of extreme importance and all units must fail safe. Where solenoid-operated spring offset valves are used to unload a pump, energizing the solenoid should put the pump on load. If there is a failure of the electrical control voltage, the pump will then automatically off-load.

Anti-burst valves can be fitted to cylinder ports so if there is a hose failure, the anti-burst valve, being subjected to an abnormal flow rate will automatically close.

Accumulators are sometimes used to provide an emergency power source to complete a cycle or return actuators to a safe condition in the event of main pump failure.

Manual overrides may be fitted into solenoid valves, so the valve can be operated manually for setting-up purposes or in the case of failure of the control electrics.

The signals from pressure switches can be used to automatically retract or reverse actuators which have jammed. Emergency stop switches should be located at strategic points. The operation of any stop switch must put the machine into a safe condition before switching off. This may involve retracting cylinders, reversing motors, returning control valves to a neutral position, discharging accumulators, etc. The exact function which the emergency stop switch performs depends upon the particular application. It may have the same function as the normal stop switch or act in a completely different mode of operation. When the machine is part of a production sequence, all the machines in that sequence may have to be stopped.

All moving parts must be guarded and where necessary electrical inter-lock switches fitted. In some cases hydraulically operated shot pins are used to lock guards in position. All guarding has to comply with the relevant safety regulations.

6.1.1 Design information required

The exact function of the system must be known, and its relationship to other units and parts of the process. The designer must obtain information or make decisions on most of the following points.

ACTUATORS

- *Cylinders*. A thrust/stroke and speed/stroke profile is required for both extend and retract strokes for each cylinder. Also, any variations in both speed and thrust, cylinder stroke and cushioning requirements, type of cylinder (double or single acting, through rod, air bleeds), construction (tie rod or non-tie rod design), special seal requirements, and cylinder control, locking in-position, and counterbalance of loads.
- *Semi-rotary actuators*. Speed, angle of rotation and torque profiles are required together with cushioning requirements, air bleeds and special seals.
- *Motors*. Speed/torque profiles are required for each direction of rotation, acceleration and retardation requirements, and special seal requirements.

SEQUENCE OF OPERATIONS
Type of sequence – event or time based or a mixture. Details of sequence include changes in pressure or flow rates.

METHOD OF CONTROL
This is manual, mechanical, hydraulic, pneumatic, electrical, microprocessor, proportional or servo.

OPERATING CONDITIONS
The location of system and environmental conditions, dirt, temperature, humidity etc. are required with limitations as to noise levels on any machines. The skill of operational personnel and maintenance facilities have also to be known.

OPERATING PRESSURES
Is there a maximum pressure or can the designer decide this?

SPECIAL REQUIREMENTS
Will the system be continually used? What are the typical hours' use per week? Is this a 'key' system? (This will determine the spares level to be recommended.) Details of any safety interlocks are required. Also, any special requirements for the working fluid? The actual fluid used will determine the material to be used for seals; also, some fluids are incompatible with certain metals. The use of water-based fluids limits the maximum operating pressure, inlet conditions and driven speed of certain pumps and motors.

COMPONENT SUPPLIERS
Is the customer using a standardized range of component suppliers, or is the designer free to select particular manufacturers of components?

6.2 SUMMARY OF BASIC FORMULAE AND RULES

Basic units and conversion factors are given in Table 1.1 in Chapter 1.

6.2.1 Fluid flow

Flow Q is measured in:

   liters per minute
   milliliters per second
   gallons per minute
   cubic inches per second

Note
1 liter = 61 in^3 = 0.22 Imperial gallons = 0.26 US gallons

1 Imperial gallon = 277.4 in^3 = 4.546 liters
1 US gallon = 231 in^3 = 3.785 liters $\simeq \frac{5}{6}$ Imperial gallon

Quantity flowing

The quantity of fluid flowing through a pipe or orifice is the product of flow area × average flow velocity. The quantity flowing through the pipe shown in Figure 6.1 is

$$Q = \frac{\pi D^2}{4} V = \frac{\pi d^2}{4} v$$

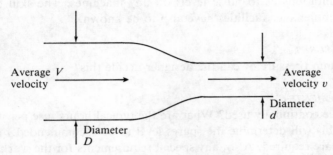

Figure 6.1 Flow through a pipe.

EXAMPLE 6.1

A pump delivery port is 30 mm in diameter and its suction port 45 mm in diameter. If the pump delivery is 120 l/min, calculate the average flow velocity at the delivery and suction ports.

    Quantity flowing = Area × Average velocity
(i) Consider delivery port:

$$\text{Average velocity} = \frac{\text{Quantity}}{\text{Area}}$$

$$= \frac{120 \times 10^{-3}}{60 \, (\pi/4) \, (0.03)^2} \left(\frac{m^3}{s} \times \frac{1}{m^2} \right)$$

$$= 2.83 \text{ m/s}$$

(ii) Consider suction port:

$$\text{Average velocity} = \frac{\text{Quantity}}{\text{Area}}$$

$$= \frac{120 \times 10^{-3}}{60 \, (\pi/4) \, (0.045)^2} \left(\frac{m^3}{s} \times \frac{1}{m^2} \right)$$

$$= 1.26 \text{ m/s}$$

Section 6.2 Summary of basic formulae and rules

Types of flow

The flow pattern can be either *laminar* or *turbulent*.

(a) Laminar flow is where the fluid particles move in straight lines in the direction of flow. This type of flow is also known as 'streamline' or 'viscous'. When laminar flow occurs over a surface, the resistance to the flow is owing to the fluid viscosity, the surface roughness having no effect. Under laminar-flow conditions, the pressure drop in a pipe is directly proportional to the viscosity of the fluid and the velocity of flow.

(b) Turbulent flow is where the fluid particles can move across the direction of flow in a random manner.

Turbulent flow occurs at higher velocities than viscous flow. The change from viscous to turbulent flow under steady-state conditions occurs when Reynolds' Number (Re) has a value of between 2000 and 3000. Under steady flow conditions, when Re < 2000, flow is viscous, and when Re > 3000, flow is turbulent. A transition state exists between these two values and flow can be in either condition.

Reynolds' Number is a non-dimensional parameter and is given by

$$\text{Re} = \frac{VD}{\nu}$$

where V is the mean flow velocity, D is a linear dimension of the flow surface which in the case of a circular pipe is the diameter, and ν is the kinematic viscosity of the fluid.

EXAMPLE 6.2

A fluid flowing through a 20-mm bore pipe has a viscosity of 40 cSt. If the fluid flow rate is 50 l/min, determine the type of flow in the pipe.

Calculate Reynolds' Number, Re:
Fluid kinematic viscosity, $\nu = 40$ cSt
$= 40$ mm^2/s

(*Note*: 1 cSt $= 40 \times 10^{-6}$ m^2/s)

$$\text{Fluid velocity} = \frac{\text{Quantity flowing}}{\text{Area of flow}}$$

$$= \frac{50 \times 10^{-3}}{60\,(\pi/4)\,(0.02^2)} \left(\frac{\text{m}^3}{\text{s m}^2}\right)$$

$$= 2.65 \text{ m/s}$$

$$\text{Re} = \frac{VD}{\nu}$$

$$= \frac{2.65 \times 20 \times 10^{-3}}{40 \times 10^{-6}} \left(\frac{\text{m}}{\text{s}} \times \text{m} \times \frac{\text{s}}{\text{m}^2}\right)$$

$$= 1325$$

As Re < 2000, flow is viscous.

6.2.2 Pressure losses

Pressure drops in pipe lines

Whenever a fluid flows there must be a corresponding pressure drop across the section in which the flow occurs. In hydraulic systems the pressure drop should be kept as low as possible to give a high transmission efficiency. It is usual to design a system so that the flow through the pipe work is viscous. In order to achieve this, the suggested maximum flow velocities in hydraulic systems are for

(i) Suction lines 1 to 1.5 m/s, 3 to 5 ft/s
(ii) Pressure and return lines 4 to 5 m/s, 13 to 16 ft/s.

When the back-pressure resulting from pipe friction in the return line has to be kept to a minimum, it will be necessary to use a pipe bore greater than that given by this formula. In such cases it may be considered as a suction line.

To calculate the actual pressure drops in the pipes, charts and tables are available in hydraulic data reference books giving pressure drop per meter run of pipe, for different flow rates and pipe bores for fluids of given viscosity and specific gravity. Allowances have to be made for the effects of bends and fittings. Pressure drops for these are obtainable from reference books and manufacturers' catalogs and are usually expressed as an equivalent length of straight pipe of the same diameter as the fittings.

Pressure drops across components

VALVES

Whenever a restriction occurs in a pipe there will be a pressure drop across that restriction. With flow-control valves, manufacturers will provide details of pressure drops at various flow rates. In the case of check valves, the pressure drop depends on the control spring in the valve and will vary with the quantity flowing. Where flow-control valves are used, the pressure drop over the valve depends on circuit conditions and the total system pressure can be dissipated as is the case in a 'bleed off' flow control. Characteristics for the pressure drop over various flow paths in a typical directional control valve are shown in Figure 3.58 in Section 3.3.3 of Chapter 3.

FILTER

The total pressure drop in a filter is normally taken as the pressure loss through the housing plus the drop over a clean filter element. Within the housing the loss is proportional to the quantity of fluid flowing and the fluid specific gravity. Whereas over the element it is dependent upon the quantity flowing and the fluid viscosity. As the element gets progressively contaminated the pressure differential across it increases. If a bypass valve is fitted the maximum pressure drop may be taken for the purposes of calculation as the setting of the valve. In a non-bypassing filter (dirt fuse) the total permissible pressure drop is theoretically only limited by the maximum rated pressure of the unit. In practice, a differential pressure of 7 bar is generally considered suitable.

OIL COOLERS

These contain a number of small bore passages through which there is consequently a

Section 6.2   Summary of basic formulae and rules

considerable pressure drop and it is proportional to the quantity flowing. There are mechanical limitations to the maximum pressure drop which is permissible and in a shell and tube cooler typical operating limits are: maximum pressure through tubes (cooling water) 7 bar; maximum pressure over tubes (hydraulic fluid) 17 bar. It is common practice to fit some form of pressure limiting valve to bypass the cooler if the pressure drop becomes excessive.

6.2.3   Cylinder formulae

A detailed treatment of cylinders is given in Chapter 4. This section is a summary of the more commonly used equations.

Consider a double-acting cylinder first under extend conditions. Let D = full bore diameter, d = rod diameter, L = stroke, P_1 = pressure at full bore end, P_2 = pressure at annulus end, Q = input flow rate, q_E = output flow rate (extending), q_R = output flow rate (retracting).

Static extend thrust is $(\pi/4)[P_1 D^2 - P_2(D^2 - d^2)]$

Dynamic thrust is less than static thrust due to seal friction, fluid friction, piston inertia, etc. As an approximation, dynamic thrust is

$0.9 \times$ Static thrust

$= 0.9 (\pi/4)[P_1 D^2 - P_2(D^2 - d^2)]$

This is the dynamic thrust developed by the cylinder and does not take the load, its inertia or friction into account.

If Q is the flow rate of fluid into the full bore end of the cylinder and q_E is the flow rate leaving the annulus end then

$$\text{Piston velocity} = \frac{\text{Flow in}}{\text{Full bore area}} = \frac{\text{Flow out}}{\text{Annulus area}}$$

$$= Q/(\pi D^2/4) = q_E/(\pi/4)(D^2 - d^2)$$

and

$$q_E = Q(D^2 - d^2)/D^2$$

If L is the piston stroke, then the time to extend is given by

$$\frac{\text{Extend volume}}{\text{Flow rate to full bore end}}$$

$$= (\pi D^2/4) L/Q$$

or

$$\text{Extend time} = \frac{\text{Stroke}}{\text{Extend velocity}}$$

$$= L/[Q/(\pi D^2/4)]$$

$$= \pi D^2 L/4Q$$

Under retract conditions the flow Q is applied to the annulus end and the flow from the full bore end is q_R.

The static retract thrust is

$$P_2\left(\frac{\pi}{4}\right)(D^2 - d^2) - P_1(\pi/4)D^2$$

Dynamic thrust is approximately

$$0.9\left[P_2(\pi/4)(D^2 - d^2) - P_1\left(\frac{\pi}{4}\right)D^2\right]$$

Piston velocity is

$$Q/(\pi/4)(D^2 - d^2)$$
$$= q_R/(\pi/4)D^2$$

Therefore,

$$q_R = QD^2/(D^2 - d^2)$$

During the retract stroke there is a greater flow of fluid leaving the cylinder than entering it. This must be taken into account when sizing components.

$$\text{Retract time} = \frac{\text{Retract volume}}{\text{Flow to annulus end}}$$
$$= (\pi/4)(D^2 - d^2)L/Q$$

Thus the time for one complete cycle of the piston assuming the flow of fluid is the same to both ends and neglecting the change-over time between strokes, is given by

$$\text{Cycle time} = \frac{\text{Total swept volume}}{\text{Quantity flowing}}$$
$$= [(\pi/4)D^2 + (\pi/4)(D^2 - d^2)]L/Q$$

PISTON ROD BUCKLING STRENGTH

The maximum safe working thrust or load which can be applied to a solid piston rod of d cm diameter is $\pi^3 E d^4/64 L^2 S$ where E is the modulus of elasticity (2.1×10^6 kg/cm^2 for steel), S is the safety factor (usually 3.5) and L is the equivalent or free buckling length (cm). The relationship between L and the piston rod stroke length which depends upon the method of fixing can be found in Figure 4.21 in Chapter 4.

6.2.4 Pump formulae

The delivery of a pump may be given in any of the following ways:

1. The volume delivered at a given speed and pressure, e.g. 5 l/min at 200 bar and 1000 rev/min.
2. The swept volume per revolution of the pump, e.g. 1.5 in^3 per revolution. This is a theoretical value and makes no allowance for slip.

3. A complete set of characteristic curves giving deliveries at various speeds and pressures.

The quantity of fluid delivered will decrease as the pump wears.

Let D_p = displacement of the pump per revolution, n_p = pump speed, P_p = pressure rise across the pump, then

Theoretical delivery = $D_p \times n_p$

Theoretical input torque = $\dfrac{D_p P_p}{2\pi}$

EXAMPLE 6.3

In a particular pump unit, D_p is 10 cm^3 = 10×10^{-6} m^3, n_p is 1500 rev/min, and P_p is 150 bar = 150×10^5 N/m^2.

Then

$$\text{Theoretical delivery} = D_p \times n_p$$
$$= 10 \times 1500 \text{ cm}^3/\text{min}$$
$$= 15 \text{ l/min}$$

$$\text{Theoretical torque} = \dfrac{D_p}{2\pi} \times P_p$$
$$= (10/2\pi) \times 10^{-6} \times 150 \times 10^5 \, (\text{m}^3 \times \text{N/m}^2)$$
$$= 23.87 \text{ Nm}$$

Pump efficiencies

(i) Volumetric efficiency $_p\eta_v$ is

$$\dfrac{\text{Actual pump delivery}}{\text{Theoretical pump delivery}} = \dfrac{Q_p}{D_p n_p} \qquad (6.1)$$

where Q_p is the actual pump delivery.

(ii) Torque or mechanical efficiency $_p\eta_t$ is

$$\dfrac{\text{Work output per revolution}}{\text{Work input per revolution}} = \dfrac{D_p P_p}{2\pi T_p} \qquad (6.2)$$

where T_p is the input torque at the pump drive shaft.

(iii) Overall pump efficiency $_p\eta_o$ is

$$\dfrac{\text{Hydraulic output power}}{\text{Input power}} = \dfrac{Q_p P_p}{2\pi n_p T_p} \qquad (6.3)$$

From equations (6.1) to (6.3), it may be shown that

$$_p\eta_o = {_p\eta_v} \times {_p\eta_t}$$

6.2.5 Hydraulic motor formulae:

(i) Volumetric efficiency $_m\eta_v$ is

$$\frac{\text{Theoretical flow rate}}{\text{Actual flow rate}} = \frac{D_m n_m}{Q_m} \qquad (6.4)$$

where D_m is the motor displacement per revolution, n_m is the motor rotational speed, and Q_m is the rate of fluid flow into the motor.

(ii) Torque or mechanical efficiency $_m\eta_t$ is

$$\frac{\text{Actual work done per revolution}}{\text{Theoretical work done per revolution}} = \frac{2\pi T}{P_m D_m} \qquad (6.5)$$

where T is the motor output torque and P_m is the pressure drop across the motor.

(iii) Overall efficiency $_m\eta_o$ is

$$\frac{\text{Actual output power}}{\text{Theoretical output power}} = \frac{2\pi n_m T}{Q_m P_m} \qquad (6.6)$$

From equations (6.4) to (6.6) it can be shown that

$_m\eta_o = {_m\eta_v} \times {_m\eta_t}$

EXAMPLE 6.4

A hydraulic motor having a displacement of 500 ml per revolution, operates at a speed of 75 rev/min and is required to develop an output torque of 1200 Nm. The volumetric and torque efficiencies of the motor are 0.9 and 0.94 respectively. Determine: (i) pressure drop over the motor, (ii) the input flow and (iii) the overall efficiency.

(i) Theoretical motor torque $= \dfrac{D_m P_m}{2\pi}$

Actual motor torque $= \dfrac{D_m P_m}{2\pi} {_m\eta_t}$

Therefore

$$P_m = \frac{1200 \times 2\pi}{500 \times 10^{-6} \times 0.94} \text{ N/m}^2$$

$= 160$ bar

(ii) Theoretical motor flow $= D_m n_m$

Actual motor flow $= \dfrac{D_m n_m}{_m\eta_v}$

$= 500 \times 10^{-3} \times 75 \times (1/0.9)$ (l/min)

$= 41.7$ l/min

(iii) Overall efficiency, $_m\eta_o$ $= 0.9 \times 0.94$

$= 0.846$

6.3 POWER-PACK DESIGN

6.3.1 Layout

Standard commercial hydraulic power packs are usually designed down to a price and therefore do not incorporate all the most desirable features. They generally take one of two forms, either *immersed pumps* or *tank-top mounting*.

Immersed pump

The electric motor is mounted vertically on top of the tank with the pump and often some of the associated valving immersed in the fluid. This is an inexpensive method of construction but servicing is difficult particularly with the larger power motors. The main advantage is that the pump suction and drain ports will be unrestricted and open to the fluid. A variation of this type has the motor mounted on the tank side where sealing of the bell housing and motor end plate can be difficult. It is necessary to drain the reservoir before maintenance work on the pump.

Tank-top mounting

This arrangement which is known as the JIC standard is neat and accessible but it has one major disadvantage in that the pump has to work with a slight negative suction head. Suction conditions are aggravated when using water-based and high-density fire-resistant fluids.

With both types of power-pack design, the reservoir top plate must have sufficient strength to carry the weight of the pump set.

Custom-built power packs

For custom-built power packs a slightly more expensive but recommended method of construction is for the pump unit to be mounted beneath, to one side of (Figure 6.2) or even remotely from the reservoir, thus providing a positive head to the pump whilst the components are still accessible for maintenance. It is usual to incorporate a full flow stop valve in the pump suction line to allow for pump removal. Unauthorized or accidental closing or partial closing of this valve can cause pump damage. Ideally it should have an electrical limit switch interlocked to the pump motor control circuit.

The features include anti-vibration mountings fitted under the electric motor feet, the pump being connected to the motor by a rigid bell housing and flexible coupling. A flexible suction hose between the pump and tank, together with the flexible motor mounts, isolates the pump unit from the tank with consequential noise reduction. Aluminum bell housings are cheap and readily available but a further reduction in noise level can be accomplished by using ferrous castings or plastics. Gaskets on the motor and pump

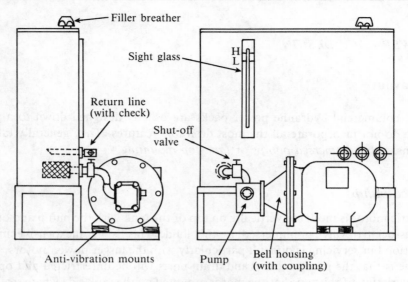

Figure 6.2 Pump mounted adjacent to reservoir.

mounting faces also help in noise reduction. Where noise is a major problem, acoustic hoods may be used.

Bell housing and coupling

These ensure correct alignment and provide a flexible connection between the pump and its drive motor (Figure 6.3). There are national and international standards for electric motor and pump mounting faces and shaft ends. A range of bell housings and couplings covers the various combinations.

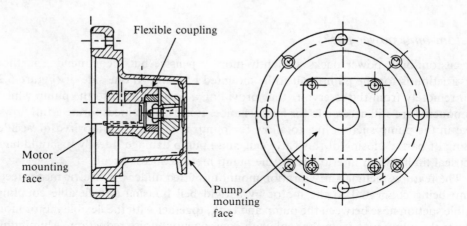

Figure 6.3 Bellhousing and coupling.

6.3.2 Reservoirs

The functions of a fluid reservoir in a power hydraulic system are the following:

1. To provide a chamber in which any change in volume of fluid in the hydraulic circuit can be accommodated. When cylinders extend there is an increased volume of fluid in the circuit and consequently there is a decrease in the reservoir level.
2. To provide a volume of fluid which is relatively stationary to allow entrained air to separate out and heavy contaminants to settle.
3. To make up any leakage occurring in the system.
4. To provide a filling point for the system.
5. To provide a radiating surface to allow the fluid to cool.

Reservoir types

NON-PRESSURIZED

The reservoir may be vented to atmosphere using an air filter or a separating diaphragm. The type most commonly used in industry, normally has an air breather filter, although in very dirty environments diaphragms or air bags are used.

PRESSURIZED

The pressurized reservoir usually operates at between 0.35 and 1.4 bar and has to be provided with some method of pressure control; this may be a small air compressor maintaining a set charge pressure. In motor circuits where there is little change in fluid volume in the reservoir, a simple relief valve may be used to limit the air pressure which will alter with changes in temperature. The advantages of a pressurized reservoir are that it provides boost pressure to the main pump and prevents the ingress of atmospheric dirt.

Reservoir size

The reservoir capacity should be adequate to cater for changes in fluid volume within the system, and with sufficient surface area to provide system cooling. An oversize reservoir can present some disadvantages such as increased cost, size and longer warming-up periods when starting from cold. There are many empirical rules for sizing reservoirs:

(a) The minimum reservoir capacity should be twice the pump delivery per minute. This must be regarded as an absolute minimum and may not be sufficient to allow for volume changes in the system.
(b) The reservoir capacity should be 3–4 times the pump delivery per minute. This may well be too high a volume for mobile applications.
(c) The reservoir capacity should be 2–15 liters per installed horsepower. This may result in very large reservoirs when high-pressure systems are used.

All the above rules are based on conventional shaped reservoirs. Special shapes require special consideration.

Where heat dissipation from the reservoir is a critical factor, this can be calculated by a basic formula:

$$H = hA\,\Delta T \times 3.6$$

where H is the heat transferred (watts), h is the transfer coefficient, A is the surface area in square meters, and ΔT is the temperature differential (°C).

For a vertical plate of height L,

$$h = 1.42\left(\frac{\Delta T}{L}\right)^{1/4}$$

For a horizontal plate of width W,

$$h = 1.32\left(\frac{\Delta T}{W}\right)^{1/4}$$

These formula apply to natural radiation; normal air circulation round the reservoir will increase the cooling considerably. For maximum heat radiation, the reservoir should be of minimum height and maximum length (see Figure 6.4).

The equations neglect the cooling effect of the pipework valves and actuators which may have a surface area comparable with that of the fluid reservoir.

The reservoir should not have any horizontal lips or angles on the vertical face as this interferes with the natural air convection.

The cooling efficiency can be increased by using a finned design (Figure 6.5); the fins should be vertical not horizontal. To assist the free circulation of external air and hence cooling, the reservoir should be mounted clear of the ground.

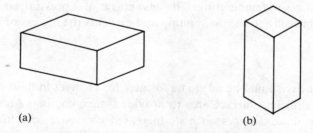

Figure 6.4 Reservoir shape. (a) Best design for heat transfer. (b) Poor heat transfer.

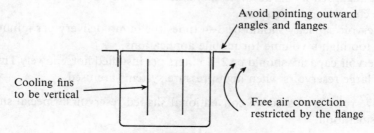

Figure 6.5 Effect of fins and flanges.

Section 6.3   Power-pack design

Table 6.1 Fluid operating temperatures.

Temperature	Mineral oil (°C)	Water in oil 60/40 (°C)	Water–glycol (°C)	Phosphate ester (°C)
Maximum local temperature	100	65	65	150
Maximum temperature for continuous operation	65	40	40	95
Maximum temperature for optimum fluid life	40	25	25	65

The operating temperature of the fluid will greatly affect its life. Table 6.1 gives an indication of operating temperatures for various types of fluid. These are typical values only and the exact value for a specific hydraulic fluid should be ascertained from the fluid supplier.

EXAMPLE 6.5

In a hydraulic system operating at 200 bar the pump delivery is 25 1/min and the input power to the pump drive is 10 kW. The system cycle is such that the pump is unloaded for 60% of the operating time. The overall efficiency of the system when it is on load is 65%. If the ambient temperature is 15°C and the maximum permissible fluid temperature in the reservoir is 50°C, calculate a suitable size for the fluid reservoir assuming (a) normal air circulation round the fluid reservoir doubles the cooling owing to natural radiation; (b) the fluid reservoir is of square section of side a with a length of $2a$.

(i) Heat dissipation from vertical plates is given by

$$H_v = h_v A \Delta T \times 3.6$$

where

$T = 35°C$ and $A = 6a^2$. Also
$h_v = 1.42 \, (35/a)^{1/4}$
$ = 3.45 \, a^{-1/4}$

and

$H_v = 3.45 \, a^{-1/4} 6a^2 \times 35 \times 3.6$ (watts)
$ = 2608 \, a^{7/4}$ watts

(ii) Heat dissipation from horizontal top plate is given by

$$H_H = h_H A \Delta T \times 3.6$$

where $\Delta T = 35°C$ and $A = 2a^2$. Also

$h_H = 1.32 \, (35/a)^{1/4}$
$ = 3.21 \, a^{-1/4}$

Therefore
$H_H = 3.21 \, a^{-1/4} \, 2a^2 \times 35 \times 3.6$
$ = 809 \, a^{7/4}$ watts

Heat dissipation owing to natural radiation is

$$H_V + H_H = 2608\,a^{7/4} + 809\,a^{7/4}$$
$$= 3417\,a^{7/4} \text{ (watts)}$$

Heat dissipation with normal air circulation may be taken as twice that owing to natural radiation and equal to $6834\,a^{7/4}$ watts.

Heat input to system during 'on load' part of cycle is $10 \times (1 - 0.65) = 3.5$ kW. But the system is only on load for 40% of the time, thus the average heat energy input to the fluid is given by

$$\text{Average heat input} = 0.4 \times 3.5 \text{ (kW)}$$
$$= 1.4 \text{ (kW)}$$
$$= 1.4 \times 10^3 \text{ watts}$$

For thermal equilibrium the heat energy entering the system must be equal to the heat energy dissipated from the system, therefore

$$6834\,a^{7/4} = 1.4 \times 10^3$$
$$a^{7/4} = 0.205$$
$$a = 0.404 \text{ m}$$

Assume tank internal measurements are 0.4-m wide by 0.4-m high by 0.8-m long, then volume of fluid contained is given by

$$Q = 0.4 \times 0.4 \times 0.8 \text{ m}^3$$
$$= 0.128 \text{ m}^3 = 128 \text{ liters}$$

In practice, the tank will have to be higher than 0.4 m as there must be a clearance volume above the oil. Sizing the reservoir from the rule of thumb formula, i.e. reservoir capacity is equal to 3–4 times the pump delivery per minute,

Reservoir capacity = 75 to 100 liters

This value is similar to the value calculated.

If a very large fluid reservoir is needed to dissipate the heat energy it may be advantageous to use a fluid cooler as described later in this chapter.

Reservoir design and construction

The reservoir must be equipped with a filling point and an air vent point. These are often combined as a filler breather unit complete with a filling mesh. Combined units of this nature tend to have a slow filling rate and the filling mesh is often deliberately punctured to speed up the inflow. This defeats its purpose and allows large dirt into the reservoir. It is also very easy to leave the air-filter cap off after filling permitting contaminants to enter. It is better to have a separate filling point complete with a quick-release coupling and integral filter unit, in which case all the fluid is pumped into the reservoir through the filter. Spin-on air breathers are available with fine filter ratings: 25 μm absolute is typical ($\beta_{25} = 75$).

In extremely dirty environments the fluid can be isolated from the atmosphere by using a sealed tank and a flexible separator to allow for variations in oil level (see Section 5.2.2 of Chapter 5). This may take the form of a bag into which air is drawn or expelled as the fluid level varies.

Section 6.3 Power-pack design

Flexible separators are usually limited to applications where there is only a small variation in fluid level, i.e. motor circuits. A special type of air breather has to be fitted. In the event of the pressure inside the reservoir falling below 0.03 bar vacuum (i.e. because of the loss of fluid from the system) air is allowed into the reservoir through a 25 μm absolute filter, thus protecting the pump from too low a suction pressure and consequential damage from cavitation. It will usually permit the reservoir to become pressurized to 0.3 bar above which pressure it opens acting as a safety valve.

The pump suction line should be fitted with an adequately sized strainer. The strainer should be located so that there is a well distributed flow and with its lowest point at least 75 mm above the bottom of the reservoir (to prevent any debris on the bottom of the tank being pulled into the suction line). The top of the strainer should be at least 150 mm below the lowest fluid level to prevent a vortex being formed and air drawn into the suction line.

System return lines and drain lines should terminate below the minimum fluid level, to prevent aeration of the fluid. The main returns should be fitted with diffusers which reduce the return oil velocity and prevent turbulence. Where diffusers are not used return line ends should be cut at 45%. Return outlets should be carefully positioned to minimize interaction between the return flow and the reservoir boundaries. If the return pipes have anti-syphon holes aeration may be caused by the jet of fluid emanating from them during normal operation. This can be reduced by fitting deflector plates.

The reservoir must be fitted with some form of level indicator, generally a sight glass which should have the maximum and minimum levels clearly marked. A float switch can be incorporated to give an alarm or shut the pump down if the oil level falls below a certain value. This is very important on large systems where there is the possibility of a pipe failure or other sudden leak. Shutting the pump down will protect the pump from cavitation and reduce the spillage of fluid.

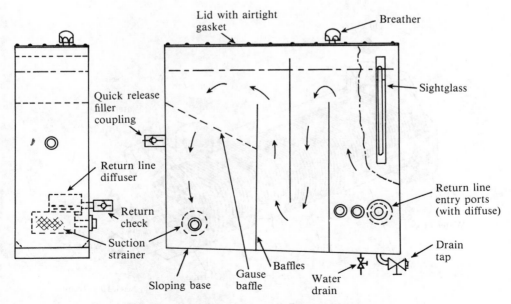

Figure 6.6 Reservoir with good design features.

Baffles are fitted in the reservoir to separate the return lines from the suction lines. They create a long flow path, which reduces surging effects, encourages dirt separation and improves cooling. A fine wire-mesh baffle of between 60 and 200 mesh size (250 to 75 μm aperture), set at an angle of 20–40° to the horizontal, with the top edge terminating below the fluid surface will considerably aid the separation of entrained air. Where the hydraulic fluid is an oil-and-water emulsion, baffles should not be used as they will tend to separate the emulsion. With these fluids good circulation in the reservoir is important to maintain the emulsion. Figure 6.6 shows a reservoir designed to include most of these features. Although it is not the best shape for heat dissipation, it has other merits. If the fluid temperature inside the reservoir varies outside specified limits, temperature switches can be used to give a warning, switch in coolers or heaters, shut down pumps, and prevent start-up etc. Where systems operate outdoors in cold temperatures or, for example, servo systems, which in order to preserve accuracy need to work at a constant temperature, heaters may be incorporated. These are a special type of immersion heater having a low wattage relative to the surface area (maximum 6 W/m^2). A domestic immersion-type water heater will 'burn' the oil.

Fluid conditioners

In the case of large reservoirs and systems where on-line filtration is impossible, separate 'clean-up' filtration loops should be fitted (see Chapter 5, Section 5.2). These may include an oil cooler. There are two types of oil cooler in use: water tube and air blast coolers.

The water-tube oil cooler consists of a series of interconnected copper tubes through which the cooling water passes surrounded by a jacket through which the hydraulic fluid passes (Figure 6.7). It is quiet in operation and can be arranged so that the oil pressure is higher than the water pressure, consequently any leakage is more likely to be of oil into

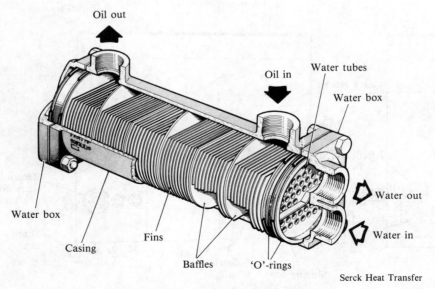

Figure 6.7 Water-tube oil cooler.

the water which is less serious than contamination of the hydraulic fluid. When a water-tube cooler is used there will be a considerable flow of water involved and a separate water-cooling tower and a circulating water supply may be necessary. Usually the water supply is thermostatically controlled so that it is only switched on when required.

The air-blast oil cooler is similar in construction to a vehicle radiator with a powered air fan. It should be situated in a cool area so that cold air is blown over the radiator. The air-blast coolers tend to be noisy but on small installations are preferable to water coolers owing to running and installation costs. Air-blast coolers are now available to fit between the pump and the electric motor as part of the bell housing and coupling. Such coolers do not need a separate electric motor drive, but it is necessary to take into account the extra power needed when sizing the pump drive.

6.3.3 Centralized hydraulic systems

The forerunners of modern gas, electricity and water authorities were a large number of local and municipal undertakings meeting the demands of large towns and industrial customers. In the late nineteenth and early twentieth century there existed several public hydraulic power companies providing water power in a similar manner.

The earliest systems were in dockyards driving machinery for cranes, dockgates and sluices, commencing with an installation at Hull docks in 1876. There were sizable undertakings in Manchester, Liverpool, Glasgow, Birmingham, Sydney and Melbourne, but the largest and most successful was the London Hydraulic Power Company which survived almost a century until 1976. At the peak of its existence in the 1920s, it was pumping over 1600 million gallons per week through its 184 miles of 7 inch diameter pipe. The pressure available was 700 lb/in^2 (48 bar), and over 8000 machines were connected. An installation in Antwerp transmitted energy hydraulically, converting it on site into electricity for lighting purposes − competitively with gas and electricity supplies. In general, the most popular employment was for lifts, hoists and presses; the normal supply pressure of 700 to 1200 lb/in^2 (48 to 83 bar) being increased by intensifiers for use in the latter.

The change to hydraulic oil as the working fluid made such large installations impractical, but there has recently been a renaissance of centralized systems operating on a smaller scale. The idea of a hydraulic ring main running round a factory in a manner similar to the pneumatic ring mains which are in everyday use has many attractions. The principal difference, apart from the higher pressures involved and the obvious seriousness of leaks, is that a return pipe has to be provided as well as a supply. Their most frequent use is in a factory or process plant where a group of related machines is powered from a central source.

Individual versus centralized systems

There is still much controversy over the use of centralized hydraulic systems as opposed to individual power packs. The features of each will now be compared. In general, what is an advantage for one is a disadvantage for the other. Both have their merits and every application must be assessed on its own particular requirements.

INDIVIDUAL POWER-PACK SYSTEMS
Advantages
- They will be completely independent of each other
- Different grades or types of fluid can be used as appropriate to each system
- Each can operate at different pressures
- If one circuit fails the others will still be operative
- Power packs can be adjacent to the machine

Disadvantages
- More power packs and components to maintain
- Increased cost
- More floor space required
- Greater total power

CENTRALIZED HYDRAULIC SYSTEMS
Advantages
- Single oil reservoir
- Standby pump(s) can be designed into the power pack
- Reduced cost
- Single unit to be maintained
- Less space required
- Fluid-conditioning can be incorporated either off-line or in the return line at much lower cost than when individual power packs are used
- Total power-pack capacity may be less than for individual power packs

Disadvantages
- If there is a failure in the power pack all the systems fail (unless standby facility included)
- Longer pipe runs are involved than with individual power packs
- If more than one operating pressure is involved, pressure-reducing valves will have to be fitted to the appropriate circuits
- Large volume of fluid in the reservoir can be a considerable fire hazard and it may be necessary to 'tank' the power pack room and install an automatic fire-extinguisher system, e.g. flooding the power pack room with carbon dioxide
- There may be interaction between the circuits.

6.4 HYDRAULIC ACCUMULATORS

An accumulator is a device used to store energy. In a hydraulic system the energy is stored in the form of fluid under pressure.

6.4.1 Types of accumulator

Dead-load accumulator

Such a unit is shown in Figure 6.8 and consists of a single-acting vertical cylinder which

Section 6.4 Hydraulic accumulators

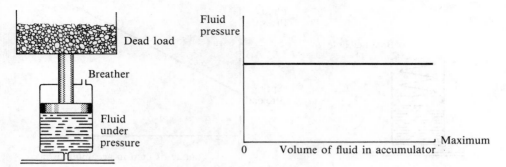

Figure 6.8 Dead-load accumulator.

raises a heavy load or weight. Large volumes of fluid can be stored but very heavy weights are needed.

EXAMPLE 6.6

A dead-load accumulator of the type shown in Figure 6.8 has a cylinder bore of 500 mm and is to operate at a system pressure of 200 bar. The dead load required will be the product of the piston area times the pressure.
Load is

$$(\pi/4) \times \left(\frac{500}{1000}\right)^2 \times 200 \times 10^5 \; (m^2 N/m^2)$$

$$= 3.93 \times 10^6 \; N \; (\text{approx. 400 tonnes})$$

The volume of fluid stored will depend upon the stroke of the cylinder.

Scrap metal, bricks, concrete, soil etc. may be used to form the dead load. The principal advantage is that the discharge pressure is constant whereas with all other types of accumulator the pressure varies as the volume of fluid stored. Disadvantages are large physical size and a slow response owing to the high inertia of the load and piston. The cylinder bore has to be machined, and seals built into the piston; wear will occur between the piston and cylinder body.

Spring-loaded accumulator

A free piston which is spring loaded moves within a cylinder as shown in Figure 6.9. The physical characteristics of the spring limit the piston stroke and hence the volume of fluid which can be stored. As the volume of fluid in the accumulator increases, the spring is compressed and the spring force increases.

Gas-loaded accumulator

In this design a compressed gas, usually nitrogen but sometimes air, is used to pressurize the stored fluid. There are three major types, which are differentiated by the method of

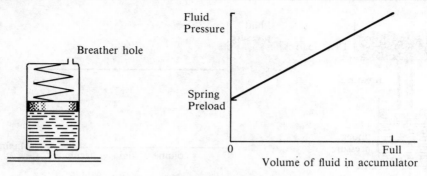

Figure 6.9 Spring-loaded accumulator.

interfacing the gas and fluid. In one type there is direct contact whereas in the other two a piston or membrane is used as the separator.

FREE-CONTACT GAS-LOADED ACCUMULATOR
This consists of a vessel with a fluid port at the bottom and a gas-charging port at the top as shown in Figure 6.10. Mounting has to be in the vertical position, as there is no piston separating the gas and fluid. The free-surface accumulator requires little machining and is consequently relatively simple to manufacture. As the gas charge is in direct contact with the hydraulic fluid, some entrainment of the gas in the fluid will occur. The degree of entrainment can be limited by restricting the flow velocity of the hydraulic fluid into and out of the accumulator so that excess agitation does not occur.

The accumulator must not be fully discharged when in service or the charge gas will flow into the hydraulic system. Float or level switches can be used to monitor the fluid level in the accumulator and to prevent complete discharge.

A feature of this design is a very fast response time.

Gas-loaded accumulators can be very large and are often used with water or high water-based fluids using air as the gas charge. Typical applications are on water turbines to absorb pressure surges owing to valve closure and on ram pumps to smooth out the delivery flow. The exact shape of the accumulator characteristic curve in Figure 6.10

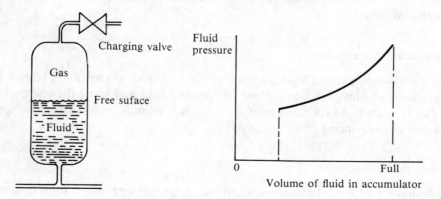

Figure 6.10 Free-contact gas-loaded accumulator.

Section 6.4 Hydraulic accumulators

depends upon the speed of compression, or expansion, of the charge gas, i.e. it is dependent upon the gas change being the following:

(i) *Isothermal* (constant temperature). This occurs when the expansion or compression of the gas is very slow. The relationship between absolute pressure P and volume V of the gas is constant, $PV = $ Constant.

(ii) *Isentropic* (adiabatic). This is where there is no flow of energy into or out of the fluid. The law which the gas obeys is given by $PV^\gamma = $ Constant, where γ is approximately 1.4.

(iii) *Polytropic*. This is somewhere between isothermal and isentropic. This gas change is governed by the law, $PV^n = $ Constant, where n is somewhere between 1 and 1.4 and is known as the polytropic coefficient.

PISTON-TYPE GAS-LOADED ACCUMULATOR

This is similar to the free-contact type except that a 'free' piston separates the gas and fluid (Figure 6.11). Although a piston-type accumulator can be used inclined to the vertical, less wear will result when it is vertical. Any failure tends to be gradual and is normally owing to deterioration of the piston seals and wear in the cylinder bore. The response time is adversely affected by the inertia of the piston and the effect of seal stiction. Discharge characteristics are similar to those of the free-surface type, although piston inertia limits the maximum discharge rate.

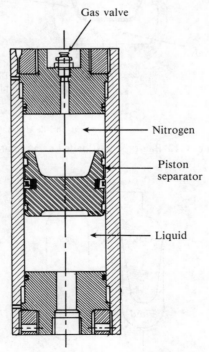

Fawcett-Christie Hydraulics

Figure 6.11 Piston-type gas-loaded accumulator.

BAG-TYPE GAS-LOADED ACCUMULATOR

Here, separation of gas and fluid is by means of a flexible membrane in the form of a bag or diaphragm (Figure 6.12). A valve fitted to the exhaust port prevents the bag being extruded. Fast-response speeds result from the relatively low inertia. Bag-type accumulators usually cost less than piston types and hold their gas precharge longer. Failure tends to be sudden owing to rupturing of the bag or diaphragm. In most designs of accumulator, the bladder is fitted through the fluid port in the base. However, the particular model illustrated in Figure 6.12 is a top serviceable construction where the bladder can be replaced by dismantling the gas valve assembly.

Although some bag-type accumulators are installed mounted on their sides, it is preferable for them to be vertical – particularly tall narrow ones. A sudden discharge can cause the poppet valve to lock closed a partially discharged unit.

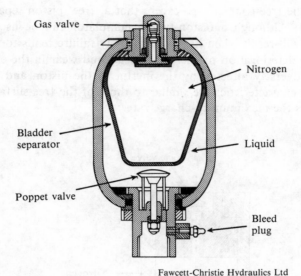

Fawcett-Christie Hydraulics Ltd

Figure 6.12 Bag-type gas-loaded accumulator.

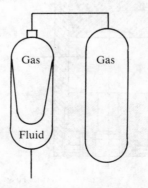

Figure 6.13 Use of gas back-up bottle.

To increase the effective hydraulic fluid capacity of an accumulator a gas back-up bottle is used as shown in Figure 6.13. The accumulator is used to store the fluid and the back-up bottle to store the additional gas charge.

6.4.2 Accumulator applications

Accumulators are used for:

- Fluid supply
- Pump delivery pulsation damping
- Pressure surge damping
- Standby or emergency power source
- Thermal expansion compensation
- Leakage compensation
- Counterbalancing
- Vehicle suspensions.

Fluid supply

One of the most frequently used applications is to provide a high flow rate of fluid over a short period of time (typical pumping circuits are in Figures 2.36 and 3.11). A pump with a low delivery rate is used to charge the accumulator over a long period of time. Stored fluid is discharged into the system when required at a high flow rate. During periods when the accumulator is charged and there is no circuit demand, the pump is unloaded or switched off until the stored fluid needs to be replenished.

To size the accumulator, consider the changes in pressure and volume of the gas charge: charging takes place slowly and the gas compression can be considered to be isothermal. Discharge is rapid and may be considered as adiabatic. The initial precharge gas pressure P_1 is normally just below the minimum working pressure P_3 of the hydraulic system. This is to prevent the accumulator bladder constantly closing the anti-extrusion check valve, typically $P_1 = 0.9 P_3$. The maximum system pressure P_2 is the fluid pressure when the accumulator is fully charged. P_2 should not be greater than three times the minimum working pressure or the elastomer material of the bag or diaphragm may be damaged. Figure 6.14 shows the accumulator in the various conditions.

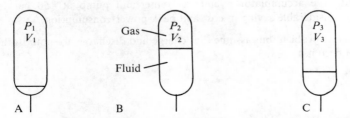

Figure 6.14 Accumulator gas charge conditions.

EXAMPLE 6.7

In a particular hydraulic system a discharge of 100 liters is required during six seconds once every minute. Noting that all calculations involving gas laws require values of pressure and temperature to be in absolute units, and that V_1, V_2 and V_3 refer to the gas volumes, calculate a suitable size of pump and accumulator if:

(a) Minimum system pressure is 100 bar gauge.
(b) Maximum system pressure is 150 bar gauge.
(c) Accumulator precharge pressure is 90% of minimum system pressure.
(d) The gas is compressed isothermally (PV = Constant) during the charging.
(e) The gas expands adiabatically ($PV^{1.4}$ = Constant) during discharge.

In Figure 6.14, the conditions are as follows: A, precharge, $P_1 = 90$ bar gauge, V_1 is unknown; B, fully charged, $P_2 = 150$ bar gauge, V_2 is unknown; C, discharged, $P_3 = 100$ bar gauge, V_3 is unknown.

Solution

Volume of oil required during six seconds is 10 liters. Considering the pump size, the pump must be capable of delivering 10 liters per minute (average demand). The volume of oil required from the accumulator is circuit demand less the volume supplied by the pump during that time, i.e. $10 - (10 \times 6/60)$ liters = 9 liters. This is the maximum volume of oil to be stored in the accumulator, i.e. $V_3 - V_2 = 9$ liters.

During discharge (gas expansion)

$$P_2 V_2^{1.4} = P_3 V_3^{1.4} \tag{6.7}$$

$$(V_3/V_2)^{1.4} = P_2/P_3 = 151/101 = 1.495$$

So

$$V_3/V_2 = 1.332,$$

then

$$1.332 V_2 - V_2 = 9$$
$$V_2 = 9/0.332 = 27.1 \text{ liters}$$

During charge (gas compression)

$$P_1 V_1 = P_2 V_2$$
$$91 V_1 = 151 \times 27.1$$
$$V_1 = (151 \times 27.1/91) = 45 \text{ liters}$$

Use a 50-liter accumulator. If a simple pump circuit is used, a 100-liter/min pump at 100 bar is necessary. With the accumulator circuit, a 10-liter/min pump at 150 bar is sufficient, representing a considerable saving in cost and peak-power consumption.

Consider the same problem but assume that charge and discharge are both isothermal, then equation (6.7) becomes

$$P_2 V_2 = P_3 V_3$$

Therefore

$$\frac{V_3}{V_2} = \frac{P_2}{P_3} = 1.495$$

so

$$1.495 V_2 - V_2 = 9$$

and therefore

$$V_2 = 9/0.495 = 18 \text{ liters}$$

During compression

$$P_1 V_1 = P_2 V_2$$

So,

$$V_1 = (151 \times 18)/91 = 29.9 \text{ liters}$$

This calculation using isothermal discharge results in a much smaller accumulator being suggested than if adiabatic discharge is assumed. In all probability the law followed will be polytropic lying somewhere between isothermal and adiabatic. If the size is found using adiabatic discharge, it will be sufficient whichever type of discharge occurs. A capacity greater than necessary is not detrimental because the maximum system pressure P_2 (when the accumulator is fully charged) can be correspondingly reduced, which will improve the overall system efficiency. Should the accumulator capacity be less than required, it will be incapable of storing sufficient fluid, lengthening the cycle time. If part of the gas charge leaks from the accumulator, there will be a reduction in the usable volume of fluid stored which also increases the cycle time. Generally, the gas law governing expansion is polytropic with an index between 1 and 1.4; however where the accumulator discharge is rapid, an index of 1.4 must be used to ensure sufficient fluid capacity.

Pump delivery pulsation damping

The delivery from most hydraulic pumps is not absolutely constant with time but is subject to pulsations or surges. For instance with a five-cylinder piston pump there will be five pulsations per pump revolution. In most applications these surges are unimportant as they are partially smoothed by the pipework upstream of the pump. When a system demands constant delivery, an electrohydraulic servo system for instance, an accumulator positioned upstream of the pump acts as a pulsation damper. A small accumulator will filter pressure pulsation to a negligible level but complete elimination is impossible (see Figure 6.15.)

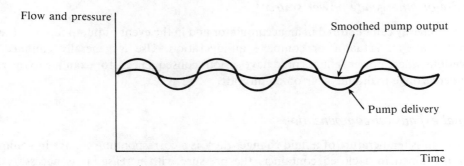

Figure 6.15 Pump delivery pulsation damping.

Pressure surge damping

Sudden closure of a valve causes pressure transients or shock waves. In a water system this phenomenon is known as 'water hammer' because it is usually accompanied by considerable noise. Initially fluid adjacent to the valve is stopped and compressed; a pressure wave travels back through the column of fluid which is being brought to a standstill. On reaching the far end of the straight pipe, a decompression wave forms travelling back to the valve. These waves travel back and forth until the energy is expended. The more rapid the valve closure, the more severe is the pressure surge generated.

The pressure generated by sudden closure is given by

$$\Delta P = \rho a \Delta V$$

where ΔP is the pressure rise, ρ is the fluid density, $a = (B/\rho)^{1/2}$ is the speed of sound through the fluid, ΔV is the change in velocity, and B is the bulk modulus of the fluid.

The derivation of these expressions can be found in most standard textbooks on hydraulics.

For example, if fluid flowing at the rate of 4 m/s is suddenly stopped there will be a pressure surge of approximately 45 bar. This surge is superimposed on the system pressure and can cause damage to components, and sudden movements of hydraulic hose and pipework. Partial damping can be effected by the judicious placing of an accumulator close to the valve to receive and absorb pressure transients (Figure 6.16).

If long pipe lines (50 m or longer) are involved, the length of the line and speed of closure must be considered. Fortunately long pipe runs are rarely encountered in power hydraulic systems. Most advanced textbooks on water hydraulics deal fully with the effects of slow closure and long pipe runs.

Pressure shocks resulting from external mechanical impacts on cylinders can also be absorbed by an accumulator.

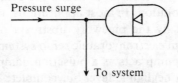

Figure 6.16 Pressure surge damping.

Standby or emergency power source

Hydraulic energy can be stored in an accumulator and in the event of pump failure, drawn on to operate an actuator or complete an operation. Use is generally confined to applications where power failure could have serious consequences, for example to operate the undercarriage hydraulic gear on an aircraft.

Thermal expansion compensation

When the mass temperature of a fluid changes there is a corresponding change in volume. If it is restrained in a closed container, the pressure will increase. It is necessary to compensate for thermal expansion in a closed system. The variation in volume of the

Section 6.4 Hydraulic accumulators

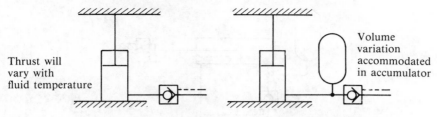

Figure 6.17 Thermal expansion compensation.

closed system is the volume which has to be stored in the accumulator (Figure 6.17). The volumetric change ΔV of a volume of fluid V subject to a change in temperature $t\,°C$ can be expressed as

$$\Delta V = V t K_v$$

where K_v is the coefficient of volumetric expansion of the fluid. (This may be taken as 0.0007 for a 1 °C change in temperature.)

The required volume of the accumulator can be calculated knowing the stored volume and the permissible pressure variation.

EXAMPLE 6.8

Calculate the increase in pressure if a cylinder 300 mm diameter × 500 mm stroke is locked in the extended condition and then subjected to a 20 °C rise in temperature. What is the volume of fluid to be stored in an accumulator fitted to compensate for the thermal expansion? (The bulk modulus of the fluid may be taken as 15 000 bar.)

Total volume of fluid contained in the cylinder is

$(\pi/4)0.3^2 \times 0.5 \text{ m}^3$
$= 35.3$ liters

Change in volume is

$35.3 \times 0.0007 \times 20$
$= 0.49$ liters

which is the additional volume to be stored.

Without an accumulator, the change in pressure in the closed system is ΔP, i.e.

$$\Delta P = \frac{B \Delta V}{V}$$

$= 15\,000 \times \left(\dfrac{0.49}{35.3}\right)$

$= 210$ bar

Leakage compensation

If there is only a very small leakage in a system containing a pump unloader valve, the pressure will fall causing the unloading valve to chatter. This chatter can be eliminated by

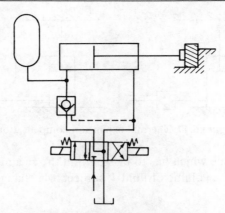

Figure 6.18 Leakage compensation.

using an accumulator to make up the leakage. In the clamping application shown in Figure 6.18 the accumulator compensates for leakage across the actuator piston or check valve seat, and maintains the clamping force.

Counterbalancing

Large masses may be counterbalanced by using a hydraulic cylinder and an accumulator in a self-contained closed loop (Figure 6.19). The quantity of fluid stored will depend on the volume of the cylinder displacement and the degree of underbalance or overbalance permissible. In counterbalancing applications, it is usual to have an accumulator with a volume of at least twice the counterbalance cylinder total displacement with a gas back-up bottle of about five times the accumulator capacity.

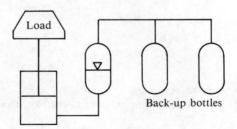

Figure 6.19 Counterbalancing.

Vehicle suspensions

A growing application of accumulators is on automobile suspensions. In the Citroën system (Figure 6.20) each wheel has a suspension strut containing an accumulator which absorbs shocks and to provide a rapid response a main accumulator augments the pressure fluid supply. Constant level control is accomplished under varying load and road conditions by mechanical servo valves linked to the front and rear anti-roll bars. These regulate the flow of fluid to and from the strut cylinders maintaining constant level. If, for

Section 6.4 Hydraulic accumulators

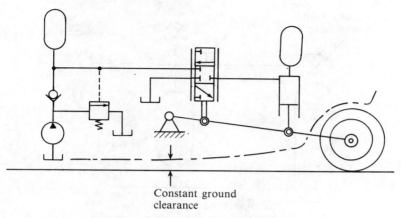

Constant ground clearance

Figure 6.20 Vehicle suspension system.

example, the vehicle bounces the chassis lifts relative to the wheels. Referring to Figure 6.20, this situation is best understood by considering the wheels dropping in which case the height corrector (servo) valve opens releasing pressure from the struts and the wheels return to their correct height thus closing the valve. If the wheels lift the valve opens supplying fluid to the strut cylinders forcing the wheels back down which closes the valve. Similarly an increase in load tends to lower the chassis (equivalent to raising the wheel) and more fluid is fed into the struts counteracting the movement. Nominal ground clearance can be varied by altering the ratios of the linkages to the servo valves.

6.4.3 Operation and safety precautions

Considerable care should be taken when working on accumulators which are pressure vessels and potentially dangerous.

Charging the accumulator with gas

It is recommended that only nitrogen be used as the charging gas and *never under any circumstances oxygen*, as there would be a serious risk of explosion. Although air has been used to charge large free-surface water-type accumulators, it is not to be advocated when the fluid used has a mineral oil content.

Normally the gas precharge pressure is 90% of the minimum system working pressure. Accumulators can be ordered from the supplier in the charged condition. Where charging takes place on site, or is undertaken by the customers in their workshops, connection between the accumulator and nitrogen cylinder has to be through a charging set as shown in Figure 6.21.

The charging instructions should be read carefully. In certain free-piston type accumulators it is recommended that 75 ml of hydraulic fluid be put into the gas side of the accumulator to lubricate the piston seals before the charging set is connected. The maximum precharge pressure generally attainable using nitrogen cylinders is 140 bar. For

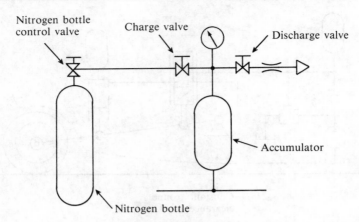

Figure 6.21 Accumulator charging.

higher precharge pressures, a boost pump is necessary. Most users purchase the accumulator precharged by the manufacturer.

The charging set is connected and the charge valve slowly opened allowing nitrogen to flow from the gas cylinder. When the pressure gauge reads the required value, the charge valve is closed. Should the precharge pressure be greater than desired, a discharge valve can be opened to bleed excess gas slowly to atmosphere through a fixed restrictor.

The accumulator precharge should be periodically checked as the gas may leak. The gas charge pressure can be measured by using a charge set or alternatively by charging the accumulator with fluid and then letting it slowly discharge, observing the fluid pressure at some convenient test point. As the fluid flows out of the accumulator, the pressure will gradually fall, until the anti-extrusion valve closes the fluid port at which point the pressure will suddenly become zero. The pressure reading immediately before the sudden pressure drop is the precharge pressure of the gas.

Any back-up bottles must be connected before the accumulator is precharged.

Pressure testing of accumulators

Hydraulic accumulators are pressure vessels and are subject to safety regulations. The test requirements of pressure vessels depend upon the maximum operating pressure and volumetric capacity. Initially the accumulator will be accompanied by a pressure test certificate issued by the manufacturer. The system should then be tested after installation and at regular prescribed intervals thereafter.

Safety regulations vary for different countries but some, or all, of the following safety devices pertain: shut off valves, pressure gauges, pressure relief valves, automatic pressure discharge valves; warning signs etc. may have to be fitted. Accumulator safety blocks are available containing these various valves. A typical installation is shown in Figure 6.22. A solenoid-controlled accumulator unloader valve coupled to the pump motor starter circuit ensures that all the pressurized fluid is automatically discharged when the pump is switched off. Maximum discharge rate from an accumulator is limited by the fluid port sizes but it is often necessary to control the flow by a restrictor. In some countries,

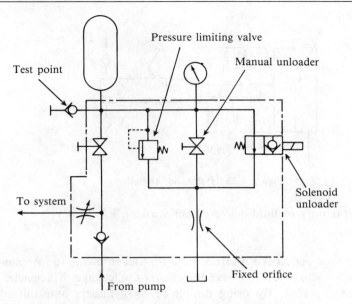

Figure 6.22 Accumulator safety block circuit.

legislation requires the pressure limiting (safety) valve to be factory preset, tested, certified and specially sealed.

Note *Before carrying out any work on a system including an accumulator, the hydraulic fluid must be fully depressurized. Before any attempt is made to dismantle an accumulator the gas pressure must be discharged.*

6.5 HYDRAULIC INTENSIFIERS

Intensifiers are used in hydraulic systems to generate a high pressure output from a low pressure input. Depending upon the design, the input source may be pneumatic or hydraulic and the output a fixed quantity of fluid for each operation (single shot) or a continuous delivery (similar to a pump output).

A simple single shot piston type intensifier shown diagrammatically in Figure 6.23 comprises a large diameter piston powered from the input source, driving a small diameter piston. Fluid is taken in and expelled with an action similar to that of a piston pump. Pressure intensification is directly proportional to the ratio of piston areas (i.e. proportional to the ratio of diameters squared).

$$PA = pa$$

so theoretical output pressure $p = PA/a = PD^2/d^2$ where P = input pressure, A = low pressure piston area, a = high pressure piston area, D = low pressure piston diameter, and d = high pressure piston diameter. Similarly the amount of fluid delivered per operation relative to the quantity of input fluid is inversely proportional to the ratio of piston areas.

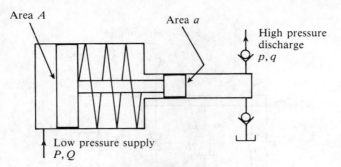

Figure 6.23 Hydraulic intensifier.

Theoretical quantity of fluid delivered per stroke q is

$$Qa/A = Qd^2/D^2$$

where $Q =$ input flow per stroke. Performance is reduced owing to internal leakage and friction – the higher the output pressure the greater the leakage. Efficiencies in the order of 80% may be expected. By using double-acting cylinders and suitable valving a continuously operating intensifier can be constructed.

Pneumatically powered intensifiers are available with ratios such that pressures in excess of 2000 bar are obtainable from a 7 bar air supply. The input air supply pressure is regulated to set the required maximum fluid pressure. When this pressure is reached the device will stall maintaining pressure and only delivering fluid when there is a demand.

Rotary type flow dividers are able to perform as pressure intensifiers. Their use for this purpose is described in Section 3.2.6 of Chapter 3.

6.6 DESIGN STUDY – A SIMPLE HYDRAULIC PRESS

The first hydraulic press patented by Joseph Bramah in 1795 was based on Pascal's Law and consisted of a single acting cylinder and a hand pump. This system is still extensively used today in small portable equipment, for example hydraulic car jacks, car lifts and portable lifting equipment. The basic circuit for all applications consists of a hand pump feeding a gravity return cylinder and a manual stop-valve from the cylinder to the reservoir. A relief valve may be built into the hand pump.

Although the hand pump is ideal for many portable applications, it is too slow and the system is not flexible enough for the majority of industrial applications. The complexity of modern press tools and components produced necessitates a large 'daylight' or press opening, so the press can be loaded and unloaded. Thus the major part of the stroke is often used to close the dies, the remainder of the stroke actually doing the work. With large presses, in order to reduce the cycle time, a rapid closing of the dies at low thrust followed by a slower speed at high thrust is required. There are several methods of achieving this type of motion hydraulically and to investigate the various systems, consider the following problem.

EXAMPLE 6.9

A single acting upstroking moulding press has a cylinder diameter of 400 mm, a stroke of 250 mm of which 225 mm is to close the dies onto the work piece at a cylinder pressure of 20 bar. The final 25 mm stroke is to be at a cylinder pressure of 350 bar. The press is to produce 60 components per hour. The breakdown of the required maximum operating time is:

Rapid approach 225 mm	= 5 seconds
Pressing over final 25 mm	= 5 seconds
Cure time (i.e. hold under full thrust)	= 25 seconds
Return time	= 10 seconds
Unload and reload press	= 15 seconds
Total cycle time	= 60 seconds

The return time is estimated for gravity return. The only times which are absolutely fixed are the cure and the load times.

Solution 1

Fixed output pump

This is the simplest possible circuit (Figure 6.24). The thrust developed by the cylinder (in kilonewtons, kN) during rapid approach is given by

Pressure × Area

$$= 20 \times 10^5 \times \left(\frac{\pi}{4}\right) \times (0.4)^2 \; (\text{Nm}^2/\text{m}^2)$$
$$= 0.251 \times 10^6 \, \text{N}$$
$$= 251 \, \text{kN}$$

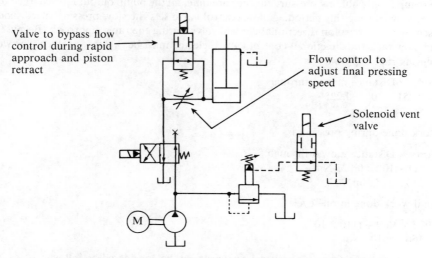

Figure 6.24 Solution 1: single fixed displacement pump circuit.

The thrust developed during final pressing (in meganewtons) is

$$350 \times 10^5 \times \left(\frac{\pi}{4}\right) \times 0.4^2 \ (Nm^2/m^2)$$

$$= 4.4 \times 10^6 \, N$$

$$= 4.4 \, MN$$

For the pump size required:

1. RAPID APPROACH

Cylinder area $= \pi \times 0.4^2/4 = 0.126 \, m^2$
Cylinder speed $= 0.225/5$ (m/s)
Flow required $= 0.126 \times (0.225/5) \ (m^3/s)$
$\qquad\qquad\quad = 0.00566 \, m^3/s$
$\qquad\qquad\quad = 5.66 \, liters/s$
$\qquad\qquad\quad = 340 \, liters/min$

2. FINAL PRESSING

Cylinder speed $= 0.025/5$ (m/s)
Flow required $= (0.126 \times 0.025)/5 \ (m^3/s)$
$\qquad\qquad\quad = 0.00063 \, m^3/s$
$\qquad\qquad\quad = 37.8 \, l/min$

In the single-pump system, the pump must deliver 340 l/min against the maximum pressure of 350 bar. The maximum theoretical input power will be

$$\frac{340}{60 \times 10^3} \times 350 \times 10^5 \ \left(\frac{m^3}{s} \times \frac{N}{m^2}\right)$$

$$= 198.3 \, kW$$

As the pump is only holding pressure for the cure time, all the pump output is discharged over the relief valve during this period. A flow-control valve sets the slow pressing speed, and is bypassed by a two-position directional-control valve during the rapid approach. Because of the high flow rates the directional-control valves selected must be two-stage valves. The work done during rapid approach is

Force × Distance moved through
$= 0.251 \times 10^6 \times 0.225$
$= 56.5 \times 10^3 \, Nm$

The work done during pressing is

Force × Distance moved through
$= 4.4 \times 10^6 \times 0.025 \, Nm$
$= 110 \times 10^3 \, Nm$

The total work done in one cycle is

$(56.5 \times 10^3) + (110 \times 10^3)$
$= 166.5 \times 10^3 \, Nm$

The energy supplied to the pump during one cycle can be considered as follows:

(a) During rapid approach with the pump operating against a pressure of 20 bar for a period of 5 seconds, the energy supplied is

$$20 \times 10^5 \times \left(\frac{340 \times 10^{-3}}{60}\right) \times 5 \left(\frac{N}{m^2} \times \frac{m^3}{s} \times s\right)$$
$$= 56.7 \times 10^3 \text{ Nm}$$

This value should agree with the actual work done during rapid approach. The slight difference in values is due to the approximations made in calculating the thrust and flow rate.

(b) Energy supplied to the pump during the 30-second period pressing and curing is given by

$$350 \times 10^5 \times 340 \times 10^{-3} \times 30 \left(\frac{N}{m^2} \times \frac{m^3}{s} \times s\right)$$
$$= 5950 \times 10^3 \text{ Nm}$$

Total energy supplied to the pump per cycle is

$(56.7 \times 10^3) + (5950 \times 10^3)$ Nm

which is approximately 6×10^6 Nm per cycle. The overall system efficiency is

$$\frac{\text{Work done}}{\text{Energy supplied}}$$
$$= \frac{166.5}{6000}$$
$$= 2.78\%$$

This is obviously a very inefficient system.

Solution 2

Multipump circuit

This incorporates three pumps — a high-flow low-pressure pump, a low-flow high-pressure unit, and a small pump to provide a pilot supply for the main directional valve (Figure 6.25). The main pumps are unloaded during part of the cycle.

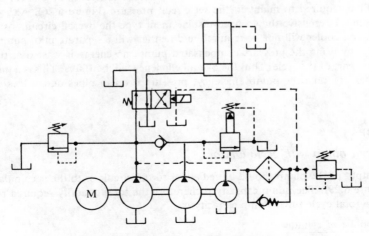

Figure 6.25 Solution 2: multi-pump circuit.

For rapid approach, the output from both main pumps, will be used. In final pressing, only the high-pressure low-volume pump is used. So, from Solution (1) pump size for final pressing is 37.8 l/min against a maximum pressure of 350 bar. The theoretical input power for final pressing is

$$\frac{37.8 \times 350 \times 10^5}{60 \times 10^3} \text{ watts}$$

$$= 22.05 \text{ kW}$$

and for rapid approach the low-pressure pump will have to deliver $340 - 37.8$, i.e. 302.2 l/min. The theoretical input power for rapid approach up to 20 bar is

$$\frac{340 \times 20 \times 10^5}{60 \times 10^3} \text{ watts}$$

$$= 11.33 \text{ kW}$$

The maximum power requirement for this system is 22.05 kW. Once again, during the cure time the high-pressure pump will be spilling-off across the relief valve and an oil cooler will be needed. The theoretical input energy during rapid approach is power (watts) × time

$$11.33 \times 10^3 \times 5 \left(\frac{\text{Nm s}}{\text{s}}\right)$$

$$= 56.7 \times 10^3 \text{ Nm}$$

The input energy during final pressing and curing is

$$22.05 \times 10^3 \times 30 \left(\frac{\text{Nm s}}{\text{s}}\right)$$

$$= 661.5 \times 10^3 \text{ Nm}$$

The total input energy is 718×10^3 Nm. As before, work done per cycle is 166.5×10^3 Nm. The theoretical system efficiency is therefore $(166.5/718) = 0.23$ or 23%.

Solution 2 (modified)

Using two pumps, the high-pressure pump has a pressure-compensator control and only delivers the fluid required to maintain the set circuit pressure (Figure 6.26). Although this pump will be more expensive than a fixed-displacement type the overall circuit cost may be reduced as the oil cooler will not be required and neither will a separate pilot pump.

In theory, by using the pressure-compensated pump, no energy is wasted at the pump during the cure part of the cycle. Thus the overall efficiency will be 100%. This is a theoretical value and does not take the pump efficiency, pressure drop in pipes and valves etc. into account.

Solution 3(a)

A fixed-output pump with accumulator

Ideally the output of the pump should be fed either to the circuit or to the accumulator (see circuit in Figure 6.29). The pump capacity is therefore the total quantity required per cycle divided by the total cycle time. Pump capacity is

$$\frac{\text{Swept volume of cylinder}}{\text{Total cycle time}}$$

Section 6.6 Design study—a simple hydraulic press

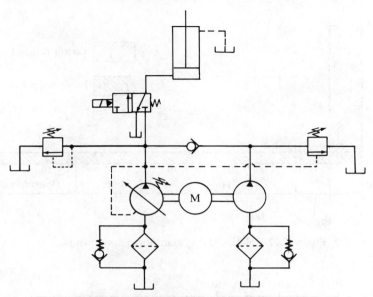

Figure 6.26 Solution 2 (modified): multi-pump circuit.

$$= \frac{\text{Area of piston} \times \text{piston stroke}}{\text{Total cycle time}}$$

$$= \frac{0.126 \times 0.25}{60} \left(\frac{m^2 \times m}{S}\right)$$

$$= 0.000525 \text{ m}^3/\text{s}$$
$$= 0.525 \text{ l/s}$$
$$= 31.5 \text{ l/min}$$

Note This is a single-acting gravity-return cylinder.

In practice, a higher-capacity pump than theoretically indicated would be used to make up for internal leakage and wear which will occur within the pump. The operating pressure of the pump would have to be greater than the maximum pressure requirement of the circuit because of the operating characteristics of gas-loaded accumulators. Assume a maximum operating pressure of 400 bar, and an accumulator gas precharge pressure of 90% of the maximum pressure required at the cylinder, i.e. $0.9 \times 350 = 315$ bar.

To determine the size of accumulator required, draw a diagram showing the flow-rate demand of the circuit for one complete cycle as shown in Figure 6.27. A second chart can be drawn showing flow into and out of the accumulator for a complete cycle. This is not necessary in a simple case such as this where the oil capacity of the accumulator can be readily found, but in complex cycles an accumulator flow chart as shown in Figure 6.28 is useful. Downward-sloping lines represent flow from the accumulator and upward-sloping lines flow into the accumulator. The chart commences at a nominal point at the start of the cycle and will finish at the same height at the end of the cycle, provided a pump of the theoretical capacity is used. The maximum overall height or amplitude of the chart gives the available oil capacity required from the accumulator. Point A in Figure 6.28 represents the quantity of oil which will be delivered from the accumulator during the rapid approach. This is made up of the circuit demand less the pump output for the rapid approach time, i.e. $(340 - 31.5)$ l/min

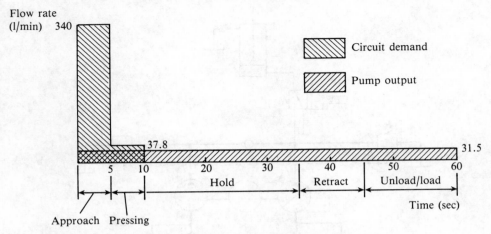

Figure 6.27 Solution 3a: system flow requirements.

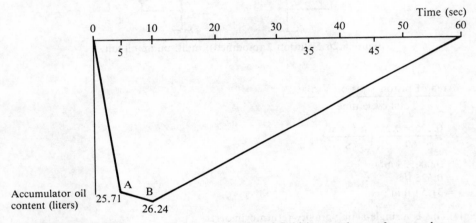

Figure 6.28 Solution 3a: chart showing flow of fluid into and out of accumulator.

for 5 seconds, i.e.

$$\frac{(340 - 31.5) \times 5}{60} = 25.71 \text{ liters}$$

Point B is the discharge from accumulator at point A plus the additional oil flow required from the accumulator during the pressing operation, i.e.

$$25.71 + \frac{(37.8 - 31.5)}{60} \times 5$$

$$= 25.71 + 0.53$$
$$= 26.24 \text{ liters}$$

For the remaining 50 seconds of the cycle, the accumulator is being charged by the pump at the rate of 31.5 l/min. The accumulator charge during this period is

$$(31.5 \times 50)/60 = 26.25 \text{ liters}$$

This is approximately the same as the quantity of fluid discharged per cycle.

Section 6.6 Design study—a simple hydraulic press

The maximum height of the chart is 26.25 liters indicating the fluid quantity which the accumulator must be capable of delivering at a circuit pressure of 350 bar. This is the pressure at the time when the accumulator is fully discharged.

For the accumulator-sizing calculation, assume adiabatic condition in the gas during discharging and isothermal during charging. Let the gas pressure and volume conditions of the accumulator be: $P_1 V_1$ when precharged, $P_2 V_2$ when fully charged and $P_3 V_3$ when discharged. These conditions are illustrated in Figure 6.14. The gas precharge pressure P_1 may be taken as 90% of the minimum working pressure P_3 of the system.

$P_1 = 0.9 \times 350 = 315$ bar gauge $= 316$ bar absolute
V_1 is accumulator size and is unknown
$P_2 = 400$ bar gauge $= 401$ bar absolute
V_2 is unknown
$P_3 = 350$ bar gauge $= 351$ bar absolute
V_3 is unknown

The difference between V_3 and V_2 is the quantity of fluid the accumulator can discharge between pressures of 400 and 350 bar. Therefore $V_3 - V_2 = 26.25$ liters.

Note In all gas-law calculations, absolute pressures must be used (in most other hydraulic-system calculations it is usual to use gauge pressure).

For initial charge, assume isothermal compression, i.e.

$$P_1 V_1 = P_2 V_2$$
$$316 \times V_1 = 401 \times V_2$$
$$V_1 = 1.269 V_2$$

For discharge assume adiabatic expansion

$$(V_3/V_2) = \left(\frac{P_2}{P_3}\right)^{1/1.4} = (401/351)^{1/1.4}$$
$$= 1.100$$

So
$$V_3 = 1.100 V_2$$
$$V_3 - V_2 = 26.25 \text{ liters}$$

Thus
$$0.1 V_2 = 26.25$$
$$V_2 = 260.25 \text{ liters}$$

Hence
$$V_1 = 1.269 \times 260.25 = 330.26 \text{ liters}$$

The accumulator volume required is thus 330 liters. This volume can be reduced by using an accumulator with back-up gas bottles. The actual volume of fluid discharged from the accumulator is 26.25 liters and an accumulator having a capacity of 30–40 liters with back-up gas bottles of 300-liter capacity could be used.

The accumulator volume could be further reduced by using a higher maximum pump pressure although this will increase the cost and reduce the choice of pump unit and components.

In order to control the speed of approach and pressing, flow-control valves must be used between the accumulator and the cylinder. As there is a considerable pressure drop over the flow-control valve for rapid approach (from 400 bar in the accumulator to 20 bar at the

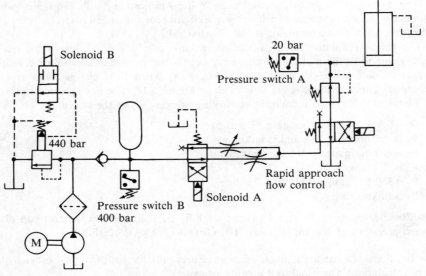

Figure 6.29 Solution 3a: accumulator circuit.

cylinder) an oil cooler will probably be needed. In Figure 6.29, pressure switch A operates solenoid A to give the different flow rates. Pressure switch B operates solenoid B to vent the relief valve at an accumulator pressure of 400 bar. In this circuit the theoretical input power will be that required to drive a pump, delivering a flow of 31.5 l/min against a maximum pressure of 400 bar. So the theoretical input power is

$$\frac{31.5}{60 \times 10^3} \times 400 \times 10^5$$

$$= 21.0 \text{ kW}$$

The pump is on-load for the whole of the 60-second cycle, so the theoretical input energy to pump per cycle is

$$21 \times 10^3 \times 60 \text{ (Nm)}$$
$$= 1260 \times 10^3 \text{ Nm}$$

The work done by the piston per cycle is as before (166.5×10^3 Nm) and the theoretical overall efficiency is $(166.5 \times 10^3)/(1260 \times 10^3) = 0.132$ or 13.2%.

Solution 3(b)

Low-pressure accumulator

An alternative would be to select a pump to satisfy the flow demand for the pressing period of the cycle and to use this pump to charge an accumulator to 20 bar for rapid approach (Figure 6.30). Although this involves a somewhat larger pump it would operate at a maximum pressure of 350 bar for only 5 seconds on each cycle and a maximum of 20 bar for the rest of the cycle. This will make a more efficient system supplying fluid at the pressure required by the circuit and will thus eliminate the need for an oil cooler. The circuit required to achieve this dual-pressure system will be more complex than the simple accumulator system and because of the cure period at full pressure, a pressure-compensated pump is shown in the circuit diagram in Figure 6.30.

Section 6.6 Design study—a simple hydraulic press

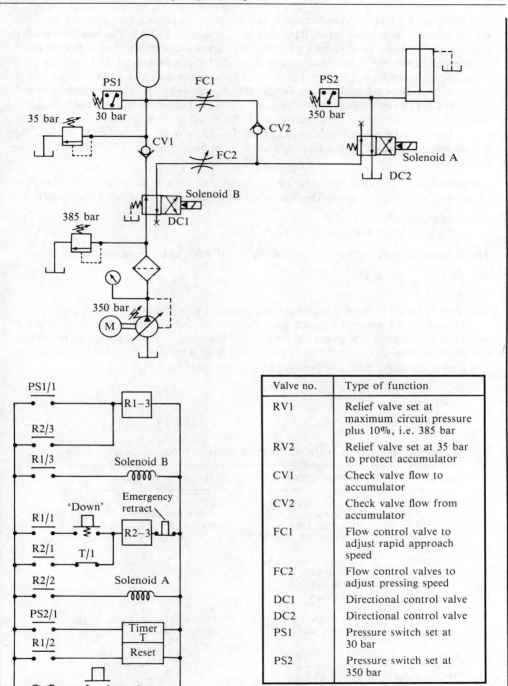

Figure 6.30 Solution 3b: low pressure accumulator circuit with pressure compensated pump.

Valve no.	Type of function
RV1	Relief valve set at maximum circuit pressure plus 10%, i.e. 385 bar
RV2	Relief valve set at 35 bar to protect accumulator
CV1	Check valve flow to accumulator
CV2	Check valve flow from accumulator
FC1	Flow control valve to adjust rapid approach speed
FC2	Flow control valves to adjust pressing speed
DC1	Directional control valve
DC2	Directional control valve
PS1	Pressure switch set at 30 bar
PS2	Pressure switch set at 350 bar

In Figure 6.30, relay nomenclature is as follows: R1–3 denotes that relay R1 has three sets of contacts designated R1/1, R1/2 and R1/3, similarly for R2. The electrical-control circuit is initiated by a spring-return push button switch which is interlocked so that it cannot be operated before the accumulator is fully charged. The pressure switch PS1 operates when the accumulator is charged to 30 bar pressure. The sequence shown in Table 6.2 can now be started by operating the push-button switch. When pressure in the cylinder attains 350 bar, PS2 operates and contact PS2/1 starts the timer which after an adjustable delay (25 seconds in this case) opens contact T/1. The time available for charging the accumulator is the cylinder return time plus the unloading/reloading time, i.e. 25 seconds. The quantity of fluid the accumulator must store for the rapid approach is the fluid required by the cylinder less the pump output.

Assume the pump output is q liters per minute maximum. Then fluid stored in accumulator in 25 seconds is $(25 \times q)/60$ liters. The fluid required by cylinder during rapid approach is

Piston area × Piston stroke
$= 0.126 \ (\text{m}^2) \times 0.225 \ (\text{m}) = 28.35$ liters

This is equal to pump delivery in the rapid approach period plus accumulator discharge

$$\left(\frac{q \times 5}{60}\right) + \left(\frac{25 \times q}{60}\right) = q/2$$

Therefore $q/2 = 28.35$ liters. The pump size required is thus $2 \times 28.35 = 56.7$ l/min.

The motor size required depends upon the delivery and pressure of the pump which is pressure-compensated. The pump delivery is 56.7 l/min up to a pressure of 30 bar during accumulator charge time, and for final pressing will be 37.8 l/min at 350 bar as in Solution (2). The theoretical power during accumulator charging period is

$$\frac{56.7 \times 30 \times 10^5}{60 \times 10^3} = 2.835 \ \text{kW}$$

Table 6.2 Sequence of operation (Reference Figure 6.30)

Signal	Electrical contact	Hydraulic operation
PS1 operates (PS1/1 closes)	Energizes relay R1–3 R1/1 closes R1/2 resets time R1/3 operates solenoid B	Pump flow deadheads (pump goes to zero swash)
Push button	Energizes relay R2–3 R2/1 closes Maintains R2–3 R2/2 closes Maintains R1–3 R2/3 closes Operates solenoid B	Accumulator and pump supply fluid to cylinder
Pressure switch PS2 (PS2/1 closes)	Timer started Time delay = 25 s T/1 opens Relay R2–3 de-energized Contact R2/2 opens Relay R1–3 de-energized R1/3 opens Solenoid B de-energized	Accumulator starts charging for next stroke. Cylinder retracts under gravity

Section 6.6 Design study—a simple hydraulic press

During the final pressing the power required will be as in Solution (2), i.e. 22.05 kW. The actual power consumption per cycle will be less than that using a high-pressure accumulator as the accumulator is only charged to 30 bar.

The pump will be operating against a maximum pressure of 30 bar during rapid approach, return and unloading time, i.e. for 30 seconds. It will operate at 350 bar during the final pressing time which is 5 seconds but at a reduced displacement with only make-up fluid being delivered during the cure period. The theoretical average power consumption per minute will be

$$\left(\frac{2.835 \times 30}{60}\right) + \left(\frac{22.05 \times 5}{60}\right) = 3.25 \text{ kW}$$

Thus theoretical input energy to pump per cycle is given by

$3.25 \times 10^3 \times 60$
$= 195 \times 10^3$ Nm

As before the work done by the piston per cycle is 166.5×10^3 Nm, so the theoretical system efficiency is

$$\frac{166.5 \times 10^3}{195 \times 10^3}$$
$= 85.4\%$

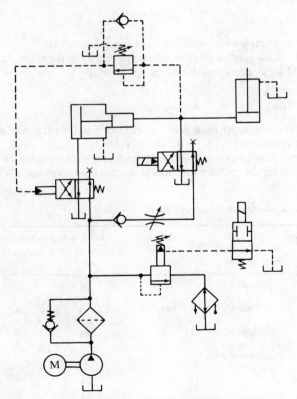

Figure 6.31 Solution 4: fixed displacement pump and intensifier.

Solution 4

A constant-delivery pump and intensifier

The pump is used directly to give the rapid approach and through an intensifier for the final pressing. The flows and pressures required are, as in Solution (1), i.e. 340 l/min at a pressure of 20 bar for rapid approach and 37.8 l/min at a pressure of 350 bar for final pressing. Consider the intensifier (Figure 6.31):

Intensification ratio = 350/20 = 17.5:1

Theoretically, the product of pressure and flow on the output side is equal to that on the input side of the intensifier. So if q is the required input flow,

$$350 \times 37.8 = q \times 20$$

Therefore $q = (350 \times 37.8)/20 = 661.5$ l/min. The theoretical delivery of the pump must be 661.5 l/min against a pressure of 20 bar for the final pressing but for the rapid approach, a flow of 340 l/min is required. A flow-control valve must be used to restrict the speed of approach and if a fixed-displacement pump is used, 321.5 l/min will pass over the relief valve during rapid approach.

The theoretical input power to the system is

$$\left(\frac{661.5}{60 \times 10^3}\right) \times (20 \times 10^5) \left(\frac{m^3}{s} \times \frac{N}{m^2}\right)$$

$$= 22\,050 \text{ watts}$$

$$= 22.05 \text{ kW}$$

This is the same maximum power requirement as for Solution (2).

The pump will be on load during the initial approach, final pressing and holding time. Thus the theoretical input energy to the pump per cycle is

$$22.05 \times 10^3 \times 35$$
$$= 771.7 \times 10^3 \text{ Nm}$$

The work done by the piston per cycle is as before (165×10^3 Nm), so the theoretical system efficiency is $(165 \times 10^3)/(771.7 \times 10^3) = 0.214$ or 21.4%.

Table 6.3 shows flow of energy during one cycle assuming a fixed-displacement pump which is unloaded during the cylinder return, unload and load operations. The average power

Table 6.3 Solution 4: summary of energy flow.

Operation	Time, t (s)	Pump outlet to circuit (l/min)	Circuit demand (l/min)	Relief valve flow, q (l/min)	Pressure p (bar)	Energy to heating fluid, $q \times p \times t \times (100/60)$ (Nm)
Rapid approach	5	661.5	340.0	321.5	20	53 580
Final pressing	5	661.5	661.5	0	20	–
Cure time	25	661.5	0	661.5	20	551 250
Return	10	0 (pump unloaded)	–	–	–	–
Unload and load	15	0	–	–	–	–
Total	60					604 830 Nm/min

Section 6.6 Design study—a simple hydraulic press

dissipated into the fluid is (604 830/60) Nm s (watts) = 10.08 kW. This is only the theoretical value. In practice the pump output would be greater than that calculated to allow for increased pump leakage with wear. The pump pressure would have to be increased to take care of pressure drops in the system. In this type of circuit which is shown in Figure 6.31, an oil cooler is essential although this can be overcome by using a pressure-compensated pump as mentioned in the modified Solution (2).

Solution 4 (modified)

A slightly more efficient alternative to Solution (4) is to use a pump calculated to meet the flow required for rapid approach with a lower ratio intensifier. This involves a smaller capacity pump working at a higher pressure and eliminates the need for a flow control valve. It also permits smaller valves and pipework in certain parts of the circuit. The intensification ratio is

$$\frac{\text{Flow rate available from pump}}{\text{Output flow required}}$$
$$= 340/37.8$$
$$= 9:1$$

Theoretical pressure at intensifier inlet to achieve 350 bar at outlet is $350/9 = 38.9$ bar. The theoretical input power during rapid approach is

$$\left(\frac{340}{60 \times 10^3}\right) \times 20 \times 10^5$$
$$= 11.33 \text{ kW (as Solution 2)}.$$

Maximum theoretical power requirement during final pressing will be

$$\frac{340}{60 \times 10^3} \times 38.9 \times 10^5$$
$$= 22.04 \text{ kW}.$$

Theoretical input energy is

$$(11.33 \times 10^3 \times 5) + (22.04 \times 10^3 \times 30)$$
$$= 717.85 \times 10^3 \text{ Nm}$$

Overall theoretical efficiency is

$$= \frac{165 \times 10^3}{717.85 \times 10^3} \times 100$$
$$= 23\%$$

This is a slight improvement on Solution (4) but the energy saved during rapid approach is insignificant compared with the huge loss during curing which is almost unchanged. Although the flow rate is approximately half, the pressure has doubled.

Solution 5 (a)

Using a prefill valve and side cylinder

The initial rapid approach is at low thrust which is developed by a small-diameter side cylinder operating at full system pressure. The side cylinder lifts the main cylinder which is filled with fluid drawn from the overhead tank via the prefill valve during the initial approach

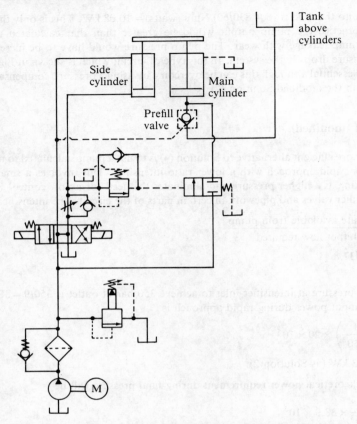

Figure 6.32 Solution 5a: using prefill valve and side cylinder.

(Figure 6.32). At the end of the approach stroke pressure rises 'firing' the sequence valve which switches the two-port directional control valve diverting most of the pump flow directly to the main cylinder. For the final pressing, the main cylinder and side cylinder are both at full system pressure. (Prefill valves are described in Section 3.3.1 of Chapter 3.)

The side cylinder must exert a thrust of 0.251×10^6 newtons at a pressure of 350 bar (from Solution 1). Thus cylinder area is

$$\frac{\text{Thrust}}{\text{Pressure}}$$

$$= \frac{0.251 \times 10^6}{350 \times 10^5} \left(\frac{\text{Nm}^2}{\text{N}}\right)$$

$$= 7171 \times 10^{-6} \text{ (m}^2\text{)}$$

$$= 7171 \text{ mm}^2$$

Cylinder diameter is

$$[(4/\pi) \times 7171]^{1/2}$$

$$= 95.5 \text{ mm (say, 100 mm)}$$

Then actual cylinder area is 0.00785 m^2. The pressure required to give a thrust of

Section 6.6 Design study—a simple hydraulic press

0.251×10^6 N is

$$\frac{0.251 \times 10^6}{0.00785}$$

$= 319.7$ bar

This pressure is too near the final pressing pressure for a sequence valve or pressure switch to satisfactorily detect the conclusion of the rapid approach. Therefore the next larger standard cylinder will be selected; this has a bore of 125 mm and a 70 or 90 mm diameter rod.

Cylinder full bore area is

0.0128 m^2.

Pressure required for rapid approach is

$$\frac{0.251 \times 10^6}{0.0128} \frac{N}{m^2}$$

$= 204$ bar

Flow for rapid approach is

$$0.0128 \times \frac{0.225}{5} \text{ m}^3/\text{s}$$

$= 34.6$ l/min

Total flow required for final pressing is the sum of the flows to the main cylinder and side cylinder. The flow to main cylinder is

37.8 l/min (from Solution 1)

Flow to side cylinder is

$$0.0128 \times \frac{0.025}{5} \text{ m}^3/\text{s} = 3.8 \text{ l/min}$$

Total flow required is

$37.8 + 3.8 = 41.6$ l/min

Pump size must be in excess of 41.6 l/min. If a fixed displacement pump is used, some of the flow will have to pass over the relief valve during rapid approach. Heat will be generated at the relief valve and the flow control valve. However, because of the side cylinder the effective piston area is increased for the pressing stroke and the system working pressure can be reduced accordingly.

Area of main cylinder $= 0.126$ m^2
Area of side cylinder $= 0.0128$ m^2
Effective area $= 0.1388$ m^2

The pressure to give a thrust of 4.4 MN is

$$\frac{4.4 \times 10^6}{0.1388} \text{ N/m}^2 = 317 \text{ bar}$$

and the energy used for final pressing is

$$\frac{41.6}{10^3} \times \frac{5}{60} \times 317 \times 10^5 = 109.8 \times 10^3 \text{ Nm}$$

During curing time virtually the full pump output flows over the relief valve, consequently an oil cooler will be needed.

Solution 5(b)

Alternatively, use can be made of a twin-pump system (Figure 6.33) with one fixed-displacement pump having a delivery of 35 l/min and the second pump being pressure-compensated with a maximum delivery of 7 l/min. The fixed-displacement pump provides for rapid approach which takes place at the load-induced pressure (204 bar). Pressure increases for the final 25 mm of stroke actuating the pressure switch which energises the two-port solenoid valve and both pumps feed the circuit. During curing the large pump flow is dumped and the pressure-compensated pump automatically regulates its output to just maintain pressure. Only the large pump is utilized on the retract stroke which takes place at low pressure; the 'meter-out' flow control is present to prevent load overrun. Pressure settings on the relief valves, pump compensator and pressure switch are critical for the circuit to function correctly. Suggested values for each component are shown on the drawing. The relief valve in the pressure-compensated pump circuit is set approximately 20% above the compensator setting and only relieves when surges or shock pressures occur.

The absence of flow controls means that full pump output is used for cylinder movements at load pressure. At other times flow is dumped or the pressure-compensated pump is 'de-swashed'. Therefore this will be a very efficient system.

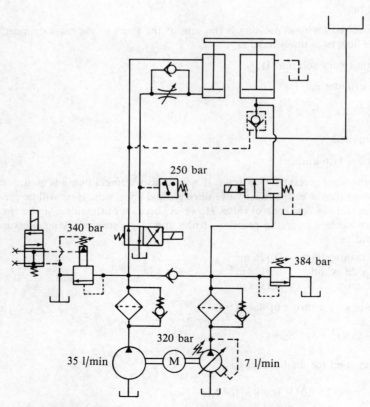

Figure 6.33 Solution 5b: twin pumps and prefill circuit.

Solution 5(c)

A further alternative is to use a prefill valve and a single pressure-compensated pump in a circuit similar to Figure 6.32 but with a two-position directional-control valve instead of the three-position valve shown. Pump delivery is 42 l/min at a maximum operating pressure of 320 bar.

Rapid approach

The pressure drop over the flow-control valve during rapid approach is given by

(Compensator setting) − (Pressure to give a side cylinder thrust of 0.251×10^6 N)
= 320 − 204 bar
= 116 bar

The flow required is as before (34.6 l/min). The energy drop over flow control is

$$\text{Quantity flowing} \times \text{Pressure drop} = \left(\frac{34.6}{10^3}\right) \times \left(\frac{5}{60}\right) \times 116 \times 10^5 \text{ Nm}$$

$$= 33.4 \times 10^3 \text{ Nm}$$

Total energy supplied during rapid approach is

$$\left(\frac{34.6}{10^3}\right) \times \left(\frac{5}{60}\right) \times 320 \times 10^5$$

$$= 92.3 \times 10^3 \text{ Nm}$$

Return stroke

Assume that a standard metric cylinder 125 mm bore and 90 mm rod has been used as the side cylinder. Take the return time as 10 seconds and as gravity return was envisaged in the original problem, assume the return thrust required at the side cylinder is zero.

Heat generated over retract flow control

Flow through retract flow control for a stroke of 250 mm in 10 s is

$$0.125^2 \left(\frac{\pi}{4}\right) \times \left(\frac{250}{10}\right) \times 10^{-3} \text{ (m}^3\text{/s)}$$

$$= 0.307 \times 10^{-3} \text{ m}^3\text{/s}$$

$$= 0.307 \text{ l/s}$$

The pressure drop across retract flow control is the pump pressure compensator setting reduced by the ratio of the annulus to full bore areas, i.e.

$$320 \times \left(\frac{0.125^2 - 0.09^2}{0.125^2}\right)$$

$$= 154.1 \text{ bar}$$

The heat generated is thus

$$0.307 \times 10^{-3} \times 10 \times 154.1 \times 10^5$$
$$= 47 \times 10^3 \text{ Nm (joules)}$$

Table 6.4 Solution 5c: summary of energy flow

Operation	Time (s)	Total pump delivery to circuit (l/min)	Circuit demand (l/min)	Energy supplied to fluid (Nm)	Energy expended in heating fluid, Nm (joules)
Rapid approach	5	34.6	34.6	92.3×10^3	33.4×10^3
Final pressing	5	41.6	41.6	112.7×10^3	
Cure time	25	nil	nil	—	—
Return	10	18.4	18.4	47×10^3	47×10^3
Unload/load	15	nil	nil	—	—
Total				252×10^3	80.4×10^3

The theoretical overall efficiency will be

$$\frac{\text{(Energy supply)} - \text{(Heat loss)}}{\text{Energy supplied}}$$

$$= \frac{252 - 80.4}{252}$$

$$= 0.68 \text{ or } 68\%$$

A summary of the energy flow is given in Table 6.4.

Solution 6

Variable-delivery pump unit

If the pump output can be matched to the circuit requirements there will be no wasted energy. The output of a variable delivery pump can be controlled by altering the eccentricity or swash-plate angle. This alteration may be effected either by an external movement or automatically by the system pressure. In some systems where precise speeds or movements are required the pump delivery can be servo-controlled or cam-controlled. The system pressure can be used to either give a constant maximum pressure or a constant power output. With the constant maximum pressure or pressure-compensated pump, the pump will be at full delivery until the set pressure is attained; the pump displacement will then be reduced to maintain that pressure. If the delivery side of a pressure-compensated pump is blocked, e.g. a cylinder fully stroked, the pump delivery will adjust to maintain the set pressure, i.e. to make up any leakage flow. With a constant-power unit the product of circuit pressure and delivery is constant within the limitations of the unit. The use of a pressure-compensated unit has already been considered in Solutions (2), (3) and (5). If it is used directly, it will require a motor capable of driving the pump at full delivery and maximum pressure so that the compensator can operate. A flow-control valve will be needed to regulate the final pressing speed.

With a constant power controlled pump the requirements are 340 l/min at 20 bar and 37.8 l/min at 350 bar. The power supplied to the pump depends upon its pressure and delivery characteristics which were discussed in Section 2.1.3 of Chapter 2. With this unit, the excess fluid must be spilled over the relief valve during the curing period and the pump unloaded at the end of the work cycle.

The purpose of this design example has been to show a variety of approaches to a particular problem. It also demonstrates that power hydraulics is not an exact science and that there are always many solutions. These are representative and do not necessarily include the best, although it is obvious that several which at first appeared feasible have proved to be very inefficient.

6.7 DESIGN STUDY – CONVEYOR FEED SYSTEM

EXAMPLE 6.10

Problem

To design the hydraulic system for a plant having three screw feed conveyors and three bunkers as shown diagrammatically in Figure 6.34. The bunker cylinders are to have a stroke of 0.5 m and exert a thrust of 2000 kg to both open and close the bunker doors. In order that the output flow from the bunkers can be regulated, the cylinders must be capable of being stopped and locked in any position between fully extended and fully retracted. If there is a failure of either the hydraulic system or the electrical control, the bunker doors must be automatically closed by the cylinders in 10 seconds. The conveyors are used as metering devices and have to be driven through a precise number of revolutions. Conveyors 1 and 2 are to be identical units having the following drive requirements:

Starting torque = 1250 Nm
Running torque = 1000 Nm
Speed range = 10–35 rev/min fully variable

The conveyor speed may have to be changed for different product mixes. This will only occur infrequently and a manually-adjustable flow control valve will be adequate.

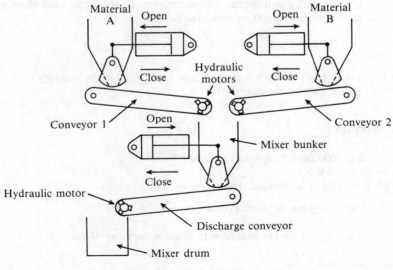

Figure 6.34 Conveyor feed system.

Conveyor No. 3 in Figure 6.34 is the mix discharge conveyor and is to have remotely variable speed control over its full speed range. Its drive specifications are:

Starting torque = 2000 Nm
Running torque = 1200 Nm
Speed range   = 2–10 rev/min

A 4:1 reduction gear-box final drive is situated between the hydraulic motor and the screw conveyor. The gear-box efficiency may be taken as 80%. All the directional control valves are to be electrically operated. The hydraulic motors must be capable of being reversed for maintenance of the screw conveyors (jamming etc.) and must have the capability of being isolated from the hydraulic circuit for maintenance or removal of the motors. The system will be in continuous use 24-hours per day, 7 days per week, apart from regular-planned maintenance shut-downs. Reliability is of paramount importance.

Solution

The merits and demerits of centralized hydraulic systems and applications using individual power packs were discussed in Section 6.3. In this problem, if one circuit fails all the others have to be stopped as they are interdependent. Therefore a solution using a centralized hydraulic system has many advantages, especially if a multi-pump system incorporating a stand-by or spare unit is used.

The design can be started by either sizing the actuators or by sketching the proposed circuit. In this case the actuator sizes will be considered first.

System working pressure

No limitations have been placed on the working pressure in the problem. Hydraulic cylinders as standard components have working pressures of 210 bar continuous (350 bar non-shock). The hydraulic motors needed are slow speed and either Gerotors with an integral reduction gear box or direct-drive radial piston units are suitable. The Gerotor motor is limited to 140 bar working and the radial piston to between 210 bar and 450 bar dependent upon design and manufacturer. Using a centralized system, a common pressure should be used where possible. Select 210 bar as the maximum system pressure.

System fluid

Again no constraints apply so select a mineral base hydraulic oil; the viscosity of the oil is to be determined by component requirements.

Cylinder selection

Thrust required is 2000 kg = 2000 × 9.81 = 19 620 N.
The cylinder stroke is 0.5 m.
The cylinder mounting: rear trunion, front eye.

Table 4.1 in Chapter 4 gives standard cylinder sizes to BS 5785:1980.

Calculate the minimum piston rod diameter to prevent buckling of the rod:

$$\text{Buckling load, } K = \frac{\pi^2 EJ}{SL^2}$$

Section 6.7 Design study—conveyor feed system

where E is the modulus of elasticity for the material of the piston rod (kg/cm^2) and is 2.1×10^6 kg/cm^2 for steel. The second moment of area of the piston rod, J, is $(\pi d^4/64)$ cm^4 for a solid circular piston rod of diameter d cm. L is the buckling length (cm), dependent upon the rod end conditions (see Figure 4.16 in Chapter 4), and S is a factor of safety which is usually taken as 3.5. In this case, $K = 2000$ kg, and L is the maximum distance between the rear trunion and the piston rod eye end. This is approximately twice the stroke length. Take $L = 2 \times 0.5$ m $= 100$ cm.
Thus

$$K = 2000 = \left(\frac{\pi^2 \times 2.1 \times 10^6}{3.5 \times 100^2}\right) \times \left(\frac{\pi d^4}{64}\right) \left(\frac{\text{kg}}{\text{cm}^2} \times \frac{\text{cm}^4}{\text{cm}^2}\right)$$

$$d^4 = \frac{2000 \times 3.5 \times 100^2 \times 64}{\pi^3 \times 2.1 \times 10^6} \text{ (cm}^4\text{)}$$

So

$$d = 2.88 \text{ cm} = 28.8 \text{ mm}.$$

The next larger standard rod diameter from Table 4.1 is 36 mm.

Assume a maximum pressure of 180 bar at the cylinder. The effective area is

$$\frac{2000 \times 9.81}{180 \times 10^5} \left(\frac{\text{Nm}^2}{\text{N}}\right)$$

$$= 0.00109 \text{ m}^2$$
$$= 1090 \text{ mm}^2$$

If a 36-mm diameter rod is used, the full bore area needed will be

$$[1090 + (\pi/4)36^2]$$
$$= 1090 + 1018$$
$$= 2108 \text{ mm}^2$$

Thus full bore diameter, D is

$$\left(\frac{2108 \times 4}{\pi}\right)^{1/2}$$

$$= 51.8 \text{ mm}$$

A standard cylinder available has a 50-mm diameter piston with a 36-mm diameter rod. The pressure required to develop 2000 kg thrust on annulus side is

$$\frac{2000 \times 9.81}{(\pi/4)(50^2 - 36^2)} \left(\frac{\text{N}}{\text{mm}^2}\right)$$

$$= 20.7 \left(\frac{\text{N}}{\text{mm}^2}\right)$$

$$= 207 \text{ bar}$$

This is far too near the maximum working pressure so choose the next cylinder in the range, i.e. a 63-mm diameter bore with either a 36-mm diameter or a 45-mm diameter rod. Consider the pressures required on annulus side for both rod diameters. The pressure required (63 mm diameter bore × 36 mm diameter rod) to give a retract thrust of 2000 kg is

$$\frac{2000 \times 9.81}{(\pi/4)(63^2 - 36^2)}$$

$$= 9.35 \left(\frac{N}{mm^2}\right)$$

$$= 94 \text{ bar}$$

The pressure required (63 mm diameter piston × 45 mm diameter rod) to give a retract thrust of 2000 kg is

$$\frac{2000 \times 9.81}{(\pi/4)(63^2 - 45^2)}$$

$$= 12.85 \left(\frac{N}{mm^2}\right)$$

$$= 129 \text{ bar}$$

The smaller rod diameter will result in a less costly cylinder, but the larger rod will be stronger.

Motor sizing

Consider conveyors 1 and 2 in Figure 6.34:

Starting torque = 1250 Nm
Running torque = 1000 Nm
Speed range = 10–35 rev/min

To select a suitable motor refer to Figure 4.43 used for the design examples in Chapter 4. The M6 motor has a start-up torque (i.e. torque at zero speed) of 1600 Nm at 207 bar and a start-up torque of 1100 Nm at 138 bar. Thus the pressure required to produce 1250 Nm start-up torque is approximately 160 bar. The running torque of 1000 Nm at 10 rev/min will require a pressure of approximately 120 bar. This motor seems satisfactory for this application. Its specification is

Displacement = 558 cc/rev
Maximum pressure = 240 bar
Volumetric efficiency = 95% or better

Consider conveyor No. 3:

Starting torque = 2000 Nm
Running torque = 1200 Nm
Speed range = 2–10 rev/min
Gear box reduction = 4 : 1
Gear box efficiency = 80%

Motor starting torque $= \dfrac{2000}{(4 \times 0.8)} = 625$ Nm

Running torque $= \dfrac{1200}{(4 \times 0.8)} = 375$ Nm

Speed range = 8–40 rev/min

The M6 motor has a starting torque of 625 Nm at approximately 75 bar and a running torque of 375 Nm at approximately 40 bar. Thus the M6 appears somewhat overpowered for this drive, but could be used if standardization is important, i.e. the three motors could be identical.

Alternatively, use the M3 motor which gives a starting torque of 625 Nm at 150 bar and a running torque of 375 Nm at approximately 75 bar. These pressures are more in line with those being considered for the rest of the system. Specification of the M3 is:

> Displacement = 280 cc/rev
> Maximum pressure = 250 bar
> Volumetric efficiency = 94% or better

Flow requirements

The next stage is to determine the flow requirements for each actuator.

CYLINDERS

Each must fully stroke (0.5 m) in 10 seconds. Velocity = $(0.5 \times 60/10) = 3$ m/min. The flow required per cylinder is

> Full bore area × speed
> = $(\pi/4) \times 63^2 \times 10^{-6} \times 3$ (m^2 × m/min)
> = 9352×10^{-6} (m^3/min)
> = 9.35 l/min

CONVEYOR DRIVES 1 AND 2

Maximum speed is 35 rev/min, using an M6 motor with a capacity of 558 cc/rev. Then the maximum flow required per motor is

> 35×558 (cc/min)
> = 19 530 (cc/min)
> = 19.53 l/min

Volumetric efficiency = 95% from data sheet. Therefore actual maximum flow is

> 19.5/0.95 = 20.5 l/min.

CONSIDER CONVEYOR 3

This uses an M3 motor, and referring to the data sheet, maximum motor speed = 40 rev/min with a motor displacement = 280 cc/rev. The maximum theoretical flow is

> $40 \times 280 = 11\,200$ cc/min
> = 11.2 l/min

The volumetric efficiency = 94%. Therefore actual maximum flow is

> 11.2/0.94 = 11.9 l/min

Assume that the worst case is when all three motors are running at maximum speed, and neglect the three cylinders. (As the cylinders have to operate, if there is a failure of hydraulic power or control electronics, accumulators will be fitted into the cylinder circuits.)

The maximum circuit flow requirement is for two M6 motors and one M3 motor at 40 rev/min = $(2 \times 20.5) + (1 \times 11.9)$, i.e. 52.9 l/min.

Pump sizing

The total flow requirement can be taken as 60 l/min at a maximum pressure of 160 bar (to give start-up torque on conveyors 1 and 2). The maximum pressure required by the cylinders is 120 bar depending upon the particular cylinder chosen. Maximum flow requirements can be met by one 60 l/min pump or two 30 l/min pumps.

Pump selection

The wide variations in flow requirements preclude the use of a fixed-displacement pump. A system pressure of 160 bar is too high for a vane pump. Therefore variable displacement axial piston pumps are indicated where the pump displacement control can automatically adjust to the circuit demand. Pressure-compensated axial pumps fulfil all the system criteria.

The maximum circuit demand is 52.9 l/min. If three pumps are used, each unit should be capable of delivering approximately 30 l/min at 160 bar, two pumps running at any time with the third pump on standby.

Consider the Vickers Systems Ltd axial piston pumps. From the data sheet in Table 2.6 the PBV10 series has a rated delivery of 21.2 liters per 1000 rev/min; this gives 30.6 l/min at 1450 rev/min. The maximum pressure rating is 210 bar.

The theoretical power input is

(Flow in cubic m/s) × (Pressure in N/m^2)
= $(30.6/60 \times 10^3) \times (160 \times 10^5) = 8.08 \times 10^3$ (Nm/s)
= 8.08 kW

Assume that the pump has an overall efficiency greater than 80%. Here, a 10 kW electric motor will be adequate. Use three of these pump units, one to be kept on standby. A circuit for the power pack is shown in Figure 6.35. The pressure compensators of the pumps in use should be set at slightly different values to prevent hunting. Suggested values are 160 and 170 bar with the relief valve set about 20% higher at say 210 bar.

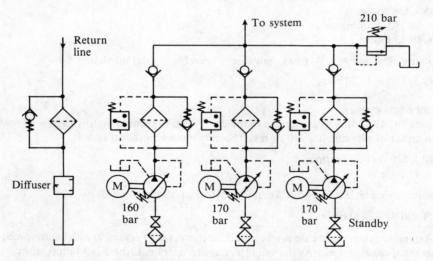

Figure 6.35 Power pack circuit.

Conveyors 1 and 2 circuits

In Figure 6.36 cross-line relief valves are used to reduce pressure surges when the hydraulic drive is stopped. The valves are to be set at 20% above the maximum pressure required, i.e. at 192 bar. The directional control valve is a two-stage valve with choke pack. A suitable valve with closed center can be selected from any valve manufacturer's catalog taking the flow

Section 6.7 Design study—conveyor feed system

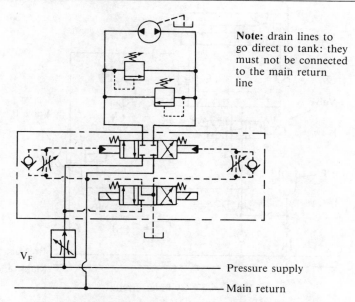

Figure 6.36 Conveyors 1 and 2 circuit.

requirements into account. A choke pack has been used to give a gradual changeover; it must be set to stop the motor as rapidly as possible without excessive hydraulic shock.

The motor speed is regulated by a pressure-compensated flow-control valve in the supply line. There is no possibility of the motor overrunning so meter-in flow control will be quite satisfactory.

Conveyor 3 circuit

The basic circuit for this conveyor will be the same as that for conveyors 1 and 2 (Figure 6.36) but the flow control valve V_F is of the proportional type. This valve is electrically modulated from the control cabin to give a remote fully-variable speed control to the mix discharge conveyor drive.

Cylinder circuit

In Figure 6.37 directional control valve A together with the pilot-operated check valves enable the cylinder to be locked in any position. Cylinder speed is controlled by a meter-in flow control valve.

If there is a failure in system pressure, PS1 will de-energize. This is arranged to automatically energize the retract solenoid on valve A. Failure of the electrical control voltage will cause the solenoid on valve B to de-energize, retracting the cylinder. Hydraulic fluid for the cylinder emergency retract is supplied by the accumulator E. A flow-control valve D adjusts the cylinder speed.

Shuttle-valve C allows the pilot check valve on the full bore side of the cylinder to be opened by a signal from valve A or valve B.

The accumulator capacity has to be sufficient to fully retract all three cylinders. Consider the accumulator gas conditions to be: pressure P_1, volume V_1 when precharged; pressure P_2, volume V_2 when fully charged; and pressure P_3, volume V_3 when discharged (see Figure 6.14).

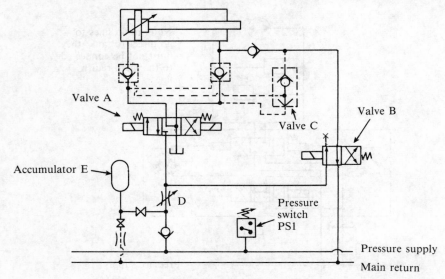

Figure 6.37 Cylinder circuit.

Let

P_1 = precharge gas pressure
V_1 = accumulator capacity
P_2 = maximum gas pressure at a volume V_2
P_3 = minimum useful gas pressure at a volume V_3

Now

P_2 = maximum system pressure = 161 bar absolute
P_3 = minimum pressure required to retract the cylinder
= 91 bar absolute if a 36 mm rod is used (worst case)
$P_1 = 0.9 P_3 = 82$ bar absolute

Volume of oil required to fully retract the three cylinders is

$$V_3 - V_2$$

$$= \left(\frac{\pi}{4}\right)(63^2 - 36^2) \times 500 \times 3 \text{ (mm}^3\text{)}$$

$$= 3.15 \text{ l}$$

Assume adiabatic charge $\gamma = 1.4$ because high pump flow rate gives a rapid change.

$$P_1 V_1^\gamma = P_2 V_2^\gamma$$

Therefore

$$\frac{V_1}{V_2} = \left(\frac{P_2}{P_1}\right)^{1/\gamma}$$

$$= (161/82)^{1/1.4}$$
$$= 1.96^{1/1.4}$$
$$= 1.61$$

So

$$V_1 = 1.61 V_2$$

Section 6.7 Design study—conveyor feed system

Assume isothermal discharge because restrictor D limits the flow rate

$$P_2 V_2 = P_3 V_3$$
$$161 V_2 = 91 V_3$$

Therefore

$$V_3 = 1.77 V_2$$

Now

$$V_1 = 1.61 V_2$$

and

$$V_3 - V_2 = 3.15 \text{ liters}$$

So

$$1.77 V_2 - V_2 = 3.15$$

i.e. $0.77 V_2 = 3.15$

$$V_2 = 4.1 \text{ liters}$$

Thus

$$V_1 = 1.61 \times 4.1$$
$$= 6.6 \text{ liters}$$

Select a suitable accumulator from catalogs. Most manufacturers have 7-liter and 10-liter models available; it is preferable to select the larger unit in order to have some spare capacity.

Oil reservoir

As pressure-compensated pumps have been chosen, the heat generated will be relatively low. Using the empirical formulae that reservoir capacity is 3–4 times pump flow rate per minute, a reservoir with an oil capacity of 200 liters should prove adequate. This can be checked by calculating the heat energy generated at maximum conveyor speeds when they are under minimum load (see Section 6.2).

CHAPTER SEVEN

HYDRAULIC SYSTEM MAINTENANCE

Many hydraulic systems are designed without any consideration being given to the maintenance of the system once it is in use. Frequently the prime requirement is minimum initial cost without thought of the running and maintenance costs. Consequently it is possible that:

- The filtration in the system will be inadequate.
- There will be a lack of test points.
- There will be no facilities for monitoring wear.
- Valves and other components will be mounted in inaccessible positions.
- It may be necessary to empty the oil reservoir before certain components can be examined or replaced.

7.1 EQUIPMENT AND PRACTICES WHICH BENEFIT MAINTENANCE

7.1.1 Good housekeeping practice

System filtration and cleanliness

It is claimed that over 80% of all breakdowns in hydraulic systems can be directly or indirectly attributed to fluid contamination. By using adequate filtration as described in Chapter 5 (Section 5.2) on contamination control the breakdown frequency can be drastically reduced.

- Ensure that the filters fitted to the system are of the recommended absolute rating, and are of a size sufficient to pass the full pump flow under cold-start conditions. If a suction line filter is used there should also be a pressure line or return line filter in the system.
- Check that the reservoir is properly sealed and all sealing gaskets are in good condition. Check that the air-filter breather is clean, of a sufficiently fine micronic rating and adequately sized to handle the air flow. A 'spin-on' type return line filter element makes an excellent air breather.
- When filling or topping up the oil reservoir, use a filter pack to transfer the oil – not an old watering can.

- The reservoir must be fitted with a drain so it can be emptied periodically and cleaned together with the suction strainers. In mineral-oil systems, the reservoir drain may be used to drain off any water which has condensed in the reservoir.

Test points and testing

To obtain an indication of the performance of any hydraulic system it is important to be able to measure the pressures at various points during the course of operation. Typical points where the pressure should be measured are at the pump output (to set the relief valve), downstream of any pressure reducing valves, upstream of sequence valves, counterbalance valves, pressure switches etc., so that these components can be accurately set. Also on both sides of all actuators so their performance can be judged, and so on. If pressure gauges are placed at all these points a large number of expensive gauges will be needed. The alternative is to use pressure test points which are small self-sealing couplings, and to connect a gauge to these points when a measurement is required. The test points can also be used for temperature measurement, air bleeding and for fluid sampling in conjunction with a sampling valve.

Condition monitoring

COMPONENTS

The condition of pumps, hydraulic motors and control valves with external drains can be judged by measuring the leakage flow in the drain lines. The drain line should be installed so that it can be easily tapped into or disconnected, the leakage collected in a measuring vessel and the leakage flow rate of the component monitored. It is essential that the leakage rate is measured under constant load or pressure conditions and the first reference leakage flow recorded when the system component is new. In a large installation it is useful to have a purpose-built flow meter but for smaller installations a measuring jar and a stop-watch are adequate for measuring leakage flow.

FLUID

Contamination monitoring of the hydraulic fluid can be useful on large or sophisticated installations. Regular checks establish that the specified level of cleanliness is being maintained and provide guidance on the correct running of the machine.

Containers, tubes and apparatus used for sampling must be scrupulously clean or the tests will be invalidated. Laboratory cleaned bottles are obtainable. British Standard BS 5540 'Evaluating particulate contamination in hydraulic fluid' specifies methods for cleaning containers, procedures for bottling the samples and methods of defining levels of contamination.

Samples which are to be compared should all be taken from the same point and after the equipment has settled at its normal running temperature. This is likely to be at least an hour after the machine has been switched on.

Ideally, sampling points should be built into the system preferably in low pressure (less than 14 bar) lines where there is relatively constant flow. Samples may be extracted from pressurized lines by using specially designed valves located at strategic points in the system. These valves contain an assembly of grooved discs which create a labyrinth-like path – thus reducing the pressure to atmospheric at low fluid velocity. In the same situation a needle valve could add contaminant owing to erosion of the valve face and seat.

A 250 ml sample is considered adequate but considerably more fluid should be bled-off prior to collecting the actual sample. Once the valve has been opened it should not be touched again until the procedure is completed.

Reservoir sampling can provide satisfactory data and the sample may be obtained by means of a small vacuum pump, syphoning or dipping. Whichever method is adopted the area surrounding the access point must first be cleaned before it is opened and the sample should not be taken from near the bottom as water or sediment may be entrained. It is best obtained from a centralized zone of the reservoir away from the pump inlet and return pipes but not from a stagnant area.

The contamination can be analyzed in two ways:

1. By particle count, giving size and distribution of dirt particles in the oil. This shows how efficient or otherwise the filtration is in the system. Particle counts may be obtained from sample analysis or inline monitoring (see Chapter 5).
2. By chemical analysis, showing the composition of the contaminant and the quantity present. This can be used to indicate the wear of various components in the system.

An inexpensive and easily applied evaluation technique called a 'patch test' gives a good indication of fluid condition and can be undertaken on-site. A set quantity of sample fluid is vacuum filtered through a membrane which collects the contaminant. The membrane is compared with master slides to give a qualitative assessment of the contamination level. Types of contaminant are often identifiable by examination through a simple microscope. For more accurate results the fluid samples should be sent to a specialist laboratory. Many oil companies and filter manufacturers operate an analysis service.

Continuous monitoring of temperature, pressure, vibration, noise levels, cycle time and so on is very expensive and is only of use if a significant deviation of any parameter from its normal value is instantly recognized and appropriate action taken.

Periodic checks should be made of the running condition of all pumps and actuators to try to locate possible breakdowns before they occur. Accelerometer transducers on bearings can detect signs of wear or bearing breakdown by measuring changes in vibration. Temperature sensors can measure increases in temperature at bearings showing increased friction forces and possibly bearing wear. The noise generated by a component gives an indication of its condition. A stethoscope can be used to listen to flow noises in components and is useful in detecting internal leakages, for example a relief valve seat may be damaged resulting in a slight leakage; this leakage will cause a flow noise in the return line from the relief valve.

The normal values of the various operating parameters such as noise levels, operating temperatures, cycle times, etc. should be measured by instruments and recorded. Do not rely on human senses for measuring or upon the memory for recording these values.

Component wear

The wear in a hydraulic system is mainly caused by contamination of the oil; as has been

Section 7.1 Equipment and practices which benefit maintenance   299

emphasised throughout this book, efficient filtration and good housekeeping will reduce wear to a minimum. The actual wear occurring in components can be determined by monitoring changes in leakage flow and by analyzing the wear material in the oil and that caught in the filter elements.

The other major causes of component wear are misalignment of pumps and actuators possibly caused by fixing bolts becoming loose, pipework and valves moving, and vibration either mechanical or hydraulic. Physical damage can occur as a result of abuse, misuse, bad positioning or inadequate protection of components.

Some of these problems can be alleviated by introducing a system of routine maintenance.

Component position

All components should be positioned so they are easily accessible for adjustment, inspection and replacement. This is usually part of the design of the system and the user can often do little or nothing to improve component accessibility.

A lockable shut-off valve in the pump suction lines and low-pressure check valves in the return lines ensures that any component in the system can be removed for maintenance without emptying the reservoir. Where expensive pumps are used, shut-off valves which incorporate electrical limit switches are available to interlock the control circuit. It ensures that the pump cannot be started unless the suction valve is fully open. This is done to prevent pump cavitation.

Records and information

As has been stressed previously it is essential to maintain up-to-date information on all systems. Necessary data and documents should comprise the following:

- A complete circuit diagram and parts list giving details of all pressure settings, operating speeds and test-point pressure readings. This must be updated every time any modification is carried out.
- A spares' stock list giving details of all spares carried for each system, and interchangeability with other hydraulic systems. Wherever possible a standardization policy should be adopted when purchasing new machines. Many valves and cylinders are made to CETOP, NFPA or international standards and so different manufacturers' components are often interchangeable.
- A record should be kept of any breakdowns giving the symptom, cause and remedy of the fault, together with the date and length of time the machine was down. This record will help the maintenance engineer when the same fault has previously occurred. It will also help the maintenance management in putting forward a scheme of preventative planned maintenance. It may also be of use to the plant management in determining the reliability of different types of equipment, and influence the purchase of new machines or systems, or highlight the necessity for a design modification.

7.1.2 Fluid storage and handling

Storage

Hydraulic fluids should preferably be stored in a clean, well-ventilated room which meets all applicable safety standards. However, if outside storage in unopened drums is unavoidable then the following precautions should be observed:

- Drums should be stored on their sides and kept clear of the ground by wooden packers or runners to prevent rusting of the undersides. They should never be stacked directly on a clinker surface which is particularly corrosive to metal.
- If for any reason drums have to be stored on their ends they should be raised off the ground and stood upside down (i.e. with their bungs at the bottom). Failing this they should be tilted so that rain water cannot collect around and submerge the bungs. Water contamination is undesirable whatever the grade of fluid and it is not generally realized that moisture can enter a drum through what appears to be a perfectly sound bung. A drum standing in the open is subjected to the heat of the day and, of course, cools down again at night. This results in expansion and contraction with the effect that the air in the space above the oil level is subjected during the day to slightly higher than atmospheric pressure and at night to a slight vacuum. These changes in pressure may be sufficiently great to initiate a breathing action whereby air is forced out of the drum by day and drawn into it by night. If, therefore, the bungs through which the breathing takes place are surrounded by water some of the water may be sucked into the drum and, with the passage of time quite considerable quantities may accumulate.
- When oil is stored in bulk it is probable that water condensation will combine with fine dust and build up a layer of sludge at the bottom of the tank. Consequently it is advisable that storage tanks should be fitted with dished or sloping bottoms provided with drain cocks which will enable dregs to be drawn off periodically. Wherever practical, a periodic cleaning routine for bulk storage tanks should be established.
- Finally, carry out a regular inspection of all storages, checking for leaks and ensuring that identification markings remain clear and legible.

Handling

A fluid drum should never be opened by cutting a large hole in it or by completely removing one end, since, even if the hole is kept covered, the chances of contamination are greatly increased. Similarly, it is extremely bad practise to dip an open container into the fluid because not only does this permit the ingress of contaminant but the outside surface of the dipper itself may be dirty.

Drums should be placed on their sides on wooden cradles of convenient height and fluid dispensed by means of a tap under which a drip tray is placed. Alternatively the drum may be stood on its end and fluid withdrawn by means of a pump, the pump intake immersed in the fluid through the bung hole.

Containers, funnels, etc. used for dispensing hydraulic fluid must be kept scrupulously clean, and reserved for that specific purpose. They should be washed periodically in paraffin or other suitable solvent and dried using lint free rags. Cotton waste or woollen

Section 7.1 Equipment and practices which benefit maintenance 301

rags must not be used as they leave behind fibers which will eventually find their way into the hydraulic system.

Remember: The benefit of good clean storage can be completely nullified if the fluid becomes contaminated in transit from the drum to the machine.

7.1.3 Installation and commissioning of hydraulic systems

Any equipment, components etc. taken on-site must be stored in a clean area. The location where the machine is to be installed must be clean and all construction work, painting etc. finished before installation is commenced. The order in which installation is undertaken depends upon the machine and the site. Each section of the machine must be sealed to prevent ingress of dirt before starting on another section. All pipework must be cleaned internally and then plugged. Any pipework which has been hot-formed or welded should be descaled, pickled and cleaned internally and externally. Protect the bore of pipework with a thin film of mineral oil or a protective fluid which is compatible with the hydraulic fluid. Clean the inside of the reservoir, bolt on reservoir access plate, checking that the sealing gaskets are in place to stop the entry of dirt and air. Plug all reservoir ports. Check that all components have plugged ports; these plugs must not be removed until just before the pipework is fitted.

All pipework must be supported in such a manner as to prevent vibration. Pipes must be separated from each other; two pipes touching will rub and be weakened, and in extreme cases may wear a hole through the pipe wall. Flexible hoses must be installed in such a manner to prevent buckling, nipping, twisting or abrasion. Check the minimum bend radii and that there is sufficient slack to allow for any movement.

Power-pack installation

Position the cleaned reservoir on suitable anti-vibration material and fix in position. Check that the pump motor assembly is correct with the pump and motor shafts connected by a suitable flexible coupling or similar device, and correctly aligned by a bell housing or similar. Locate the pump motor set on suitable anti-vibration mounts and fix in position. Connect the pump suction to the reservoir via any shut-off valves, suction filters etc., using a flexible suction hose. There will be some relative movement between the pump and reservoir which will transmit vibration and may damage a rigid pipe. Suction hose is internally reinforced to prevent collapse when the inside is subject to negative pressure. It is advisable to use a lockable shut-off valve in the pump suction line if one is needed; this prevents unauthorized closure of the pump suction.

Cylinders, valves, manifolds and motors should be mounted in position before attempting to connect them. Pipework should be installed so that no stress is transmitted to the component. For further details see Chapter 5, Section 5.4 on pipes and fittings.

Filling the system

Make sure that all components have been correctly installed and fully piped with all joints tightened. Check that air filters are in position and tank drain valves closed. Fill the reservoir with fluid taken from new drums. A part-used drum may be contaminated by dirt or other fluids. The fluid should be pumped from the drum through a suitably rated filter into the reservoir. Fill with fluid to the upper fluid level as shown on the sight glass.

Any accumulator in the system should have been charged with gas according to manufacturer's instructions.

Open the pump suction valve if fitted; switch on the tank heaters if fitted to raise the fluid temperature to the minimum operating temperature.

Initial starting and commissioning

Under ideal conditions a system should be flushed with fluid velocities higher than those likely to be encountered under normal running conditions. The direction of flow should be reversed as often as possible. Also fluid temperature should be as high as can be achieved without damage to the structure of the fluid. To achieve these conditions will involve the use of special flushing rigs; however the majority of systems are flushed using their own pumps. The following procedures assume this to be the case.

- Loop pump output through a pressure or a return line filter back to reservoir. It may be possible to do this with the existing valving or it may be necessary to fit extra pipework. Adjust the relief valve to its minimum pressure setting.
- If a drain is fitted to the pump check that the pump casing is filled with fluid. Ensure that the direction of rotation of the pump is correct. If the pump is a manually adjustable variable-delivery unit, set it at half delivery.
- Start the pump, watch for leaks, check the pump for correct running paying particular attention to vibration noises and temperature. For systems containing relatively large volumes of pipework and cylinders it will probably be necessary to top up the reservoir with fluid. Check filter indicators and change filter elements as required. (For a variable pump increase the delivery to maximum after four minutes' running.)
- Run the pump for several hours to flush the fluid reservoir and power pack. Top up the reservoir if required.
- Return the circuit to normal – remove temporary pipework linking pressure and return lines, if this was installed. Loop out all actuators and sensitive components such as servo valves, precision flow control valves etc., substituting lengths of flexible hose for the actuators and flushing blocks for the valves.
- Start up the system with the relief valve still set at minimum pressure. Run the system operating control valves to get fluid flow through all pipework and valves. Top up reservoir and change filter elements as required. Run under these conditions for one to two hours depending on size of system. The larger the system, the longer should be the running time. Fifty theoretical recirculations of the fluid can be taken as a minimum flushing time.
- Ideally, fluid samples should be taken at regular intervals and the contaminant level checked with flushing continuing until the desired cleanliness level has been obtained.
- Reconnect actuators, servo valves etc. and run system under no load conditions. It may be necessary to bleed air out of the system at some cylinders which are above the reservoir level. Gradually increase the relief valve setting until the operating pressure is reached.

Section 7.1 Equipment and practices which benefit maintenance   303

- Adjust and lock all adjustable pressure and flow control valves.
- Check all pressure gauge readings for whole of system cycle.
- Check all bolts are tightened to the correct torque.
- Check all safety circuits are operational and all guards are in place.

7.1.4 Routine maintenance

Operator tasks (to be undertaken during operation of plant):

- Visual examination of system for damaged or leaking pipes, fittings and components.
- Visual examination of fluid level in reservoir and fluid condition.
- Visual check of operating pressures, filter condition indicators.
- Check guards are in place.
- Check operation of system and work produced.

Periodic maintenance (weekly or monthly etc., dependent upon operating conditions):

- Carry out operator tasks.
- Check fixing of all units.
- Check pressure readings at test points in system.
- Check pumps for noise level and operating temperature.
- Check all actuators for damage, noise level, operating temperature, output speeds and forces.
- Check precharge of any accumulators.
- Check correct operation of any interlocks.

Annual maintenance

- Empty fluid reservoir, check fluid condition.
- Clean reservoir internally and externally, examine for rust.
- Clean strainers.
- Clean air passages in air blast coolers.
- Examine all hoses, pipework and fittings for damage, wear or leaks. Replace as required.
- Examine electric motor, clean air passages.
- Examine flexible coupling between pump and motor.
- Check filter elements, replace any which have been in service for twelve months.
- Clean filter bowl.
- Check filter condition indicators for correct operation.
- Check leakage of pumps and motors by running under normal conditions and comparing leakage rates with that of new unit or manufacturer's recommendations. If leakage is excessive return to manufacturer for overhaul.
- Check leakage across piston seals of cylinders, re-seal as required. If replacing cylinder seals obtain a full seal kit from the manufacturer and change all seals.

When restarting after a major overhaul follow recommendations for initial starting given in the section on Installation and Commissioning.

It is difficult to lay down a set of checks which will cover all applications. For example, it may be advantageous to carry out pump and motor leakage tests on a more frequent basis.

7.2 TROUBLE SHOOTING IN HYDRAULIC SYSTEMS

In far too many instances the method of fault-finding in hydraulic systems is by trial and error. The maintenance engineer removes components in a random manner and replaces them with new ones hoping to cure the fault. This can be a very time-consuming and expensive exercise, and can indeed introduce additional faults into the system. Every time a joint is broken, contaminants are introduced.

Using a simple logical fault-finding technique can considerably reduce down-time and ease the task of the maintenance engineer. To use this technique it is essential to have the following:

(a) A good understanding of the function of all the hydraulic components in the system.
(b) A complete and up-to-date circuit diagram. All too often it is found that machines have been modified and that these alterations have not been shown on the circuit diagrams. Ideally an up-to-date circuit diagram and copies of all relevant handbooks, parts list, maintenance schedules and spares' stock list should be kept near the machine. If this information is kept in an office, it is inevitably the case that breakdowns occur when the office is locked.
(c) A parts list showing full part number and manufacturer of each component.
(d) An operational schedule giving details of sequence of operations, cylinder speeds, motor speeds, setting pressure of relief valves, pressure-reducing valves etc.

7.2.1 Test equipment

Ideally every user of hydraulically-operated machines should have the following equipment available to the maintenance engineer:

(a) A flow meter, or range of flow meters, capable of measuring the flow from the largest pump on the plant and from the smallest drain line. The flow meter should have a 'built-in' pressure-compensated flow-control valve and pressure gauge, so that components such as pumps can be tested under load conditions: a badly worn or damaged pump can deliver rated flow at low pressures but fail completely at normal system pressure.
(b) Test gauges with wandering leads to plug into pressure test points strategically positioned in the circuit, so that critical pressures can easily be monitored. Pressure gauges permanently fitted on machines tend to be very unreliable and vulnerable to mechanical damage.
(c) A hydraulic test unit on which any component which has been repaired can be fully tested before being replaced in the machine.
(d) A contamination measuring device for monitoring the hydraulic fluid for dirt particle content, type of particle and chemical deterioration of the fluid.

Unfortunately only a very few maintenance departments will have this equipment available. Most engineers have to operate with the aid of a pressure gauge only.

Section 7.2 Trouble shooting in hydraulic systems

7.2.2 General rules for hydraulic maintenance engineers

- Before working on a machine check the effect on interlocked parts or machines.
- Chock up all cylinders and parts which may fall under gravity.
- Isolate electrical supply and lock control cabinet.
- Isolate pump and ensure it cannot be started accidentally.
- Bleed fluid to relieve any pressure in system by cracking fittings – cover with cloth to prevent oil spray. **Particular care must be taken in the case of accumulator circuits.**
- Plug all pipe ends and ports of components to keep out contaminants.
- Ensure that components stripped are marked to facilitate correct assembly.
- Wash components in the correct fluid. If in doubt use clean hydraulic fluid as used on the machine.
- Use torque wrenches for tightening components. **Do not over tighten.**
- Use extreme care when starting machine for first time after overhaul – a pipe left off can cause a flood of oil; a valve spool reversed may cause a cylinder to fall instead of rise; actuators may operate out of sequence.

7.2.3 The concept of logical fault finding

This is a diagnostic technique in which all the symptoms have to be carefully considered so the fault can be localized to one section of the circuit. If the fault affects all cylinders and motors it must be something common to them all, i.e. a failure in supply either of hydraulic fluid – a fault in the power pack – or in the electrical control system. Should the fault be unique to one actuator, start by considering the components associated with that actuator.

    To determine the possible cause of a fault, the exact function and effect of each component associated with the system must be known. Consider the action of each component in turn, with respect to the symptoms. Make a note of the component or components which if maladjusted or malfunctioning could result in all these symptoms. When this has been done select the component which can be most easily tested, for example adjusting flow or pressure settings, manually operating solenoid valves and limit switches. Only when all the simple tests have been completed should any attempts be made to disconnect pipework or remove components. Every time a joint is broken, contamination will enter the system; always plug pipes and manifolds when components are removed.

Fault finding procedure

INITIAL INFORMATION

This is from the machine operator

1. Where and when does fault occur?
    (a) On all cylinders and motors.
    (b) On one only.
    (c) Under all load conditions.

2. Type of fault
   (a) Complete stoppage.
   (b) Reduced speed or thrust.

3. How soon did fault occur?
   (a) Suddenly ... (breakdown)
   (b) Gradually ... (general wear)
   (c) Periodically ... (intermittent fault)

4. Unauthorized adjustment
   (a) Has someone adjusted the machine?
   (b) Has any attempt been made to repair unit?
   (c) Has machine recently been modified or repaired?

INITIAL CHECK ON MACHINE
- Check electrical supply is switched on to both power and control circuits.
- Check oil level, condition, temperature.
- Check pumps for correct running.
- Check filters.
- Check pressures.
- Check visually for broken or burst pipes, leaks from components.

FAULT DIAGNOSIS
If fault is localized to part of circuit, only that area need be considered. If fault is general to all the circuit the cause must be something common to whole circuit, e.g. pump, main relief valve, suction line. Initially trace the fault to a particular part of the circuit so that only components in that part need to be considered. Next consider each component and determine which components could cause the fault. Only when possible causes of the fault have been determined should any components be tested. Carry out the easiest tests on suspect components first, i.e. tests which can be done *in situ*, e.g. manual operation of solenoid valves, pressure settings, flow settings, and oil leaks and damaged pipes. If this fails to locate the fault, remove the component *most likely* to be faulty and check that first.

EXAMPLE 7.1

System specification:
The cylinder shown in Figure 7.1 is to traverse under load in both directions. Cycle duty 6–8 cycles per hour, 24 hours per day.

Fault occurs

Initial information

Cylinder extends but will not retract. Can be forced back using separate jacking cylinder.

Section 7.2 Trouble shooting in hydraulic systems

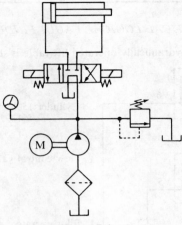

Figure 7.1

Initial check

(a) Steady pressure on extend stroke, pressure rises to relief valve setting at the end of stroke prior to solenoid being de-energized.
(b) On energizing retract solenoid pressure rises to relief valve setting after a short time delay.
(c) Condition normal in center condition.

As this is a simple circuit, the whole circuit can be considered.

Unit	Possible fault	Comments
Pump	No	Fault is one direction only Pump would affect both directions
Suction filter	No	As for pump
Relief valve	No	If the pressure setting is too low for retract, the cylinder will stall. This would not cause the delay in reaching relief valve pressure on retract
Solenoid valve	No	Retract solenoid may not be moving spool fully over, causing a restriction to flow. Will not cause time delay
Cylinder	Yes	Seals could be faulty. External guides loose. Piston detached from piston rod

Possible cause is: Piston detached from piston rod. (The symptoms assume there is no hole through the piston.)

EXAMPLE 7.2: THEORETICAL LOGICAL FAULT FINDING

The hydraulic circuit for a hydraulically operated lift table is shown in Figure 7.2.

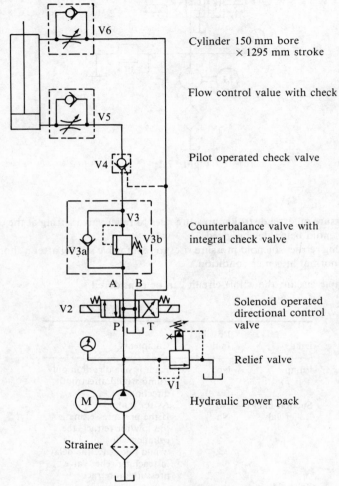

Figure 7.2 Hydraulic circuit for lift table.

Description of hydraulics

The hydraulic power pack consists of a 75-liter reservoir containing class HM mineral oil hydraulic fluid, a 2-kW motor driving a vane type pump delivering 32 l/min. A relief valve V1 sets the maximum circuit pressure which is indicated on a pressure gauge.

A four-port, three-position solenoid operated, spring-centered, directional control valve V2 directs the flow of oil to the annulus or full bore ends of the cylinder. In the mid-position all ports are connected, i.e. open center, and pump flow will be passed to tank at a low pressure.

The valve V3 is a counterbalance valve with integral check valve. The check valve V3A

Section 7.2 Trouble shooting in hydraulic systems

allows free flow of oil to the full bore end of the cylinder. When the cylinder rod is retracting the counterbalance section V3B sets up a back-pressure which prevents the cylinder running away under load.

The pilot-operated check valve V4 permits free flow to the full bore end of the cylinder. It will allow free flow from the full bore end if the annulus side is pressurized to pilot the check valve open. When the directional valve is in the center condition the pilot check valve V4 locks the oil in the full bore end of the cylinder preventing it from moving.

The flow control valves with check valves V5 and V6 meter the quantity of oil leaving the full bore end and annulus end of the cylinder respectively. V6 controls the extend speed of the cylinder and V5 the retract speed.

Example

EFFECTS OF COMPONENT MALFUNCTIONS

If the suction strainer blocks or clogs it can cause pump cavitation. Suggest annual inspection.

With no filtration the *pump* will tend to wear which will result in increased leakage and reduced pump delivery. If the operating speed of the cylinder decreases it may be as a result of pump wear.

The pilot stage of the *relief valve* V1 sets the operating pressure. If the spring in the pilot stage breaks or weakens it will be impossible to attain maximum system pressure. If the pilot stage valve seat is damaged it will leak causing the main spool to open slightly. With this circuit V5 and V6 restrict the flow to the cylinder; excess oil flows over the relief valve all the time.

The *directional control valve* V2 directs the flow of oil to or from the cylinder. If a solenoid malfunctions the valve will not switch. This fault can be checked by manually operating the valve. If a centering spring breaks, the valve will remain in an operated position, which can be checked by manually actuating the valve. Dirt in the system can cause the spool to jam. If this occurs the valve will have to be stripped and cleaned, the system flushed and a pressure line filter introduced to prevent recurrence.

The *counterbalance valve* V3 is similar in construction to the relief valve, and sets up a back-pressure when the cylinder is retracting. If this is incorrectly set or malfunctioning it may prevent operation of the pilot check valve or cause it to judder. If set too high it will prevent the piston rod being retracted.

The function of the *pilot-operated check* V4 is to lock the cylinder in any position against a downward load. V3 can cause a malfunction. If the check valve chatters it can damage the valve seat, allowing a leak of oil, which will let the cylinder creep down under external load.

In the case of *flow control valves* V5 & V6 dirt in the oil can erode the control orifice causing a loss of speed control. If the check valve fails to seat this will also give a loss of speed control. It is unusual for a check valve to jam closed but if this does occur the flow control valve will affect the flow in both directions.

A general fault in the *cylinder* is seal wear caused by contamination, excessive temperature, excessive pressures or load, or damage to piston rod. External leakage at the piston rod gland indicates that the rod or its seal is damaged. Leaking piston seals will result in a reduction of piston speed. Check by moving piston to one end of stroke, removing the pipe at that end, pressuring other end and checking for flow from the cylinder port.

Exercise

Faults which could occur on the lift table have been tabulated in Figure 7.3. List the possible causes of each fault and any checks to be carried out to determine which is the actual fault. (A suggested solution is given in Figure 7.14).

Symptom	Cause	Initial checks
Cylinder extends at normal speed, retract speed is slow		
Cylinder movement is erratic and noisy		
Cylinder movement is jerky when retracting		
Cylinder extends but will not retract. System pressure is up to the relief valve setting		
Cylinder will not extend. System pressure is low and cannot be adjusted by the relief valve		

Figure 7.3 Lift table – fault finding exercise (a suggested solution is given in Figure 7.14)

FAULT-FINDING USING FUNCTIONAL BLOCK DIAGRAMS AND TROUBLE-SHOOTING CHARTS

A hydraulic circuit can look very complex to the uninitiated and make fault-finding almost impossible. Although it is preferable for maintenance engineers to have a working knowledge of power hydraulics it is possible by using charts and block diagrams to simplify trouble shooting. There is considerable work involved in designing and drawing up the appropriate charts, so this method may only be viable if a number of similar machines are involved.

The complete system should be broken down into sections which can be considered individually, for example an actuator or group of actuators and the associated control valves.

EXAMPLE 7.3

A hydraulic power pack will be considered to demonstrate the method but this could be applied to any section of a complex hydraulic system. The power pack considered has two main pump sets each with identical motors and pressure-compensated pumps. The circuit demand is such that it can be met by one pump for the majority of the time, the second pump

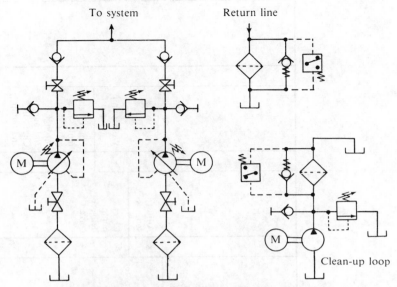

Figure 7.4 Hydraulic power pack (symbolic circuit).

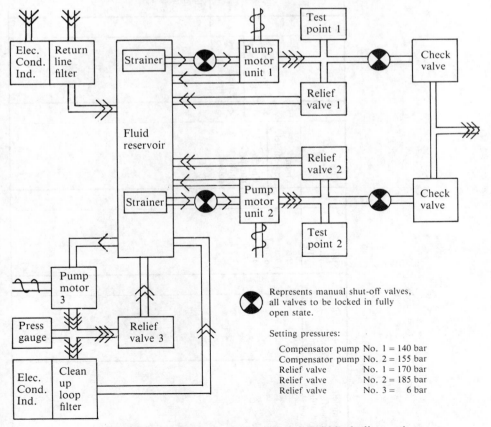

Figure 7.5 Hydraulic power pack (functional block diagram).

Complete loss of hydraulic power	Check electric motor	Start electric motor	Check interlocks	Open immediately	Check coupling	Check direction of rotation correct	Reverse immediately	Check pump compensator settings	Adjust settings	Check pump delivery correct	Worn pump – replace	Check valve faulty flow going through 2nd pump. Check if 2nd pump rotating when electric motor off
	×	Check pump suction valve ✓	×	Check pump shaft rotating ✓	×	✓	×	✓	×	✓	×	×

Hydraulic pump noisy	Pipework vibrating after pump	Pump running hot – higher than oil temperature							
	×		Damaged pump – replace						
	✓	Cavitation check suction valve open							
		×	Open immediately						
		✓	Fluid level correct						
			×	Top up with correct oil					
			✓	Oil temperature high		Check oil cooler			
			✓	Oil temperature low		Check oil heater			
					×	Oil temperature correct			
					✓	Air leaks on pump suction	Tighten or change fitting		
						×	Check oil tank for oil frothing		
							✓	Find cause of foaming and eliminate	

Figure 7.6 Hydraulic power pack (trouble shooting chart). ✓ = Yes, × = No.

only being required under peak-operating conditions. This is achieved by arranging the pressure compensators on the pumps to operate at different pressures, these pressures being sufficiently different to prevent any hunting of the pumps. As the pumps are pressure-compensated and so deliver a variable flow, a separate clean-up loop is used to condition the fluid, together with a return-line filter.

The hydraulic circuit for the power pack is shown in Figure 7.4. This circuit can be redrawn in the form of a functional block diagram (Figure 7.5) each block representing a unit of the system. Useful information such as flow rates and pressure settings can be included.

Notes To check the setting of No. 1 pump and relief valve plug a 200-bar pressure gauge into test point No. 1. Start pump motor unit No. 1. Close shut off valve 2. Increase compensator setting of pump No. 1 until relief valve operates – pressure gauge should indicate 170 bar – adjust if necessary. When relief valve No. 1 is set, reduce the compensator setting until the pressure gauge indicates 140 bar. Repeat the procedure to set the compensator on pump set No. 2 and relief valve No. 2 to the stated pressure valves. To counteract gauge inaccuracies *Use the same pressure gauge in test point 2 as was used for setting up No. 1 main pump.*

The electrical switches on the return-line filter and on the clean-up loop filter operate lights on the control panel to monitor element condition: green light for correct function, red light shows filter element needs changing.

On the block diagram full details of pressures at all test points under relevant circuit conditions should be given with instructions as to the method of obtaining these operating conditions, i.e. which solenoids have to be energized, the shut-off valves which are open, etc. When actuators are shown on the block diagram, the cylinder loads and motor torques can be specified together with speeds so that the flow control valves can be set.

The block diagrams should be accompanied by a parts' list. This should list each component, its part number, manufacturer, size of actuators, and full details of spares held in stock.

A trouble-shooting chart similar to that shown in Figure 7.6 can be drawn up for the circuit. This chart should pose questions to which there is a 'yes/no' type of answer. Dependent on the answer, the chart will show the next step in locating a fault. These charts can only be built up based on experience of the machine or similar machine, and will not cover every conceivable fault or breakdown liable to occur. As further experience is gained on the operation of a machine, the fault-finding chart can be extended.

TREE-BRANCHING METHOD OF FAULT-FINDING

This is basically another method of presenting the fault-finding chart. It asks a question which has only two possible answers, 'yes' or 'no'; the answer determines the next step to be taken.

It is said that on average it takes four times as long to locate a fault as it does to rectify it and this technique helps develop a logical and rapid approach to fault diagnosis.

EXAMPLE 7.4
Assume a cylinder is moving erratically. Figure 7.7 could be used to determine the cause. A similar chart can be made to cover complete system malfunction or for a particular valve or component. In some cases a logical sequence of questions has to be followed, in other cases the sequence is unimportant.

Section 7.2 Trouble shooting in hydraulic systems

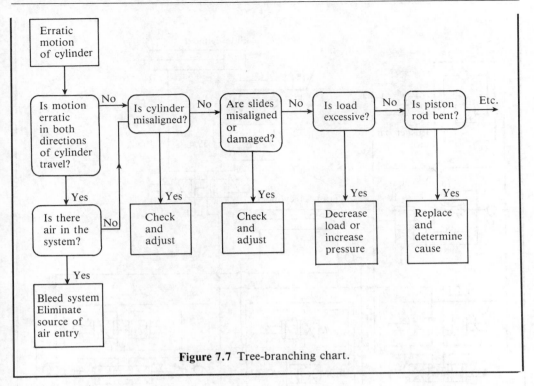

Figure 7.7 Tree-branching chart.

EXAMPLE AND EXERCISE 7.5: LOGICAL FAULT FINDING

Refer to hydraulic circuit in Figure 7.8.

In this example we are considering only part of the complete circuitry for a packaging machine. These particular circuits control a side cylinder and pusher bar which are used to position a number of small packages in a set pattern. A transfer cylinder moves the group of packages to another part of the machine where they are wrapped as a single large package.

The pressure in parts of the side cylinder and pusher bar circuits has to be limited to prevent the packages being crushed.

Oil is fed from the reservoir via a suction strainer (28) and stop-valve (33) to the suction side of a double pump unit. This comprises a 10-liter per minute vane pump (1) supplying the requirements of circuit No. 1 and a 40-liter per minute vane pump (2) feeding circuit No. 2. The maximum system pressures for circuit Nos 1 and 2 are set by relief valves (18/1 and 18/2).

Hydraulic circuit No. 1

The output from the pump (1) feeds a spring-centered, solenoid-operated, three-position valve (9) which controls the direction of motion of the side cylinder (31). A combined non-compensated flow control and check valve (12) controls the cylinder extend speed. When the cylinder (31) is retracting counterbalance valve (39) maintains the back-pressure in the full

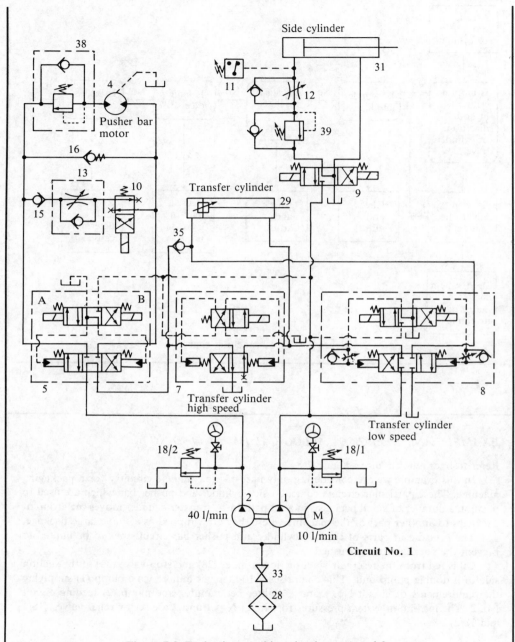

Figure 7.8 Packaging machine-circuit nos 1 and 2.

bore end. An adjustable electrical pressure switch (11) indicates the pressure in the full bore end of the cylinder. The pump (1) also supplies a two-stage solenoid-controlled pilot-operated directional-control valve (8) which gives slow speed operation of the transfer cylinder (29). This valve is fitted with a choke pack to regulate the change-over speed of the main spool.

Section 7.2 Trouble shooting in hydraulic systems

Symptom	Possible fault
Side cylinder will not retract system pressure correct	
Side cylinder extend speed low	
System pressure low. Cannot be adjusted by relief valve even when valves (8) and (9) in de-energized state	
Transfer cylinder starts very slowly when it is extended at low speed	
System will not operate	

Figure 7.9 Packaging machine-circuit no. 1 – fault-finding exercise (a suggested solution is given in Figure 7.15).

Hydraulic circuit No. 2

This circuit provides power to the pusher bar hydraulic motor (4), and for the transfer cylinder (29) high-speed travel. The maximum pressure at the pump is limited by relief valve (18/2) set at approximately 37 bar. This pressure setting can be checked using the pressure gauge and shut-off valve associated with the relief valve. When not required output from the pump will flow directly to tank at very low pressure through the valves (5) and (7). Pilot oil pressure for the two-stage valves is obtained from circuit No 1. By energizing the solenoid A on valve (5) pressure is applied to the main spool switching it to 'crossover' driving the pusher bar motor forward at high speed. If the solenoid on valve (10) is energized part of the flow bypasses the motor through flow control valve (13) which reduces the motor speed. The counterbalance valve (38) is used to create a back-pressure of approximately 4 bar. This assists in obtaining more accurate stopping of the pusher bar. Check valve (16) has a 5-bar cracking pressure. Thus when valve (5) is switched to drive the motor in reverse, pressure is limited to 5 bar. If the pressure required for the motor is in excess of this, check valve (16) will operate bypassing the flow across the motor. If valve (10) is energized while the motor is in reverse rotation the check valve (15) prevents any flow occurring through the slow speed control loop.

Valve (7) is used to control the major flow for high-speed transfer cylinder movement. With the valve in the de-energized position, the output from the pump is directed to tank. For rapid extend valve (7) is energized and oil is fed via check valve (35) to the full bore end of the transfer cylinder (29), the annulus end being connected through valve (7) to tank. At the same time additional flow is provided from circuit No. 1 through valve (8) which is energized with the main spool in crossover condition. When the transfer cylinder approaches the limit of its stroke, the high-speed valve (7) is de-energized and only the low-speed valve (8) continues to

Symptom	Possible fault
Pusher bar motor stalls under load in forward direction	
Pusher bar motor will not reverse. Functions normally in forward	
Pusher bar motor speed low in 'fast forward'	
Pump (2) noisy. Lines after pump vibrating. Pressure gauge reading erratic	
Transfer cylinder high-speed normal. Slow speed extend very slow. Retract speed normal	

Figure 7.10 Packaging machine-circuit no.2 — fault finding exercise (a suggested solution is given in Figure 7.16).

Section 7.2 Trouble shooting in hydraulic systems

supply oil to the cylinder. This retards the transfer cylinder, consequently reducing the shock loads which would be induced by stopping the cylinder suddenly. For the reverse stroke only the low speed valve is used with flow from circuit No. 1.

Exercise

A list of fault symptoms for circuit Nos. 1 and 2 are shown in Figures 7.9 and 7.10. Determine the possible cause of these faults. Where there are four boxes shown under the heading 'possible fault', this indicates four likely causes.

EXAMPLE 7.6: FAULT-FINDING

Hydrostatic transmission

The hydraulic circuit for this unidirectional transmission is shown in Figure 7.11. The oil reservoir (8) has a capacity of 40 liters and is equipped with a sight glass and a filler breather. The make-up pump (7) is a gear pump having a rated delivery of 30 l/min, and is driven by a 1.5-kW electric motor at 1440 rev/min. The oil reservoir can be isolated from the boost pump

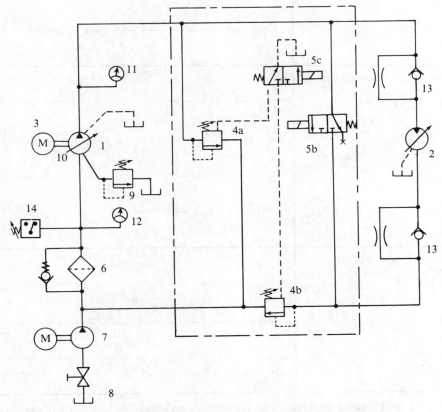

Figure 7.11 Hydraulic circuit for hydrostatic transmission.

by a shut off valve in the suction line. The output from the make-up is fed into the main return line through a filter rated at 125 l/min and into the main axial piston pump (1). A low-pressure relief valve (9) set at 1.5 bar limits the make-up pressure. This is connected to the main pump casing which forms part of the make-up circuit.

A pressure switch (14) set to operate at 0.8 bar falling, is used to indicate a failure in boost pressure in the main pump (1) which has a maximum delivery of 109 l/min at 1440 rev/min against a maximum pressure of 275 bar. This pump is fitted with a manual servo

Symptom	Possible faults and checks
Excessive main pump noise	
Oil overheating	
Loss of pressure, motor (2) stalls	
Loss of flow, motor (2) runs slow	

Figure 7.12 Fault finding on hydrostatic transmission Example 7.6, Exercise 1 (a suggested solution is given in Figure 7.17).

Section 7.2 *Trouble shooting in hydraulic systems* **321**

stroke control (10) which adjusts the swash plate angle of the pump and hence the delivery. The manual servo has a remote electrical control unit attached. The main pump drive is by a 55 kW electric motor (3).

A two-stage relief valve (4a) is connected in parallel with the hydraulic motor (2) and

Symptom	Possible causes and checks
Motor cannot be rotated by hand when hand turning facility engaged	
Motor speed low, this has occurred over a period of time	
Boost pump starts, main pump does not start, low boost pressure warning light comes on	
When stop button operated hydraulic motor stops very abruptly	
Circuit runs normally but oil reservoir temperature high	

Figure 7.13 Fault finding on hydrostatic transmission – Example 7.6, Exercise 2 (a suggested solution is given in Figure 7.18).

limits the maximum pressure at the motor to 275 bar. A second relief valve (4b) is positioned in the hydraulic motor return line, and can be used to set up a back pressure acting as a brake or retardation valve. A two-position directional-control valve (5c) is used to vent either the brake valve (4b) or the cross-line relief (4a).

The hydraulic motor develops a torque of 300 Nm at a maximum speed of 3000 rev/min with a pressure drop over the motor of 250 bar. The maximum leakage flow from either the motor or main pump should not exceed 10 l/min when they are operating at 275 bar. If this value is exceeded it indicates that excessive wear has taken place and the unit requires a complete overhaul which is best done by the manufacturer.

When starting up the main pump the cross-line relief valve is vented by valve 5c which is unenergized so that the pump is started under no-load conditions. In normal running valve 5c is energized, hence the brake valve is vented so that it does not create a back-pressure and the cross-line relief valve acts normally limiting the maximum circuit pressure.

Under stop, or emergency stop, conditions the cross-line relief valve is vented ensuring that there is no hydraulic pressure on the motor. The brake valve is put on load and sets up a back-pressure which rapidly slows down the hydraulic motor and its load. The rate of deceleration is determined by the spring setting of the brake valve. In order that the hydraulic motor can be turned by hand, a solenoid-operated directional-control valve (5b) is connected across the hydraulic motor. When the solenoid is energized a flow path is opened, bypassing the motor which can then be manually rotated.

Check valves (13) are located upstream and downstream of the hydraulic motor. The function of the checks is to prevent the motor being driven in reverse if the main pump is accidently reversed whilst it is being adjusted. The check valves have a 1 mm diameter hole drilled through the poppet allowing a small reverse flow. This enables the motor to be turned in reverse by hand.

Pump control

The pump delivery is varied by adjusting the pump swash plate angle. The forces acting on the swash plate are very large so a servo hydraulic system is built into the pump for this purpose. A servo rod protruding through the casing controls the swash-plate servo system, the servo rod being moved remotely by an electrically-driven screw (10).

Exercises

Bearing in mind how the components affect the function of the system, logically determine possible faults causing the various malfunctions listed in Figures 7.12 and 7.13. For example, where the symptom is given as 'excessive pump noise', it is suggested that there are three possible major causes. Indicate, where appropriate, checks to verify the fault.

7.3 SOLUTIONS TO FAULT-FINDING EXERCISES

The following are possible solutions to the exercises contained in Section 7.2. They indicate some of the most likely causes but do not necessarily cover all eventualities.

Symptom	Cause	Initial checks
Cylinder extends at normal speed, retract speed slow	V5 flow control set too low	Adjust
	Internal leaks across piston seal (retract)	Retract piston. Disconnect full bore port. Apply pressure to annulus and check for leaks
	V6 check valve jammed closed	Adjust V6 flow control and check if it affects retract speed
Cylinder movement erratic and noisy	Air in cylinder	
Cylinder movement jerky when retracting	V3b set too low and cylinder overrunning	Adjust V3b
Cylinder extends but will not retract. System pressure up to relief valve setting	V5 flow control closed	Adjust V5
	V3b set too high	Adjust V3b
	V4 not opening	Check pilot to V4
Cylinder will not extend. System pressure low cannot be adjusted by relief valve	V1 damaged	Check control spring
	V2 not operating	Manually operate V2
	Pump malfunction	Check pump

Figure 7.14 Suggested solution to the Exercise in Example 7.2 (Figure 7.3) – lift table.

Symptom	Possible fault
Side cylinder will not retract system pressure correct	Counterbalance value (39) set at too high a pressure
	Directional valve not moving over – manually over-ride valve, check solenoid
	Cylinder rod jammed in extend condition
	Piston has become detached from rod
Side cylinder extend speed low	Flow control value (12) misadjusted
	Piston extend seals leaking
	Directional control valves not going fully over – setting up a restriction. **Note**: If the solenoids are AC they will burn out if not going fully over
	Relief valve (18/1) set low – this reduces pressure drop across valve (12) which is not pressure-compensated, therefore the pressure drop affects the flow
	Pump delivery low – pump worn – system will not reach relief valve setting as no oil flowing over relief valve
System pressure low, cannot be adjusted by relief valve even when valves (8) and (9) in de-energized state	Pump badly worn – this would have shown in a gradual reduction in system performance
	Spring in pilot section of relief valve broken
Transfer cylinder starts very slowly when it is extended at low speed	Choke pack restrictor set too low
	Pilot valve not moving fully over
	Restriction in pilot valve or pilot lines
System will not operate	Electric motor not driving pump – coupling slipping
	Relief valve fully open – check pressure gauge reading
	Failure of electric supply to solenoid valves
	Pump broken

Figure 7.15 Suggested solution to exercise in Example 7.5 (Figure 7.9) – packaging machine – circuit no. 1.

Section 7.3 Solutions to fault finding exercises

Symptom	Possible fault
Pusher bar motor stalls under load in forward direction	Counterbalance valve (38) set at too high a pressure
	Relief valve (18/2) set at too low a pressure
	Motor (4) badly worn leaking all flow at full load
	Drive coupling from motor (4) slipping at full load
Pusher bar motor will not reverse. Functions normally in forward	Reversing solenoid on valve (15) does not operate. Main spool on valve (15) not moving over to 'tram line' condition
	Spring on valve (16) weakened, this would affect cushioning of pusher bar motor in forward direction
	Load on motor (4) high in reverse
	Check valve on valve (38) jammed shut
Pusher bar motor speed low in 'fast forward'	Check valve (16) leaking
	Solenoid valve (10) not in closed condition or leaking
	Wear in motor (4)
	Pump (2) worn – reduced delivery
Pump (2) noisy. Lines after pump vibrating. Pressure gauge reading erratic	Pump cavitation. Valve (34) partly closed. Filter (27) clogged. Low oil level, high or low oil temperature
Transfer cylinder high-speed normal. Slow speed extend very slow. Retract speed normal	Check valve (35) leaking

Figure 7.16 Suggested solution to exercise in Example 7.5 (Figure 7.10) – packaging machine – circuit no. 2.

Symptom	Possible faults and checks	
Excessive main pump noise	Oil level in tank low	Check oil level
	Restriction in suction lines, shut off valve not fully open	Check tank strainer Check valve in boost Pump suction line
	Mechanical failure of pump	Check hydraulic motor speed
Oil overheating	Relief valve (4a) set at too low a pressure	If relief valve blowing off the outlet line will be hot. Check pressure gauge reading – this will not read relief valve setting unless relief valve blowing off
	Brake valve (4b) not fully open, setting up a back pressure	Check temperature of pipes entering and leaving brake valve
	Excessive leakage across pump (1) or motor (2). *Note* Falling of pump inlet pressure is an indication of this fault	Check temperature of tank lines. Check leakage flow rates
Loss of pressure, motor (2) stalls	Relief valve (4a) set at too low a pressure	Check pressure gauge (11)
	Brake valve setting up a high back-pressure	Check pressure downstream of hydraulic motor (2)
	Check valves (13) not opening fully	Check valves (13)
	Motor (2) fault	Check pressures downstream and downstream of motor
	Mechanical fault at motor (2) drive	Examine motor drive
Loss of flow, motor (2) runs slow	Setting of pump (1) low	Check pump setting
	Motor (2) setting high	Check
	Value (5b) leaking	Check pipe temperature downstream of value (5b)
	Relief value (4a) leaking	Check temperature of line downstream of (4a)

Figure 7.17 Suggested solution to Exercise 1 in Example 7.6 (Figure 7.12) – hydrostatic transmission.

Section 7.3 Solutions to fault finding exercises

Symptom	Possible causes and checks
Motor cannot be rotated by hand when hand turning facility engaged	Solenoid valve (5b) malfunctioning
	Solenoid burnt out – manual override
	Spool jammed due to dirt
	Failure of electrical supply to solenoid
Motor speed low, this has occurred over a period of time	Excessive leakage at hydraulic motor or pump gradually increasing
	Pump servo control setting incorrect i.e. working loose
	Motor displacement increased. Check locking method
	Part of flow leaking over relief valve (4a) This would result in heating up of the fluid
Boost pump starts, main pump does not start, low boost pressure warning light comes on	Low boost pressure. Check setting of pressure switch (14) by means of pressure gauge (12). Check setting and operation of boost pressure relief valve (9)
	Check shut off valve in suction line of boost pump
	Check operation of boost bump; if it is very badly worn it may not be able to reach required pressure (about 50 psi to allow for pressure drop across filter)
When stop button operated hydraulic motor stops very abruptly	Setting on (4b) brake valve too high
Circuit runs normally but oil reservoir temperature high	Pump or motor leakage high
	Slight leakage over main relief valve (4a) or through hand turning valve (5a)
	Brake valve (4b) setting up a back pressure
	Boost pressure relief valve (9) set high

Figure 7.18 Suggested solution to Exercise 2 in Example 7.6 (Figure 7.13) – hydrostatic transmission.

CHAPTER EIGHT
CONTROL SYSTEMS

Control systems fall into two main groups – digital and analog.

In a *digital system* the control signal can only have two states, on or off. An example of such a control method is a hydraulic cylinder reciprocating between two limit switches (Figure 8.1). When the piston reaches the end of its stroke it operates a limit switch sending a signal to the directional control valve which reverses the direction of motion of the piston.

The signal from an *analog device* is dependent upon the quantity being measured. The signal can have any value between the limits set by the control device. A simple weighing machine is an example of an analog system in which the movement of a pointer is proportional to the applied load. When the output of a circuit is continuously monitored and the information used to adjust the input, it is termed 'feedback'. The control module may respond automatically to this information or it may be displayed for an operator to adjust the input command accordingly.

A thermostat in an electrical heating system measures the output and compares it with the desired temperature. When the temperature reaches the required value the electrical heating system is switched off; if the temperature falls the heater is switched on again. This is a digital 'on or off' control method with feedback. In the case of a gas control system for an oven or furnace, the thermostat senses the temperature and modulates the gas flow to the burners to give the desired value. In this case gas control is by continuous adjustment, i.e. an analog system.

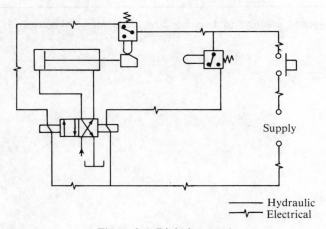

Figure 8.1 Digital control.

8.1 SERVO CONTROL

A servo control system is one in which (a) the input signal is amplified using a power source to give the output and (b) there is a feedback signal from the output. This may be shown in block schematic form in Figure 8.2.

In an open-loop servo the operator compares the actual output with desired value and then corrects any error. An example of this is the 'power-assisted steering' of a vehicle (Figure 8.3). The operator turns the steering wheel, visually observes the course of the vehicle and corrects any deviation from the desired course. A closed-loop servo includes a device which compares the input signal θ_i with the system output θ_o and feeds an error or difference signal ε into the power valve (Figure 8.4). A closed-loop servo can either be a 'regulator' or a 'follow-up' system. In the regulator type of system the object of the control loop is to maintain the output at a given value independent of all system

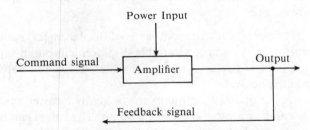

Figure 8.2 Servo control (block diagram).

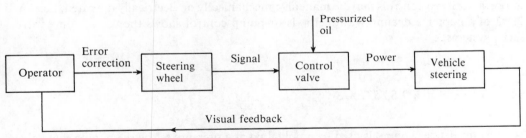

Figure 8.3 Servo system with feedback via operator (block diagram).

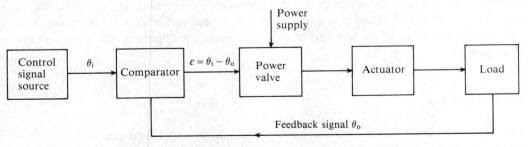

Figure 8.4 Closed-loop servo system (block diagram).

disturbances. The thermostat loop on a gas oven is a regulator system where the function of the loop is to maintain the oven setting at the desired value independent of oven load and external cooling. A follow-up system is one in which the input function is constantly changing and the output is controlled to follow the input. A typical example of a follow-up control is a copying lathe in which the cutting tool position (which is the system output) follows a path traced from a template.

Hydraulic servo systems generally fall into two groups – valve-operated or pump-operated servos.

The *valve-operated servo* controls the amount of fluid flowing into the system from the pump, the valve acting as a variable orifice. This can generate considerable heat resulting in an inefficient system. The *pump-operated servo* adjusts the delivery of a variable-displacement pump to the required flow rate, giving a more efficient system with less energy loss. The majority of pump servos employ a valve servo to control the pump displacement; the control system becomes more complex than a simple valve system.

Generally, the valve-operated servos are used on lower power applications as they offer the following advantages:

1. Components are of simpler construction, a fixed-displacement pump being used, whereas with a pump-operated servo, a variable-displacement pump supplies the main system and a second fixed-displacement pump is used for the valve servo which controls the pump displacement.
2. Valve servos have a lower inertia than pump servos giving a faster dynamic response.
3. Several valve servos can be operated off a single pump. The valves can be situated near the actuator again giving a faster response and greater flexibility than pump servos.

Pump servos are used when the power to be controlled is large and system efficiency is important. Servo controls may be manually, mechanically or electrically operated. Section 2.1.3 of Chapter 2 dealing with methods of pump control shows the use of some valve servo systems.

8.2 VALVE SERVO SYSTEMS

A constant-delivery pump is used to supply fluid at a pressure set by the pump relief valve to the servo valve. The spool in the valve is used to throttle or adjust the quantity flowing by masking or unmasking an orifice (see Figure 3.20 in Chapter 3). If the pressure drop over the valve is δP and the orifice area x then the quantity q flowing through the orifice is given by $q = cx(\delta P)^{1/2}$ where c is a constant. The actual orifice area x may not be the same as the movement of the spool from the zero position; the opening will depend on the valve geometry.

8.2.1 Valve lap

The valve in Figure 8.5(a) is shown as having a zero lap spool. This means that the spool lands are exactly the same size as the ports in the valve body. To achieve this is virtually

Section 8.1 Servo control

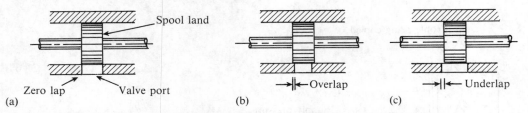

Figure 8.5 Valve lap. (a) Zero lap spool. (b) Overlap spool. (c) Underlap spool.

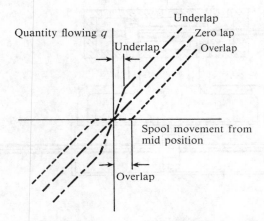

Figure 8.6 Valve lap – flow characteristics.

impossible even using sophisticated modern machining techniques. Overlap spools (Figure 8.5b) are used in ordinary hydraulic directional control valves to facilitate good sealing. Underlap spools (Figure 8.5c) are generally used in hydraulic servo valves.

Figure 8.6 shows the quantity of fluid flowing through a valve with various types of spool lap assuming the valve opening has a linear characteristic:

(a) The flow from the zero lap spool is directly proportional to the spool movement.
(b) The overlap spool has a dead zone in which no flow occurs; this dead zone is equal to the total valve overlap. If an overlap spool is used in a servo valve it will result in an unstable condition when the actuator is near the desired position, the spool having to move through the dead zone before any flow can occur.
(c) An underlap valve gives a high flow rate when the spool is in the underlap area near zero displacement or null point. A servo valve usually incorporates an underlap spool to achieve high response rates near the null point.

8.2.2 Mechanical feedback

A valve-type hydraulic servo-copying device for a machine tool is shown in Figure 8.7. If the stylus moves to the left by a distance Z it will cause the valve spool to move to the left by an amount x as the control lever pivots about the piston rod pin. The valve spool movement directs a flow of fluid to the left-hand side of the piston in the cylinder. This

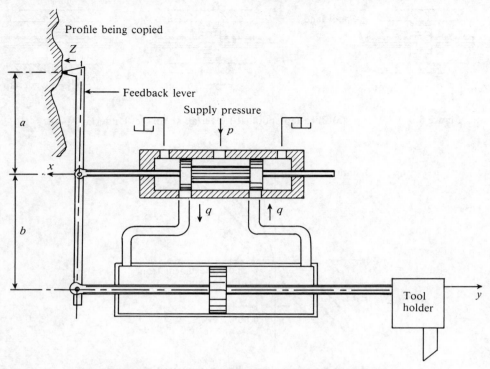

Figure 8.7 Valve-type hydraulic servo-copying device.

causes the piston rod to move to the right a distance y taking the feedback link with it, so reducing the valve spool displacement.

Consider the movement x of the valve spool. When the stylus moves a distance Z to the left, the feedback lever will pivot about the piston rod end. This will result in a movement of the valve spool to the left equal to $Zb/(a+b)$. In turn the piston will move to the right; the feedback link now pivots about the stylus. The movement y of the piston will cause the valve spool to move to the right a distance $Ya/(a+b)$.

Thus, the actual movement of the valve x is given by

$$x = [Zb/(a+b)] - [Ya/(a+b)]$$

If $a = b$, then $x = (Z - Y)/2$. Assuming a zero-lap servo valve, the quantity of fluid flowing q is given by

$q = cx\sqrt{p}$. If p is constant,

$q = K_v x$

where K_v is the valve constant at that particular pressure.

The quantity flowing into the cylinder is given by the product of the piston area A and the piston velocity dY/dt.

$$q = A \frac{dY}{dt}$$

Section 8.2 Valve servo systems

Therefore

$$K_v \, x = A \frac{dY}{dt}$$

$$K_v \frac{(Z-Y)}{2} = A \frac{dY}{dt}$$

This is a first-order differential equation and may be solved by one of the conventional operator substitutions. The Laplace transform will be used, details of which can be found in any mathematical text book covering the solution of differential equations. By applying Laplace transform assuming initial conditions are zero,

$$K_v \, (Z(s) - Y(s))/2 = AsY(s)$$

Separate the variables $Z(s)K_v = 2AsY(s) + K_v Y(s) = Y(s)(K_v + 2As)$

$$Y(s)/Z(s) = K_v/(K_v + 2As) = 1/[1 + (2As/K_v)]$$
$$Y(s)/Z(s) = 1/(1 + \tau s)$$

where τ is known as the time constant and $\tau = 2A/K_v$

8.2.3 System response

The performance of a system is expressed in terms of its response to various types of input.

Response to unit step input

The input to the system will be unity so $Z = 1$. Apply Laplace transform assuming initial conditions are zero.

$$Z(s) = 1/s$$

$$\frac{Y(s)}{Z(s)} = 1/(1 + \tau s)$$

$$Y(s) = (1/s) \, [1/(1 + \tau s)]$$
$$= (1/s)(1/\tau)/[s + (1/\tau)] \tag{8.1}$$

Using partial fractions

$$Y(s) = (1/\tau) \, [(A/s) + (B/(s + 1/\tau)]$$

$$Y(s) = (1/\tau) \frac{As + (A/\tau) + Bs}{s[s + (1/\tau)]} \tag{8.2}$$

Equating coefficients in equations (8.1) and (8.2)

$$A/\tau = 1$$

Therefore,

$$A = \tau$$
$$AS + Bs = 0$$

Therefore

$$B = -A = -\tau$$

$$Y(s) = (1/\tau)\,[(\tau/s) - \tau/(s + 1/\tau)]$$

$$= (1/s) - \left[\frac{1}{s + (1/\tau)}\right]$$

Apply inverse transform

$$Y = 1 - e^{t/\tau}$$

where Y is the output to unity step input (Table 8.1).

The value of the time constant gives an indication of the response time of a first order system. When the value of the time is that of three time constants, the output has reached 95% of its final value.

This can be expressed as a graph (Figure 8.8).

Table 8.1 Step input response.

τ	e^{-t}/τ	$Y = 1 - e^{-t/\tau}$
1	0.368	0.632
2	0.135	0.865
3	0.050	0.950
4	0.018	0.982

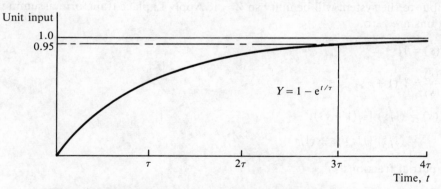

Figure 8.8 Response to step input of a first-order servo system.

Response to a ramp input

This will be a constant velocity input having zero value when the time is zero. Then

Input $Z = Vt$

Section 8.2 Valve servo systems

Using Laplace transform assuming initial conditions are zero:

$$Z(s) = V/s^2$$
$$Y(s)(1 + \tau s) = Z(s) = V/s^2$$
$$Y(s) = V/[s^2(1 + \tau s)] = (V/\tau)/s^2\,[s + (1/\tau)]$$

By partial fractions

$$\frac{1}{s^2[s + (1/\tau)]}$$
$$= A/s^2 + \left[\frac{B/s}{(s + (1/\tau))}\right]$$
$$= \frac{A[s + (1/\tau)] + Bs}{s^2[s + (1/\tau)]}$$

Equating coefficients: $A/\tau = 1$; $A = \tau$; $A + B = 0$; $B = -\tau$.

$$\frac{1}{s^2[s + (1/\tau)]} = (\tau/s^2) - \tau/[s(s + 1/\tau)]$$

$$Y(s) = (V/\tau)\left[(\tau/s^2) - \frac{\tau}{s[1 + (1/\tau)]}\right]$$

$$= V\left[\left(\frac{1}{s^2}\right) - \frac{1}{s[s + (1/\tau)]}\right]$$

From previous work,

$$\frac{1}{s[s + (1/\tau)]} = \tau\left[(1/s) - \frac{1}{[s + (1/\tau)]}\right]$$

$$Y(s) = (V/s^2) - V\tau\left[(1/s) - \frac{1}{[s + (1/\tau)]}\right]$$

Using inverse Laplace

$$Y = Vt - V\tau(1 - e^{-t/\tau})$$

This is shown graphically in Figure 8.9 indicating that in a first-order servo system there

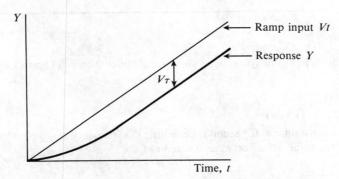

Figure 8.9 Response to ramp (velocity input) of a first-order servo system.

Response to a sinusoidal input (frequency response)

The frequency response of a system is the steady state output of that system when it is subjected to a sinusoidal input of fixed amplitude but variable frequency. The output will be sinusoidal but of a different amplitude and out of phase to the input.

Let the input be $a \sin \omega t$. Then the output will be $b \sin(\omega t + \phi)$. This can be represented as shown in Figure 8.10.

The output lags behind the input on the time axis by an angle $\phi(\omega)$ which is known as the phase angle or phase shift. The value of the phase angle is dependent upon the input frequency. The ratio of the output to the input amplitudes b/a is known as the magnitude ratio $M(\omega)$ and is also dependent upon input frequency. To obtain the frequency response first determine the Laplace transform of system output to system input. This ratio is known as the transfer function for the system $T(s)$. Next substitute $j\omega$ for each value of s in the transfer function where $j = (-1)^{1/2}$. Assuming that the input has an amplitude of unity determine the magnitude ratio and phase angle for various values of frequency.

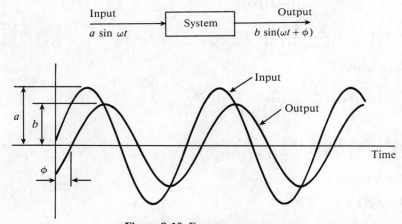

Figure 8.10 Frequency response.

EXAMPLE 8.1

The transfer function for the hydraulic servo copying device shown in Figure 8.8 was found to be

$$T(s) = \frac{Y(s)}{Z(s)} = \frac{1}{(1 + \tau s)}$$

Taking the time constant τ as 0.5 seconds determine the frequency response when the system input is $\sin \omega t$ for values of ωt between 0 and 5 rad/s.

$$\frac{Y(s)}{Z(s)} = \frac{1}{1 + 0.5s}$$

Section 8.2 Valve servo systems

Substitute $j\omega$ for each value of s:

$$\frac{Y(j\omega)}{Z(j\omega)} = \frac{1}{1 + j0.5\omega}$$

Separating the expression into the real and imaginary parts:

$$\frac{Y(j\omega)}{Z(j\omega)} = \frac{1 - j0.5\omega}{(1 + j0.5\omega)(1 - j0.5\omega)}$$

$$= \frac{1}{1 + (0.5\omega)^2} - \frac{j0.5\omega}{1 + (0.5\omega)^2}$$

This can be represented in rectangular form as Figure 8.11a. Changing the expression to polar form:

$$\frac{Y(j\omega)}{Z(j\omega)} = M(\omega)\underline{/-\phi(\omega)}$$

where

$$M(\omega) = \left[\left(\frac{1}{1 + (0.5\omega)^2}\right)^2 + \left(\frac{0.5\omega}{1 + (0.5\omega)^2}\right)^2\right]^{1/2}$$

$$= \frac{(1 + (0.5\omega)^2)^{1/2}}{1 + (0.5\omega)^2}$$

$$\phi(\omega) = -\tan^{-1}\frac{0.5\omega}{1}$$

Table 8.2 Magnitude ratio and phase angle.

ω (rad/s)	$M(\omega)$	$\phi(\omega)$
0	1.000	0.0
1	0.991	−26.6
2	0.705	−45.0
3	0.555	−56.4
4	0.447	−63.4
5	0.372	−68.2
∞	0	∞

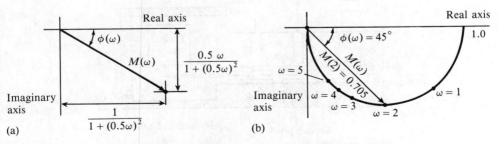

Figure 8.11 (a) Vector diagram. (b) Nyquist diagram.

Taking $\omega = 2$ then

$$M(\omega) = \frac{(1 + (0.5 \times 2)^2)^{1/2}}{1 + (0.5 \times 2)}$$

$$= \frac{1.41}{2} = 0.705$$

and $\phi(\omega) = -\tan^{-1}\frac{1}{1} = -45°$

Calculate values for $M(\omega)$ and $\phi(\omega)$ over the range of values given for ω (Table 8.2). These values can be represented as a polar plot or Nyquist diagram as shown in Figure 8.11(b).

8.2.4 Electro-hydraulic servo valves

The simplest electro-hydraulic servo valve is the single-stage spool type (Figure 8.12) where spool position is directly controlled by an electric torque motor. The spool movement is limited by the torque motor and so the valve is only suitable for small flow rates only.

In a two-stage spool-type servo valve, a directly operated valve is used to control the main spool. A feedback linkage from the main spool to the pilot spool acts in a way similar to the tracer mechanism in Figure 8.7.

To overcome the effects of static spool friction and improve valve response, the spool is kept constantly moving by superimposing a dither signal upon the control signal. Usually this dither signal is a very low-amplitude alternating current at approximately 100 Hz.

In the 'flapper valve' servo, the spool is positioned by fluid pressure on the ends of the spool. Control pressure is varied by a flapper valve opening or blocking nozzles as shown in Figure 8.13.

A 'jet-pipe' servo is similar to the flapper valve. Fluid from a jet is directed into pipes connected to the ends of the spool. A torque motor deflects the jet to vary the pressure on the spool ends, hence controlling the spool position. In the diagram of the jet pipe servo (Figure 8.14) the feedback link from the valve spool to the jet is not shown.

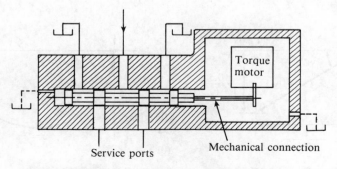

Figure 8.12 Directly operated spool-type servo valve.

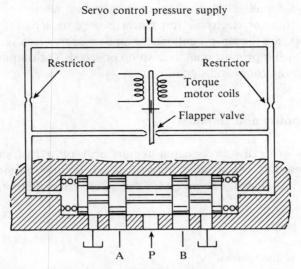

Figure 8.13 Flapper valve servo.

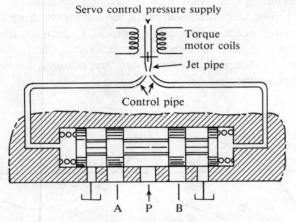

Figure 8.14 Jet pipe servo.

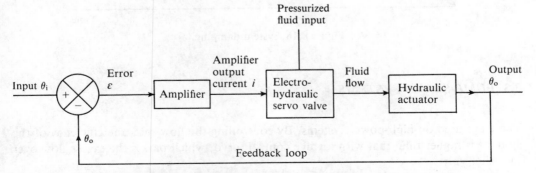

Figure 8.15 Electro-hydraulic servo valve system (block diagram).

Figure 8.15 is a simplified block diagram of a typical electro-hydraulic servo unit. Some form of electrical or electronic transducer is used to derive the feedback signal. Typical transducers are linear potentiometer and synchro resolvers. These give an electrical output which is proportional to position or speed. The output signal is fed back and compared with the command signal.

8.2.5 System response and stability

Rapid response is generally an essential requirement of a hydraulic servo system. However, if the response is too rapid the system can become unstable, in which case damping has to be introduced. The necessary degree of system damping for stability will depend upon:

- The system response
- The load inertia
- The back-lash of the system.

In electrical control systems the amplifier gain can be adjusted to vary the system response. Various damping conditions are shown in Figure 8.16.

A system is said to be 'underdamped' when the output resulting from a step input is oscillatory but ultimately reaching a steady condition. Critical damping is that at which the system just ceases to oscillate and the output achieves a steady state in the minimum possible time. When the damping is greater than that required for critical damping the system is said to be 'overdamped'.

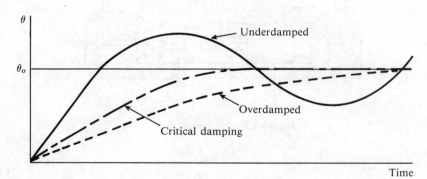

Figure 8.16 System damping.

8.3 PUMP SERVO SYSTEMS

These are used on high-power systems. By controlling the flow rate, the efficiency of the systems is higher than that with a valve control servo, which passes the excess flow over the relief valve.

The majority of pump control servo systems are used to drive hydraulic motors. A

Section 8.3 Pump servo systems

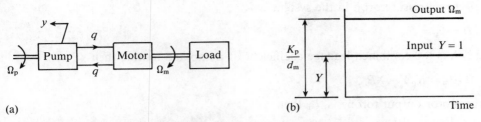

(a) (b)

Figure 8.17 Hydrostatic transmission. (a) Block diagram. (b) Ideal characteristics.

hydrostatic transmission can be represented as a block diagram as shown in Figure 8.17(a).

Assuming ideal conditions for the hydrostatic transmission (that there is no leakage, compressibility etc.),

Flow from pump q_p = Flow to motor $q_m = q$

$$\Omega_p d_p = \Omega_m d_m$$

If Ω_p the pump rotational speed is fixed and d_m the motor displacement per radian is fixed then Ω_m the motor speed is proportional to d_p the pump displacement. **Note:** displacement per revolution D_p and D_m is used elsewhere in this book. Therefore for pumps $D_p = 2\pi d_p$ and for motors $D_m = 2\pi d_m$.

Let the pump control move a distance Y and let the flow from the pump be q_p. Then $q_p = K_p Y$ where K_p is the pump flow constant for a given speed.

Then

$$q_p = q_m = K_p Y = \Omega_m d_m$$

or

$$\Omega_m / Y = K_p / d_m$$

(which can be shown graphically in Figure 8.17(b)). This is obviously unreal because the output cannot instantaneously achieve maximum speed.

8.3.1 Effect of leakage

Consider the effect of leakage at the pump and motor. Let the leakage coefficients for pump and motor be λ_p and λ_m. Let the pressure at the pump and motor be P_p and P_m.

Then pump leakage $= \lambda_p P_p$
Actual flow from pump $= K_p Y - \lambda_p P_p = q$
Motor leakage $= \lambda_m P_m$
Actual flow used in motor $= q - \lambda_m P_m = \Omega_m d_m$

Therefore

$$\Omega_m d_m = K_p Y - \lambda_p P_p - \lambda_m P_m$$

Neglecting pressure drop in the system

$$P = P_p = P_m$$

and if λ is the combined leakage coefficient of pump and motor, then

$$\Omega_m d_m = K_p Y - \lambda P$$

If the motor output torque is T_m then

$$T_m = d_m P = I \frac{d}{dt} \Omega_m$$

where I is the equivalent load inertia at the motor output shaft.

Substituting for P

$$I(d/dt)\Omega_m = d_m (K_p Y - \Omega_m d_m)/\lambda$$

Using Laplace transform

$$Is\Omega_m(s) = \frac{d_m}{\lambda} (K_p Y(s) - \Omega_m(s) d_m)$$

$$\Omega_m(s) \left(Is + \frac{d_m^2}{\lambda} \right) = \frac{d_m}{\lambda} K_p Y(s)$$

$$\frac{\Omega_m}{Y(s)}(s) = (d_m K_p/\lambda)/[Is + (d_m^2/\lambda)] = \frac{K_p}{d_m} (1/(1 + \tau s))$$

where the time constant $\tau = (\lambda I)/d_m^2$.

The response of the system to a step input Y is shown graphically in Figure 8.18.

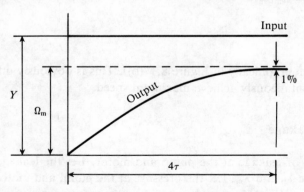

Figure 8.18 Characteristics for hydrostatic transmission considering the effect of pump and motor leakage.

8.3.2 Effect of compressibility

Bulk modulus $B =$ (Volumetric stress)/(Volumetric strain) $= P/(\Delta V/V)$ where P is the pressure, ΔV is the change in volume and V is the original volume of fluid between the

Section 8.3 Pump servo systems

pump and motor. The rate of reduction of the volume is the flow loss caused by compressibility q_c,
where

$$q_c = \frac{d}{dt} \Delta V = \frac{d}{dt}\left(\frac{VP}{B}\right) = \frac{V}{B}\frac{d}{dt}P$$

Actual flow used in the motor is

$$\Omega_m d_m = \text{(Pump delivery)} - \text{(Leakage flows)}$$

$$= K_p Y - \lambda P - V/B \frac{d}{dt} P$$

But

$$P d_m = T_m = I \frac{d}{dt} \Omega_m$$

Therefore

$$P = \frac{I}{d_m} \frac{d\Omega_m}{dt}$$

$$\Omega_m d_m = K_p Y - \lambda \left(\frac{I}{d_m}\frac{d}{dt}\Omega_m\right) - (V/B)\frac{d}{dt}\left(\frac{I}{d_m}\frac{d}{dt}\Omega_m\right)$$

$$= K_p Y - \lambda \left(\frac{I}{d_m}\frac{d}{dt}\Omega_m\right) - \frac{VI}{Bd_m}\left(\frac{d}{dt}\right)^2 \Omega_m$$

Using Laplace transform

$$\Omega_m(s) d_m = (K_p Y(s)) - \left(\frac{\lambda I}{d_m} s\Omega_m(s)\right) - \left(\frac{VI}{Bd_m} s^2 \Omega_m(s)\right)$$

$$\Omega_m(s)\left[d_m + \left(\frac{\lambda I}{d_m} s\right) + \left(\frac{VI}{Bd_m} s^2\right)\right] = K_p Y(s)$$

$$\Omega_m(s)/Y(s) = K_p/[d_m + (\lambda I s/d_m) + (VI s^2/B d_m)]$$

$$\Omega_m(s)/Y(s) = (K_p/d_m)[1/(1 + \lambda I s/d_m^2 + VI s^2/B d_m^2)]$$

$$= \frac{K_p}{d_m}\left[\frac{Bd_m^2/VI}{s^2 + (\lambda Bs/V) + (Bd_m^2/VI)}\right]$$

The characteristic equation can be shown as

$$\frac{\Omega_m(s)}{\lambda(s)} = \frac{K_p}{d_m}[\omega_0^2/(s^2 + 2\zeta\omega_0 s + \omega_0^2)]$$

Thus

$$\omega_0^2 = \frac{Bd_m^2}{VI}$$

and

$$2\zeta\omega_0 = \frac{\lambda B}{V}$$

where ω_0 is the undamped natural frequency of the system and ζ is the damping ratio.

8.3.3 Natural frequency

ω_0 is the undamped natural frequency of the system and from the characteristic equation

$$\omega_0^2 = Bd_m^2/VI$$

So the undamped natural frequency can be increased by the following:

(a) Increasing B the bulk modulus of the fluid. This is normally impossible.
(b) Increasing motor displacement d_m. This is usually selected according to torque and speed requirements. Increasing the value of d_m necessitates an increase in pump delivery to give the same motor speed.
(c) Decreasing the volume of fluid under compression. The hydraulic pump should be placed as near to the hydraulic motor as possible, preferably mounted as an integral unit. Rigid pipework rather than flexible hose should be used between the units to reduce compressibility effect.
(d) The inertia of the load should be reduced to a minimum.

8.3.4 Hydraulic stiffness

The natural frequency can be expressed in terms of the hydraulic stiffness of the system, H_s, defined as the torque required to turn the motor shaft through 1 radian when the pump swash plate is locked to give no output.

$$\text{Bulk modulus of fluid, } B = \frac{\text{Volumetric stress}}{\text{Volumetric strain}}$$
$$= P/(\Delta V/V)$$

where V is the volume of fluid between the pump and the motor and P is the pressure induced when V is increased by ΔV.

The change in volume ΔV when the motor shaft is turned through 1 radian is the motor displacement d_m

$$B = P/(d_m/V)$$

so the induced pressure P is

$$P = (Bd_m)/V$$

The motor torque $T_m = d_m P_m$ where P_m is the pressure difference across the motor. When the motor has been rotated through 1 radian the torque is

$$H_s = d_m P$$

Section 8.3 Pump servo systems

or

$$H_s = \frac{Bd_m^2}{V}$$

but

$$\omega_0^2 = Bd_m^2/VI = H_s/I$$

8.3.5 Damping ratio

The term ζ in the general equation for a second-order system is called the 'damping ratio', which is the ratio of the actual damping coefficient to that required to give critical damping.

From the characteristic equation

$$2\zeta\omega_0 = \lambda(B/V)$$

the damping ratio is dependent on the system's leakage λ and is a function of the natural frequency of the system. When ζ is less than unity the system will be underdamped as shown in Figure 8.19.

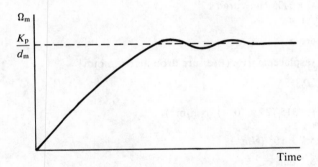

Figure 8.19 Response of system to unit step input.

EXAMPLE 8.2

In a closed-loop reversible hydrostatic transmission having a variable-displacement pump and a fixed-displacement motor, the combined leakage coefficient of the pump and motor is 10×10^{-3} l/min/bar. The total load inertia at the motor shaft is 300 Nm s^2 and the motor displacement is 25 ml/radian. The maximum motor speed is 200 rev/min and the motor is to be capable of accelerating from rest to maximum speed in 20 seconds.

(a) Determine the system working pressure neglecting any pressure drops in the pipework.
(b) Calculate the pump capacity if the pump speed is 1400 rev/min.
(c) If the overall pump efficiency is 85%, what power electric motor is required to drive the pump?
(d) Draw a suitable circuit giving suggested pump sizes and settings of relief valves.

(e) If the maximum movement of the stroke control on the pump is 0.1 m, determine the response of the system (i.e. speed of hydraulic motor) to unit step input after a time equal to 2τ, where τ is the time constant of the system.

(f) Calculate the value of the time constant τ.

Solutions

(a) Consider the motor accelerated to 200 rev/min in 20 seconds, and assuming constant acceleration, then motor acceleration

$$\alpha = \frac{\text{Speed}}{\text{Time}}$$

$$= \left(\frac{200 \times 2}{60}\right) \times \left(\frac{1}{20}\right) \left(\frac{\text{rad}}{\text{s}^2}\right)$$

$$= 1.05 \text{ rad/s}^2.$$

Accelerating torque at motor, T_m is given by

$$T_m = I\alpha$$

where I is the load inertia of 300 Nm s².

$T_m = 300 \times 1.05$ Nm s² rad s⁻²
$= 315$ Nm.

But motor torque is

(Motor displacement) × (Pressure drop across motor)

$$T_m = d_m P_m$$

$$P_m = \frac{T_m}{d_m} = 315/(25 \times 10^{-6}) \text{ (Nm/m}^3\text{)}$$

$$= 125.7 \times 10^5 \text{ (Nm}^2\text{)}$$

$$= 125.7 \text{ bar}.$$

(b) Theoretical pump delivery is

(Motor speed) × (Motor capacity) $(N_m \times d_m \times 2\pi)$
$= (200) \times (25 \times 10^{-6}) \times (2\pi) \text{ (m}^3/\text{min)}$
$= 31.4 \times 10^{-3} \text{ (m}^3/\text{min)}$
$= 31.4$ l/min

Total pump and motor leakage is combined leakage coefficient times system pressure, i.e.

$$10 \times 10^{-3} \times 125.7 \left(\frac{1}{\text{min}} \times \frac{\text{bar}}{\text{bar}}\right)$$

$= 1.26$ l/min.

Actual pump capacity is $31.4 + 1.26 = 32.66$ l/min.

(c) Electric motor power (kW) is

$$\frac{\text{Flow (l/min)} \times \text{Pressure (bar)}}{600 \times \text{Overall pump efficiency}}$$

Section 8.3 Pump servo systems

$$= \frac{32.66 \times 125.7}{600 \times 0.85}$$

$$= 8.04 \text{ kW}$$

(d) A suggested circuit diagram is shown in Figure 8.20(a). Capacity of make-up pump is to be at least twice the combined leakage rate of 1.26 l/min, preferably much more as it is common practise to bleed some of the fluid from the low-pressure side of the circuit back to tank so as to constantly introduce cool clean fluid into the loop. The make-up pump has to overcome the pressure drop in the pipework and the check valve so a make-up relief valve setting of 3 bar is usually adequate. The main relief valve would be set at approximately 10% above the maximum system pressure depending upon its reset characteristics.

(e) The circuit can be represented in block diagram form by Figure 8.20(b).
Let K_p be the pump flow constant, then

$$K_p = \frac{32.66}{60 \times 10^3} \times \frac{1}{0.1} \left(\frac{\text{liters}}{\text{min}} \times \frac{\text{m}^3}{\text{liter}} \times \frac{1}{\text{m}} \times \frac{\text{min}}{\text{s}} \right)$$

$$= 5.44 \times 10^{-3} \text{ m}^3 \text{ s}^{-1} \text{ m}^{-1}.$$

Therefore the pump flow constant K_p is 5.44×10^{-3} cubic metres per second per meter movement of the pump stroke control. Let λ_p and λ_m be the pump and motor leakage coefficients respectively in $\text{m}^3 \text{s}^{-1} \text{bar}^{-1}$.

Pump leakage $= \lambda_p P_p$
Flow from pump $= (K_p Y) - \lambda_p P_p = q$
Motor leakage $= \lambda_m P_m$
Flow used in motor $= q - \lambda_m P_m$
$= (K_p Y) - (\lambda_p P_p) - \lambda_m P_m$

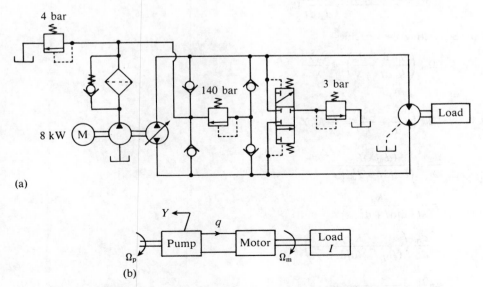

Figure 8.20 Hydrostatic transmission. (a) Circuit. (b) Block diagram.

Neglecting any pressure drops in the pipework,

$$P_m = P_p = P$$

Flow used in motor $= K_p Y - (\lambda_p + \lambda_m)P$. But flow in motor $= \Omega_m d_m$ where d_m is the motor displacement in m^3/rad and Ω_m is the motor speed in rad/s. Thus

$$\Omega_m d_m = K_p Y - \lambda P \tag{8.3}$$

where λ is the combined leakage coefficient of pump and motor.
From equation (8.3),

$$P = \frac{K_p Y - \Omega_m d_m}{\lambda} \tag{8.4}$$

Consider the acceleration α of the motor load having an inertia I when the motor torque is T_m

$$T_m = d_m P$$

Also

$$T_m = I\alpha$$

But

$$\alpha = \frac{d}{dt}\Omega_m$$

$$T_m = d_m P = I \frac{d}{dt}\Omega_m$$

Hence

$$P = \frac{I}{d_m}\frac{d\Omega_m}{dt} \tag{8.5}$$

From equations (8.4) and (8.5)

$$P = \frac{K_p Y - \Omega_m d_m}{\lambda} = \frac{I}{d_m}\frac{d}{dt}\Omega_m$$

Separating the variables

$$\Omega_m \left(\frac{I}{d_m}\frac{d}{dt} + \frac{d_m}{\lambda}\right) = \frac{K_p Y}{\lambda}$$

or

$$\Omega_m \left(I\frac{d}{dt} + \frac{d_m^2}{\lambda}\right) = \frac{d_m}{\lambda} K_p Y$$

$$\frac{\Omega_m}{Y} = \frac{(d_m/\lambda)K_p}{(I\,d/dt + d_m^2/\lambda)}$$

$$= \frac{d_m K_p}{\lambda(I\,d/dt + d_m^2/\lambda)}$$

$$= \frac{d_m}{d_m^2}\left(\frac{K_p}{(I\lambda/d_m^2)(d/dt) + 1}\right)$$

$$= \frac{K_p}{d_m}\frac{1}{[1 + (I\lambda/d_m^2)\,d/dt]}$$

Section 8.3 Pump servo systems

Using Laplace transform

$$\frac{\Omega(s)}{Y(s)} = \frac{K_p}{d_m}(1/(1+\tau s))$$

where τ the time constant is given by

$$\tau = \frac{I\lambda}{d_m^2}$$

thus

$$\Omega(s) = \frac{K_p}{d_m}[1/(1+\tau s)]Y(s)$$

Consider the response to unit step input, i.e. $Y = 1$. By Laplace transform if $Y = 1$,

$$Y(s) = \frac{1}{s}$$

$$\Omega_m(s) = \frac{K_p}{d_m}\left[\frac{1}{(1+\tau s)}\right](1/s)$$

Applying partial fraction to

$$\frac{1}{(1+\tau s)}(1/s)$$

gives

$$\frac{1}{(1+\tau s)s} = \frac{A'}{(1+\tau s)} + (B'/s)$$

$$= \frac{A's + B'(1+\tau s)}{(1+\tau s)s}$$

Equate coefficients A' and B', $B' = 1$ and $0 = A' + B'\tau$, $A' = -\tau$

$$\Omega_m(s) = \frac{K_p}{d_m}\left[(1/s) - \frac{\tau}{(1+\tau s)}\right]$$

$$= \frac{K_p}{d_m}\left[(1/s) - \frac{1}{(1/\tau + s)}\right]$$

Apply inverse Laplace transform standard forms which gives

$$1/s = 1$$

$$\frac{1}{(A+s)} = e^{-At}$$

where $A = \tau^{-1}$

$$\Omega_m = \frac{K_p}{d_m}(1 - e^{-t/\tau})$$

When $t = 2\tau$

$$\Omega = \frac{K_p}{d_m}(1 - e^{-2\tau/\tau})$$

and

$$K_p = 5.44 \times 10^{-3}\,\text{m}^3\,\text{s}^{-1}\,\text{m}^{-1}$$

This is the value of K_p for a movement of the control lever of 1 m; as the control lever has a maximum travel of 0.1 m, the maximum value K_p can have is

$$K_p(\text{max}) = 5.44 \times 10^{-3} \times 0.1 \text{ m}^3 \text{s}^{-1}$$

(Note as K_p has been multiplied by the maximum value of Y the units of K_p (max) become m^3s^{-1}.)

The motor displacement d_m is given as 25 ml/rad = $25 \times 10^{-6} \text{m}^3 \text{rad}^{-1}$.
The motor speed Ω_m after a time 2τ is given by

$$\Omega_m = \left[\frac{5.44 \times 10^{-3} \times 0.1}{25 \times 10^{-6}}\right] (1 - e^{-2}) \left(\frac{\text{m}^3}{\text{s}} \times \frac{\text{rad}}{\text{m}^3}\right)$$

$$= 18.8 \text{ rad s}^{-1}$$

$$= \frac{18.8}{2\pi} \times 60 \text{ rev/min}$$

$$= 179.6 \text{ rev/min}$$

(f) The time constant τ for the system is given by

$$\frac{I\lambda}{d_m^2}$$

$I = 300 \text{ Nm s}^2$, $\lambda = 10 \times 10^{-3}$ (1 $\text{min}^{-1} \text{bar}^{-1}$), i.e.

$$\frac{10 \times 10^{-3}}{10^3} \times \frac{1}{60} \times \frac{1}{10^5} \text{ (m}^3\text{s}^{-1}\text{N}^{-1}\text{m}^2\text{)}$$

so

$$\lambda = \frac{1}{60} \times 10^{-10} \text{ (m}^5\text{s}^{-1}\text{N}^{-1}\text{)}$$

$d_m = 25 \times 10^{-6} \text{ m}^3$

$$\tau = 300 \times \frac{1}{60} \times 10^{-10} \times 1/(25 \times 10^{-6})^2 \text{ (Nm s}^2\text{m}^5\text{s}^{-1}\text{Nm}^{-6}\text{)}$$

$$= \left(\frac{300}{60}\right) \times \left(\frac{1}{625}\right) \times \left(\frac{10^{-10}}{10^{-12}}\right) \text{ (seconds)}$$

$$= 0.8 \text{ s}$$

τ is therefore 0.8 s.

8.4 PROPORTIONAL VALVES

8.4.1 Force control

Conventional electro-hydraulic valves have digital control systems, in so far as a solenoid is either on or off, the valve spool moving to one of a discrete number of positions. A two-position, two-port, solenoid-operated directional control valve is either fully open, or, when the solenoid is energized, fully closed. This 'bang-bang' or digital operation gives rise to flow and pressure surges in the hydraulic circuit with all the resultant problems.

Section 8.4 Proportional valves

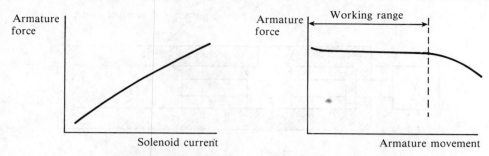

Figure 8.21 Proportional solenoid characteristics.

If the valve can be gradually closed – as can, for example, a manually-operated gate valve – a gradual transition between open and closed results. In process control plant, the openings of valves in large pipe lines are adjusted using semi-rotary actuators, electric motor and gear box drives, stepping motors etc. thus obtaining gradual flow changes and eliminating shocks. Electric motor-driven valves have been used in hydraulic systems, but they are expensive and very limited in application.

The output force exerted by the armature of a DC solenoid is dependent upon the current flowing through it. This can be utilized in the design of a proportional DC solenoid in which the force exerted by the armature is proportional to the current flowing and independent of the armature movement over the working range of the solenoid. Typical characteristics of a proportional solenoid are shown in Figure 8.21.

8.4.2 Force position control

The electrical control to the proportional valve normally uses a variable current rather than a variable voltage. Should a voltage control system be adopted, any variation in coil resistance, caused by a temperature change, will result in a change of current although the voltage remains fixed. This problem is eliminated by using a current control system.

Thus it is possible to control a force electrically and by applying the force to a compression spring its deflection can be controlled. If the spool in a valve as shown in Figure 8.22(a) is acted on by a spring at one end and a proportional solenoid at the other, the orifice size A can be varied in accordance with the control current.

The flow from the valve will be proportional to the current flowing through the

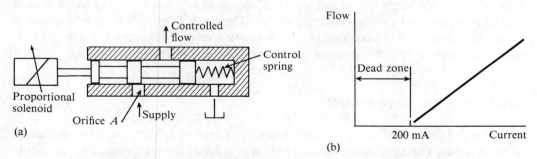

Figure 8.22 Proportional control valve. (a) Diagrammatic section. (b) Flow current characteristics.

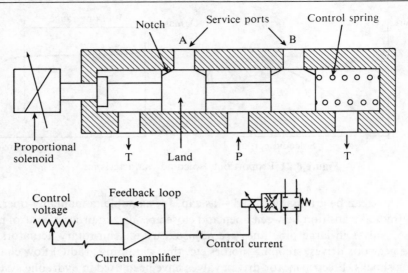

Figure 8.23 Notched spool proportional valve.

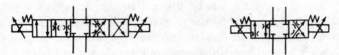

Figure 8.24 Proportional directional control valve symbols.

solenoid. Because of difficulties in manufacturing a zero lap spool, i.e. one in which the land on the spool is exactly the same length as the port in the valve body, overlapped spools are used in proportional spool valves. This means that the spool has to move a distance equal to the overlap before any flow occurs through the valve; this gives rise to a 'dead zone' in the valve characteristic as shown in Figure 8.22(b)

Using notched spools in the proportional valves gives better control of the flow rate as the orifice is progressively opened, the orifice opening and hence the maximum flow rate through the valve being determined by the notch shape. A diagrammatic sketch of the notched spool valve with its symbolic representation is shown in Figure 8.23 together with an electrical control diagram.

A proportional directional-control valve with double solenoid and spring-centered is very similar to the valve shown in Figure 8.23 except that it will have a solenoid at each end of the spool and a spring-centering device. The symbol for such a valve is shown in Figure 8.24 as either a five-position or a three-position valve; both symbols are in common use. The extremes on a five-position valve represent fully-operated conditions.

8.4.3 Spool positional control

In order to increase the accuracy, and extend the range of applications of proportional control valves, a linear transducer may be fitted to measure the spool position. The output

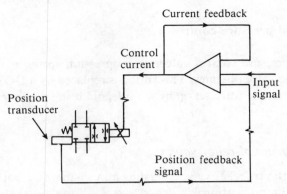

Figure 8.25 Position feedback.

from the transducer is a voltage which is proportional to the spool displacement, continuously varying through the total spool movement. The actual position of the spool is fed back via the transducer to the electrical control system and compared with the required position, the control current being adjusted accordingly. Such a system is shown in Figure 8.25.

In both the force position control of directional valves, the spool opening and hence the flow rate is controlled. The transducer used for position feedback of the spool does not monitor the quantity of fluid flowing through the valve so it is essentially an open-loop control system. Should additional accuracy be necessary it is possible to use a transducer to measure the system output and feed this back to the control circuit. In the speed-control circuit for the hydraulic motor shown in Figure 8.26 a tachogenerator or similar device is used to measure the motor speed. If such a circuit is used, the effect of the 'dead zone' in the characteristics of the proportional valve must be taken into account. This will be more critical in the case of position control rather than speed control.

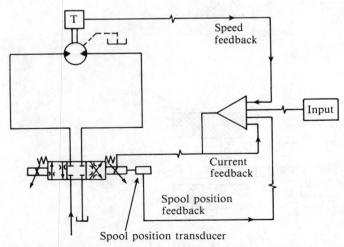

Figure 8.26 Closed-loop speed control with a proportional valve (electrical circuit is shown to one solenoid only).

8.4.4 Proportional pressure control

In a conventional pressure control valve, a compression spring is used to control the pressure at which the valve operates. This spring is replaced by a DC solenoid in the case of proportional valves, the force set up by the solenoid being dependent upon the current flowing through it.

Single-stage proportional-relief valves

Direct-acting proportional-relief valves are shown in Figure 8.27 both diagrammatically and symbolically. The proportional solenoid exerts a force on the poppet keeping the valve closed, until the hydraulic pressure at port P overcomes this force and opens the valve.

In the design of relief valve shown in Figure 8.27(a), the proportional solenoid acts directly on the valve poppet. An alternative configuration in Figure 8.27(b) uses the solenoid to tension the control spring. The force exerted by the proportional solenoid has an upper limit owing to the physical size limitations. So to increase the operating pressure of the valve the size of the orifice in the valve seat is decreased and vice versa. The operating pressure of the valve will depend on the current in the solenoid and the quantity

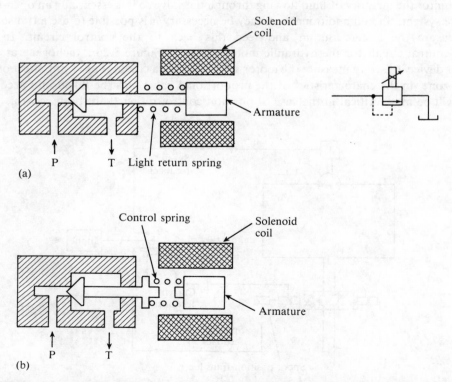

Figure 8.27 Direct acting proportional relief valves (pilot stage). (a) Solenoid acts on valve poppet. (b) Solenoid tensions the control spring.

Section 8.4 Proportional valves

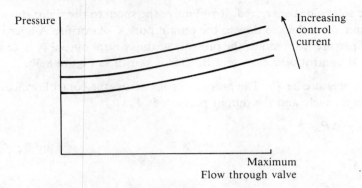

Figure 8.28 Pressure/flow characteristics for direct acting proportional relief valve.

of fluid flowing through the valve. Figure 8.28 shows the relationship between pressure set by the valve and the flow through the valve for three different control currents.

Proportional pressure-reducing valves

This operates in a similar manner to a conventional pressure regulating valve, the control spring being replaced by a proportional solenoid. However, when this solenoid is not energized, the proportional valve is closed unlike the conventional pressure reducing valve which is normally open. The output pressure of the valve shown diagrammatically in Figure 8.29 is proportional to the current flowing through the solenoid.

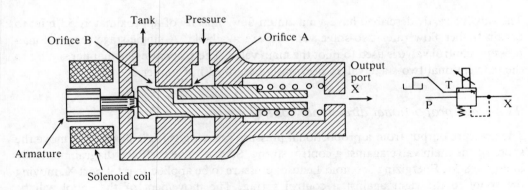

Figure 8.29 Proportional pressure-reducing valve.

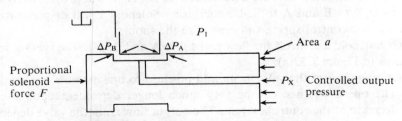

Figure 8.30 Principle of pressure-reducing valve.

When the solenoid is energized, it will move the spool to the right; the control orifice A will open and allow fluid to flow to the output port X. As orifice A aperture increases, the orifice B aperture will reduce; the pressure at the control output X is dependent upon the openings of control orifices A and B. This is shown in Figure 8.30.

Let the supply pressure be P_1. The pressure drops across the control orifices A and B are P_A and P_B respectively and the output pressure is P_X.

$$P_1 = \Delta P_A + \Delta P_B$$

and

$$P_X = \Delta P_B$$

If the control orifice B is fully closed, then P_X will equal the supply pressure P_1. The output pressure is applied to the right-hand end of the spool and if this is greater than the equivalent pressure exerted by the proportional solenoid, the spool will move to the left. This increases the opening of orifice B and reduces orifice A so reducing the output pressure. Therefore this is a relieving-type pressure-reducing valve. For equilibrium, $P_X a = F$. The output pressure is proportional to the current flowing in the proportional solenoid. There will always be a flow to tank from this type of valve if the output pressure P_X is less than the supply pressure P_1. It is essential that there is no back-pressure in the tank line if the valve is to function correctly.

8.4.5 Two-stage proportional valves

The valves already described have a maximum flow capacity of approximately 5 l/min; to obtain higher flow rates two-stage versions are available. A single-stage proportional-pressure control valve is used to pilot the main valve. These operate in a manner similar to the conventional two-stage valves previously described in Chapter 3.

Two-stage proportional directional control valve

The pressure output from a proportional pressure-reducing valve is directed to move the spool of the main valve against a control spring. Such a valve is shown diagrammatically in Figure 8.31. Energizing Solenoid 1 causes pressure to be applied to pilot port X, moving the spool to the right against a control spring. The movement of the spool will be proportional to the pressure applied to pilot port X and hence to the current in Solenoid 1. As the main spool lands are notched a movement to the right will progressively open the flow paths from P to B and A to T. De-energizing Solenoid 1 will de-pressurize spring chamber C and the control spring will centralize the spool.

Similarly Solenoid 2 controls the flow paths P to A and B to T. The symbol for such a valve is shown in Figure 8.32(a).

The operating time of the valve from mid-position to one extreme is a minimum of 40–60 ms. The operating time can be very much longer dependent upon the rate of increase or decrease of the control current. The output flow from the valve depends upon

Section 8.4 Proportional valves

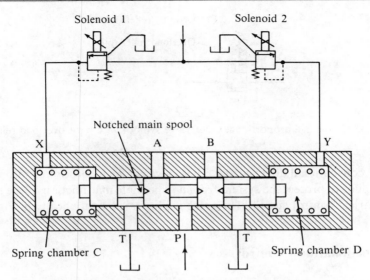

Figure 8.31 Two-stage proportional directional control valve.

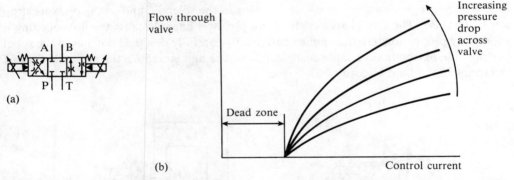

Figure 8.32 Two-stage directional control. (a) Symbol. (b) Typical flow characteristics (one path).

the pressure drop across the valve and the control current in the solenoid; a typical characteristic is shown in Figure 8.32(b) for one flow path through the valve.

Two-stage proportional relief valve

This is similar to a conventional two-stage relief valve which is described in detail in Sections 3.1.1 and 3.4.1, but with a proportional pilot relief valve controlling the main spool.

System pressure is applied via the control orifice to the pilot stage and when the pressure exceeds the force generated by the proportional solenoid, the pilot stage opens. This causes a flow across the control orifice with a resultant pressure drop. The pressures on the main spool are out of balance and so the spool lifts relieving the fluid.

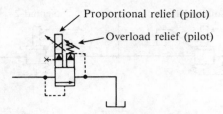

Figure 8.33 Two-stage proportional relief valve symbol with pilot overload relief valve.

A small, conventional direct-acting relief valve can be incorporated into the design as an overload pilot to protect the system from any possible malfunction of the proportional valve or electrical control circuit. The valve is illustrated symbolically in Figure 8.33.

8.4.6 Proportional flow control

Small flows may be controlled by using one pair of ports of a four-port, two-position proportional directional-control valve. For higher flows, two ports may be coupled together. These methods of connection are illustrated in Figure 8.34. The flow through the valve will be proportional to the current flowing in the solenoid and to the pressure drop across the valve. The flow characteristic is not precisely linear because the flow opening is not exactly proportional to the applied current. By carefully designed notches in the spool it is possible to obtain a variable sharp-edged orifice and so reduce the effect on flow of variations in the fluid viscosity.

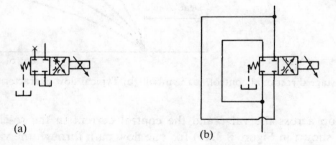

Figure 8.34 Connection of four-port proportional directional-control valve. (a) Single flow path. (b) Double flow capacity.

Pressure-compensated proportional-flow control

If a constant pressure drop is maintained across the flow control valve orifice, the flow through the valve will be independent of any upstream or downstream pressure variations. This is achieved by using a pressure-compensating cartridge as employed in conventional flow-control valves. The compensator can be considered as a remotely-operated pressure-control valve, which continuously varies its orifice opening to maintain a fixed pressure drop over the flow-control orifice. Such an arrangement is shown as a graphical symbol in

Section 8.4 Proportional valves

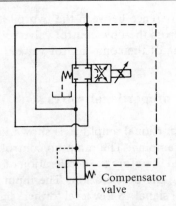

Figure 8.35 Pressure-compensated proportional flow control.

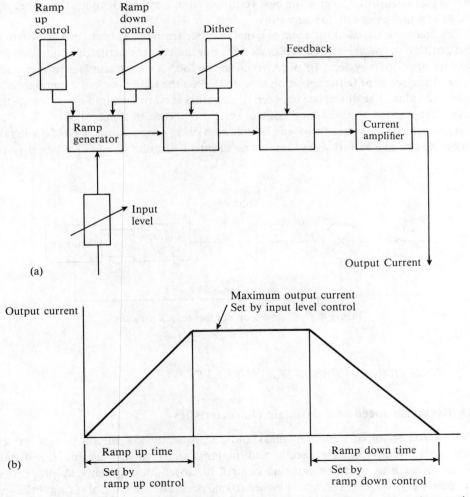

Figure 8.36 Proportional amplifier. (a) Block schematic diagram. (b) Current–time characteristics.

Figure 8.35. The pressure difference between the two pilots on the compensator valve, which is the pressure drop across the flow control valve orifice, is equivalent to the fixed force set by the control spring of the compensator valve.

8.4.7 Electrical control of proportional valves

A block schematic for a proportional amplifier is shown in Figure 8.36 together with a current/time graph showing ramping. The ramp-up control determines the rate at which the control signal increases and therefore the acceleration of the actuator. The ramp-down control is similar, controlling the deceleration. The input level control determines the maximum value of the control signal. A low level 'dither' signal is superimposed on to the control signal. The dither signal is an AC level at approximately 100 Hz. Its function is to keep the spool oscillating to overcome the effect of static friction. The amplitude of the dither signal is adjusted to give the best response from a system without the dither signal causing any movement of the actuator.

The feedback signal to the control panel is either from the output current or from the spool position in the valve. The feedback does not indicate the actuator conditions and it is still an open-loop system. In order to close the loop a transducer has to measure the output of the actuator (either position or speed), feed the signal back and compare it with the desired value. The difference between these values is converted into a new input signal. A block diagram of a closed-loop control system is shown in Figure 8.37.

Although closed-loop control can be achieved using proportional valves, it will not be as accurate or have as fast a response as an electro-hydraulic servo-valve based system.

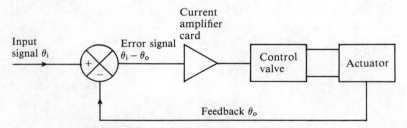

Figure 8.37 Closed-loop control (block diagram).

8.5 PROPORTIONAL VERSUS SERVO VALVES

8.5.1 Response speed and dynamic characteristics

A short travel spool of minimum mass and consequently low inertia is used in servo valves, giving high response speeds and making servo valves suitable for dynamic applications such as speed or position control in closed-loop systems. A proportional valve spool has a longer travel and is biased to one position by a control spring – the spool and spring combination having a much higher inertia than the equivalent servo valve. Application of a dither signal will reduce the effects of spool 'stiction' and inertia in both proportional and servo valves. Its value is usually adjustable and is set to give maximum response speed without any flow or pressure fluctuations being set up by the dither current.

8.5.2 Hysteresis effect

Spool position in a servo valve is controlled by a nozzle and flapper or a jet pipe system with a feedback link correcting for the spool position. A proportional valve relies on the force exerted by a DC coil acting against a spring to position the spool. There is a considerable difference in the valve output depending upon whether the current is increasing or decreasing. Proportional valves have a much higher hysteresis than servo valves.

8.5.3 Null position

Because of the underlap spool, a very slight change in control current will vary the output of a servo valve about the zero flow position. In a proportional valve there will be no output until the control current exceeds about 200 mA which is required to overcome the spring preload and the effect of using an overlap spool.

A comparison between proportional and hydraulic servo valves is given in Table 8.3. Where fast response and accurate control are required, servo valves are best suited. However proportional valves which are much more dirt-tolerant provide an economical and satisfactory alternative for many applications.

Table 8.3 Comparison of proportional and servo valves.

Parameter	Proportional hydraulic valve	Electrohydraulic servo valve
Valve lap	Overlap spool, causing a 'dead zone' on either side of the null point	Zero or underlap valve spool. No dead zone
Response time for valve spool to move fully over	40–60 ms	5–10 ms
Maximum operating frequency	Approx. 10 Hz	Approx. 100 Hz
Hysteresis	Without armature feedback approx. 5%	Approx. 0.1%
	With armature position feedback approx. 1%	

8.6 SOME APPLICATIONS OF PROPORTIONAL CONTROL VALVES

8.6.1 Control of actuators

Speed control of cylinders

The conventional speed control of cylinders is by meter-in, meter-out or spill-off flow control valves. This sets a cylinder speed which can only be manually varied. Alternatively, a cam drive can progressively close or open an adjustable orifice in accordance with a pre-set speed profile which is only changed by altering the cam profile. Examples of these types of circuit are shown in Figure 8.38 and are described in detail in Chapter 3. The acceleration and retardation of a cylinder can be controlled by:

1. Relief valves limiting the maximum pressure available to accelerate the load.
2. Using a two-stage directional-control valve with a choke pack to control the speed of movement of the main spool.
3. Using a variable-displacement pump.
4. Internal cylinder cushions or external shock absorbers can be used to decelerate the cylinder.
5. Brake valves, deceleration valves and counterbalance valves can be built into the circuit to control the deceleration and sometimes the acceleration of the actuator.

All these methods require manual adjustment and are incapable of continuous variation whilst the system is operating.

A proportional directional-control valve in the cylinder circuit enables continuous regulation of speed, acceleration and deceleration. If a control card is used to drive the valve, any adjustments to maximum current, ramp-up and ramp-down have to be carried out by adjusting potentiometers on the card. However, a microprocessor or minicomputer may be employed to control the proportional valve by varying solenoid current over different parts of the cycle.

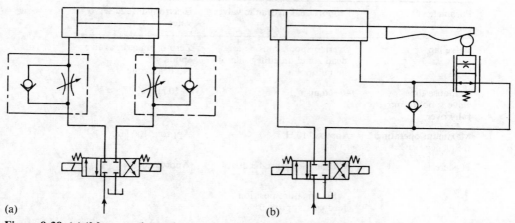

Figure 8.38 (a) 'Meter-out' speed control. (b) Cam-operated speed control (on piston extend stroke only).

Speed control of hydraulic motors

This is very similar to the speed control of cylinders, but it is relatively simple to monitor the motor speed and use a feedback system to control the proportional solenoid as was seen in Figure 8.26. The speed of response to changes in load or command will limit the applications. For high-response speeds and accurate control, servo valves must be used.

Position control of hydraulic cylinders

In order to control the position of the piston rod, a transducer has to be used to monitor the actual position. The output of the transducer is compared with the desired piston rod position and the difference fed to the current amplifier and then to the proportional solenoid. The output of the current amplifier has to be biassed so that any error signal which it receives will provide a sufficiently large output to drive the proportional valve out of the dead zone. Otherwise the system becomes unstable.

8.6.2 Pump control systems

In an ideal hydraulic system the output of the power pack in terms of quantity flowing and maximum pressure is matched exactly to the system demand. This situation is rarely ever achieved. A proportional relief valve can be used as the main relief valve to set the maximum pressure and the setting remotely varied with the system sequence. A secondary relief valve should be fitted into the system as a safety feature in case of failure of the proportional relief valve. Supply pressure is matched to system demand by controlling the proportional relief valve. It will only be useful if the flow requirement of the system is almost constant.

In order to match the pump delivery to system demand a variable-displacement pump has to be used. One possibility is to use a pressure-compensated pump with a proportional relief valve acting as the compensator (Figure 8.39). This will give the characteristics of a pressure-compensated pump with its compensator meeting the system flow demand at the set pressure. The circuit will have to include flow-control valves to give the actual flow requirement and it therefore has a restricted versatility. This method may be adopted

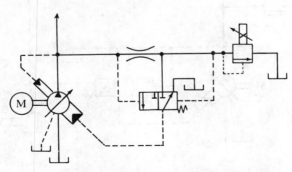

Figure 8.39 Pressure compensated pump with proportional pressure control.

where a series of actuators demanding different pressure and flows are operated sequentially.

In any system using flow-control valves there will be an associated pressure drop which can only be eliminated by removing the valves. In this case, the full pump output has to be utilized at all times. Where the flow demand varies within different parts of the sequence, this can be achieved by varying the pump displacement. Electric motor drives and cam drives are used to remotely vary the pump displacement, but both have limitations – the motor drive in response speed and the cam drive in versatility. An alternative is a flow-controlled pump with a proportional flow control valve and a proportional pressure control (Figure 8.40). The pump output can therefore be exactly matched to system demand in terms of both pressure and flow. This results in a particularly power-efficient system with little heat generation.

With valve A in the closed position as shown in Figure 8.40, the directional valve C will be piloted to the right, opening the large pump control piston to tank. The small piston causes the pump to move to zero displacement. When a current is applied to the proportional solenoid on valve A, the orifice partially opens and fluid flows providing the system pressure is below the setting of valve B. The pressures occurring across the orifice formed are applied to the pilots of valve C causing the pump displacement to be increased. When the pressure difference across the orifice in valve A balances the control spring pressure of valve C, valve C spool centralizes locking the displacement of the pump.

System pressure is set by the proportional relief valve B. If the system pressure is greater than the setting of valve B, the valve opens causing a reduced pressure on the right-hand pilot of valve C which moves to the right opening the large pump control piston to tank. Pump displacement reduces until the system pressure matches that set by the proportional control valve B, which balances valve C locking the pump displacement.

Thus in this system, the pump delivery and pressure can be matched to the system by remotely operating the proportional flow and proportional pressure-control valves A and B.

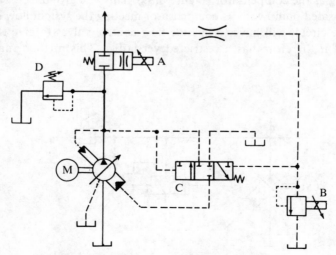

Figure 8.40 Pump with proportional pressure and flow control.

Section 8.6 Some applications of proportional control valves

The response time of the pump from maximum to zero flow will be in the order of 50–100 ms. If a pressure surge occurs in the system the pump may not respond quickly enough to reduce the surge. It is therefore prudent to fit a conventional pressure relief valve D to cater for pressure surges. This relief valve should be set approximately 20% above the maximum operating pressure of the system.

The precise continuous regulation of flow, pressure and displacement with consequential control of speed, thrust, position, acceleration and retardation achievable from modern servo and proportional systems has made hydraulics indispensable in the field of modern drive and control techniques. The special characteristics of this sophisticated equipment involves electronic circuitry matched to individual components, therefore system design and construction is naturally much more complex than for electro-hydraulic digital control. Commissioning a prototype can be time consuming and in order to carry out repairs maintenance personnel must be better trained than was the case with conventional hydraulic equipment. Nevertheless these developments have presented the hydraulic design engineer with many exciting opportunities, and extended application of the subject into new areas.

APPENDIX
EXERCISES AND SOLUTIONS

This appendix is designed to test your understanding of power hydraulics. It is divided into three sections

A.1 Circuitry questions
A.2 Hydraulic calculations
A.3 Design problems requiring calculations and circuit design.

Answers to the calculations in Section A.2 may be found in Section A.4 at the end of the appendix.

 A list of standard metric cylinders appears as Table 4.1 in Chapter 4. In all piston rod strut strength calculations assume a factor of safety of 3.5 and take the modulus of elasticity as 2.1×10^6 kg cm^{-2}. Effective strut lengths for the various types of cylinder mounting may be obtained from Figure 4.16 also in Chapter 4.

A.1 *CIRCUITRY QUESTIONS*

Examine the following circuits. Try to understand their operation and the function of each component. Some of the circuits contain deliberate draughting errors – see if you can find and correct them. Where appropriate suggest simple modifications which could improve the performance of any of the circuits. Answer the additional questions included in some of the problems. To simplify the circuit filters have been omitted from most of the examples. There is generally insufficient detail of the particular application to be precise but where you feel that filtration would be essential make recommendations. It is assumed that in all cases there will be a suction strainer on the pump.

A.1.1

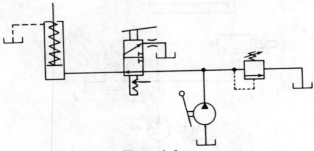

Figure A.1

Correct the draughting error.

A.1.2

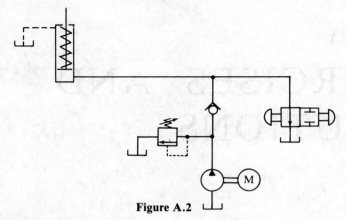

Figure A.2

Correct the draughting error.

A.1.3

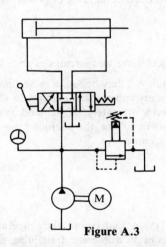

Figure A.3

In the circuit shown in Figure A.3, given that the load induced pressure on the extend stroke is 150 bar and on the retract stroke is 70 bar, and the relief valve pressure setting is 165 bar, draw a pressure–time graph for a complete extend and retract cycle. The graph should start and end with the directional control valve in the mid position. Assume the extend stroke takes 12 seconds and the retract stroke 9 seconds.

A.1.4

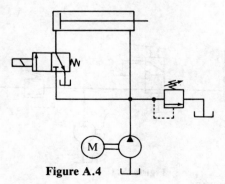

Figure A.4

Section A.1 Circuitry questions

What are the characteristics of this circuit and in what circumstances is it most usually employed?

A.1.5

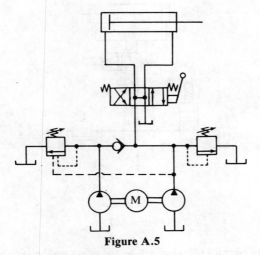

Figure A.5

Correct the draughting error in this figure.

What advantage has this two-pump circuit over a single pump system? Name a particular application where it is frequently used.

A.1.6

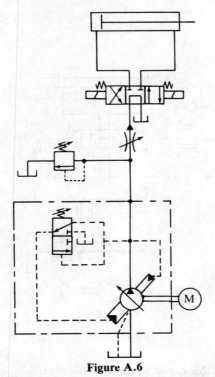

Figure A.6

What advantage has a pressure compensated pump in this application? How can the efficiency of the circuit be improved?

A.1.7

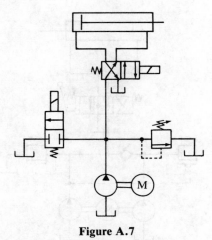

Figure A.7

Should the control circuit fail, what will be the effect on the system? Modify the circuit to 'fail safe'.

A.1.8

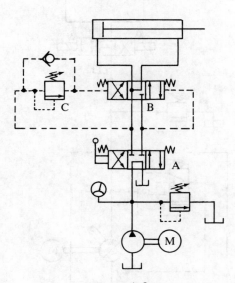

Figure A.8

The cylinder shown in this circuit which contains draughting errors cannot be retracted. Using standard valves modify the circuit so that the cylinder can be retracted by the manual directional control valve. The extend function must, however, be unchanged.

Section A.1 Circuitry questions

A.1.9

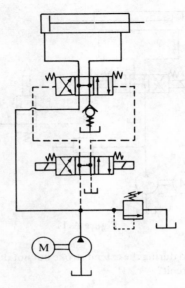

Figure A.9

What is special about the check valve in this circuit and what is its purpose?

A.1.10

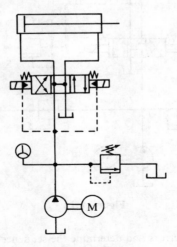

Figure A.10

When this circuit was built it would not operate. Why? Suggest a simple modification to correct this design fault.

A.1.11

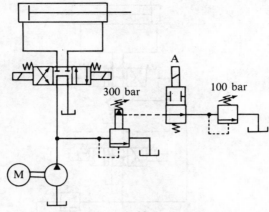

Figure A.11

Solenoid A is energized during the extend stroke but not during retract. How does this affect the function of the circuit?

A.1.12

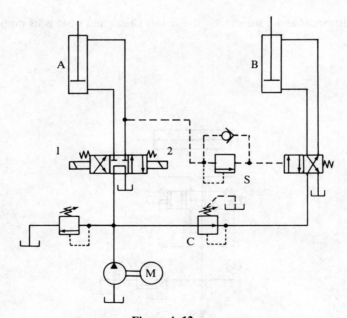

Figure A.12

Correct the draughting errors and determine the sequence of operation of the cylinders:

(a) When solenoid 1 is energized.
(b) When solenoid 2 is energized.

What is the purpose of valve C?

Section A.1 Circuitry questions

A.1.13

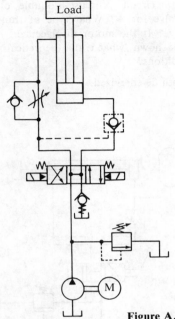

Figure A.13

This circuit is intended to hydraulically lock a loaded cylinder in any position and to control the retract speed. Comment on its suitability and modify if necessary. Suggest a pressure setting for the relief valve if the load is 10 tonnes and the cylinder bore 100 mm diameter.

A.1.14

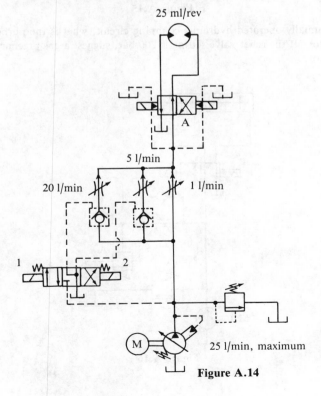

Figure A.14

Explain the function of the circuit. Why use a double solenoid valve rather than a single solenoid spring offset valve for A? What type of transition centre condition do you recommend for this valve? If the motor displacement is 25 ml/rev, and with the flow controls set at the values shown, what is the theoretical maximum motor speed possible under the following conditions?

(a) Solenoids 1 and 2 both de-energized.
(b) Solenoid 1 energized.
(c) Solenoid 2 energized.

A.1.15

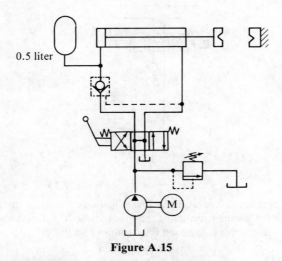

Figure A.15

In this manually-operated hydraulic clamping circuit, what is the purpose of the 0.5-liter accumulator? If the relief valve setting is 200 bar, suggest a gas precharge pressure for the accumulator.

A.1.16

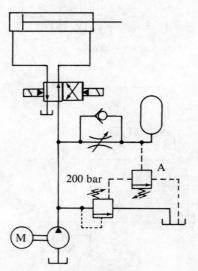

Figure A.16

Section A.1 Circuitry questions

The cylinder in the circuit is required to give a single high-speed stroke at infrequent intervals. Comment on its suitability and suggest any necessary modification. What is the function of the flow control valve? Suggest a pressure setting for valve A.

A.1.17

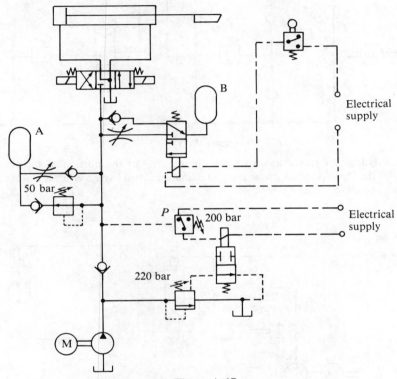

Figure A.17

A cylinder is required to extend initially at high speed but under low load. When the limit switch is actuated the stroke is completed at low speed and high load. Explain how this is achieved by the above circuit. Why are two accumulators used? What is the purpose of the pressure switch P?

A.1.18

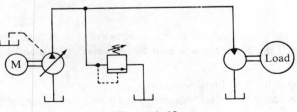

Figure A.18

This non-reversing hydrostatic transmission has a high inertia load and is found to over-run as the flow rate is adjusted to reduce speed. Suggest a suitable modification.

A.1.19

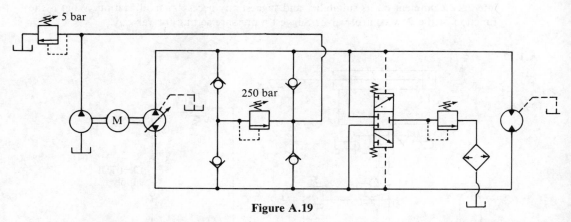

Figure A.19

In this hydrostatic transmission what is the purpose of the portion of the circuit which includes the three-position, three-port directional control valve? What is a suitable setting for the associated relief valve?

A.1.20

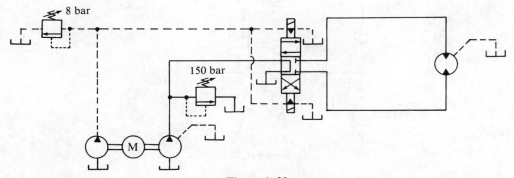

Figure A.20

This hydrostatic transmission is subject to pressure surges when the directional control valve is switched. Suggest modifications to alleviate these surges.

A.2 HYDRAULIC CALCULATIONS

A.2.1 A pump having a displacement (swept volume) of 1.7 cm^3 per revolution is driven at 1500 rev/min. If the pump has a volumetric efficiency of 87% and an overall efficiency of 76% calculate:

(a) The pump delivery in liters per minute.
(b) The power required to drive the pump when it is operating against a pressure of 150 bar.

A.2.2 A gear pump delivers 15 l/min against a system pressure of 200 bar at a driven speed of 1430 rev/min. If the input is 6.8 kW and the torque efficiency of the pump is 87% calculate the pump displacement in cm^3 per revolution.

Section A.2 Hydraulic calculations

A.2.3 A hydraulic system requires 32 l/min of fluid at a pressure of 260 bar. The pump to be used is a manually-variable axial piston pump having a maximum displacement per revolution of 28 cm^3. The pump is driven at 1430 rev/min and has an overall efficiency of 0.85 and a volumetric efficiency of 0.90. Calculate

(a) At what percentage of maximum displacement the pump has to be set.
(b) What power is needed to drive the pump.

A.2.4 A hydraulic pump having a displacement of 8.8 ml/rev runs at 2880 rev/min. If its volumetric efficiency and torque efficiency are 93% and 91% respectively determine:

(a) The actual pump delivery.
(b) The input power to the pump when it operates against a pressure of 350 bar.

A.2.5 A hydraulic circuit using 25 liters of fluid per minute is supplied by a pump having a fixed displacement of 12.5 cm^3/rev driven at 2880 rev/min. The pump has a volumetric efficiency of 0.85 and an overall efficiency of 0.75. If the system pressure is set at 180 bar by the relief valve calculate:

(a) The quantity of fluid delivered by the pump.
(b) The power required to drive the pump.
(c) The heat generated owing to the excess flow passing over the relief valve.

A.2.6 A hydraulic circuit consists of a fixed-displacement gear pump supplying hydraulic fluid to a cylinder which has a bore of 100-mm diameter, a rod of 56-mm diameter and a stroke of 400 mm. Pumps are available with displacement increasing in steps of 1 ml/rev from 5 ml; the volumetric efficiency is 88% and its overall efficiency is 80%. The pump is driven directly from an electric motor with an on-load speed of 1430 rev/min. Select a suitable pump so that the cylinder can be reciprocated through a complete cycle once every 12 seconds.

A.2.7 A pump having a displacement of 25 ml/rev is driven at 1440 rev/min by a 10 kW electric motor. If the pump's overall and torque efficiencies are 85% and 90% respectively determine:

(a) The quantity delivered by the pump.
(b) The maximum pressure the pump can operate against without overloading the motor.

A.2.8 A pump having a theoretical delivery of 35 l/min and a volumetric efficiency of 90% drives a cylinder having a bore of 110 mm, a rod diameter of 65 mm and a stroke of 700 mm. Determine:

(a) The extend and retract velocities of the cylinder.
(b) The time for one complete cycle.

A.2.9 A pump/accumulator power pack is to supply the fluid flow demanded by a hydraulic system as shown in Figure A.21. The system working pressure is 125 bar and the maximum pressure at the accumulator is 200 bar. Assuming the accumulator pre-charge pressure is 90% of its maximum working pressure. Determine:

(a) The actual pump delivery required.
(b) The maximum volume of fluid to be stored in the accumulator.
(c) The accumulator volume assuming isothermal charge and discharge of the accumulator.

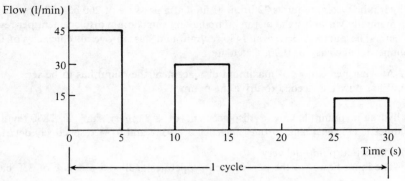

Figure A.21

A.2.10 A press cylinder having a bore of 140 mm and a 100-mm diameter rod is to have an initial approach speed of 5 m/min and a final pressing speed of 0.5 m/min. The system pressure for rapid approach is 40 bar and for final pressing 350 bar. A two-pump, high-low system is to be used; both pumps may be assumed to have volumetric and overall efficiencies of 0.95 and 0.85 respectively. Determine:

(a) The flows to the cylinder for rapid approach and final pressing.
(b) Suitable deliveries for each pump.
(c) The displacement of each pump if the drive speed is 1720 rev/min.
(d) The pump motor power required during rapid approach and during final pressing.
(e) The retract speed if the pressure required for retraction is 25 bar maximum.

A.2.11 A pump driven at 1440 rev/min having a displacement of 12.5 ml/rev and a volumetric efficiency of 87% is used to supply fluid to a circuit with two cylinders. If the cylinder dimensions are 63 mm bore × 35 mm rod × 250 mm stroke, and 80 mm bore × 55 mm rod × 150 mm stroke, find the minimum cycle time for both cylinders to extend and retract fully.

A.2.12 A hydraulic cylinder is required to exert a minimum forward dynamic thrust of 25 tonnes and a minimum return dynamic thrust of 15 tonnes. Determine a suitable standard size metric cylinder if the maximum system pressure is 200 bar. Assume that the dynamic thrust is 0.9 times the static thrust. What pressure is required at the cylinder to give the desired thrusts if the effect of back-pressure is neglected?

A.2.13 A hydraulic cylinder as shown in Figure A.22 is to accelerate a load of 50 tonnes horizontally from rest to a velocity of 10 m/min in 50 mm. Take the coefficient of friction (μ) between the load and the guides as 0.1; assume zero back-pressure. Determine:

(a) A suitable size of standard metric cylinder if the maximum allowable pressure at the cylinder is 180 bar.
(b) The fluid flow rate required to drive the piston forward at 3 m/min.

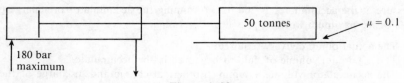

Figure A.22

Section A.2 Hydraulic calculations

A.2.14 A cylinder having a bore of 50 mm with a rod diameter of 32 mm is used to lift a vertical load of 3 tonnes. The circuit used is shown in A.23. The quantity delivered by the pump is 8 l/min, the relief valve is set at 180 bar and the cylinder extend meter-out flow-control valve is set at 4 l/min. Calculate:

(a) The extend speed.
(b) The pressure reading at gauge P_1 when the cylinder is extending at a steady speed.
(c) The pressure registered on gauge P_1 if the cylinder is extended under no load conditions, i.e. load removed.
(d) The setting of the retract flow control valve in l/min if the retract speed is the same as the extend speed.

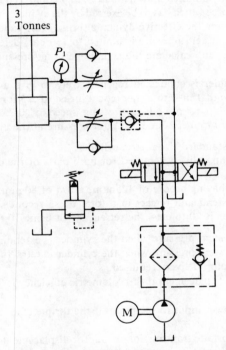

Figure A.23

A.2.15 A hydraulic cylinder is required to lift a 6000-kg load through a vertical height of 4 m. The working pressure is not to exceed 100 bar at the cylinder inlet port; neglect the effect of back pressure. The cylinder is to be front flange mounted, the load fully guided with the piston rod connection being pivoted.

Determine the sizes of a suitable standard piston rod and cylinder bore.

A.2.16 A hydraulic cylinder having a bore of 125 mm and a rod diameter of 90 mm is used to lift a load of 20 tonnes at a maximum speed of 5 m/min. The load has to be brought to rest in the cushion length of 50 mm at the end of the stroke, system pressure being applied to the full bore end of the cylinder during retardation. Determine:

(a) The pressure required to lift the load assuming zero back pressure at the annulus side of the cylinder.
(b) The fluid flow rate to raise the load at 5 m/min.
(c) The average pressure in the cushion during retardation.

A.2.17 A centre lathe head stock is to be driven by a 'constant power' hydrostatic transmission. The variable-displacement non-reversing hydraulic motor is to have a speed range of 300 to 2500 rev/min. The maximum power required at the output of the hydraulic motor is 6 kW. The maximum pressure available at the pump delivery is 125 bar and the pressure drop between the pump and motor is 5 bar. The torque and volumetric efficiencies of both the pump and motor may be taken as 0.85. Assuming an open loop transmission, determine:

(a) The ideal motor displacement and the actual pump delivery required.
(b) The input power required by the pump.

A.2.18 A hydraulic cylinder has to move a load horizontally through a distance of 3 m. The cylinder is front flange mounted and the load rigidly connected to the piston rod and fully guided. The extend force which has to be exerted by the cylinder is 1.6 tonnes and the retract force 0.7 tonne. Assume the effective dynamic thrust is 0.9 times the static thrust.

If the maximum system pressure is limited to 150 bar, determine a suitable standard metric size of cylinder and calculate the actual operating pressure.

A.2.19 A machine tool cylinder is connected regeneratively to give a rapid approach speed of 10 m/min for a stroke of 1 m with a theoretical thrust of 2.5 tonnes. It is then switched to conventional connection to provide a pressing speed of 0.25 m/min for 0.5 m with a theoretical thrust of 10 tonnes. The maximum pressure at the cylinder is to be 200 bar.

(a) Select a suitable standard size metric cylinder.
(b) Calculate the pump delivery required for both parts of the extend stroke.

A.2.20 A hydraulic cylinder having a bore of 125 mm, a rod of 80 mm diameter and a stroke of 350 mm is to fully extend and retract in a total of 15 seconds. The extend thrust to be exerted by the cylinder is 20 tonnes, the retract thrust being 10 tonnes. Determine:

(a) The theoretical system pressure when the cylinder is extending.
(b) The theoretical system pressure when the cylinder is retracting.
(c) The theoretical pump delivery required.
(d) The actual pump displacement if the volumetric efficiency is 90% and the pump driven at 1440 rev/min.
(e) The maximum power input to the pump if the torque efficiency is 85%.

A.2.21 In an open-loop hydrostatic transmission the motor displacement is 0.5 liter/rev and it is to run at 65 rev/min. The pump supplying fluid to the motor is driven at 1440 rev/min. If the torque and overall efficiencies of both the pump and motor are 95% and 85% respectively, determine:

(a) A suitable pump displacement.
(b) The pressure at the motor if the torque required at the motor is 1000 Nm.
(c) The input power to the pump if the pressure drop in the pipework etc. between the pump and motor is 5 bar.

A.2.22 A hydraulic motor is required to develop a torque of 100 Nm at a maximum speed of 600 rev/min. The maximum pressure drop across the motor is to be 150 bar. The torque and volumetric efficiencies are both 0.9. Determine:

(a) A suitable motor displacement.
(b) The flow required to the motor.

A.2.23 A lathe head stock is directly driven by a hydraulic motor. The lathe is used to turn a bar with a maximum diameter of 60 mm. The maximum tangential cutting force on the lathe tool is 2 kN and the maximum rotational speed of the head stock is 700 rev/min. The maximum pressure set by the relief valve is 200 bar and the total pressure drop between the relief valve and the hydraulic motor is 10 bar, the back-pressure at the motor being 5 bar. The overall and volumetric efficiencies of the motor are 0.85 and 0.9 respectively. Determine:

(a) The minimum displacement in cm^3/rad.
(b) The flow rate to the motor at maximum speed.
(c) The maximum hydraulic input power to the motor.

A.2.24 A hydraulic motor with a displacement of 475 cm^3/rev is used to drive directly a conveyor drum having a diameter of 0.7 m. The pressure drop over the motor is 140 bar and the actual flow into the motor 48 l/min. The overall and torque efficiency of the motor are 0.9 and 0.94 respectively. Determine:

(a) The torque at the conveyor drum.
(b) The power in kilowatts supplied to the conveyor drum.
(c) The linear speed of the conveyor belt.

A.2.25 A vehicle weighing 2 tonnes is to be driven up a slope of 1 in 10 (1 vertical in 10 measured along the slope) at a speed of 20 km/h. The coefficient of rolling resistance may be taken as 0.1. The vehicle is driven hydraulically by two fixed-displacement motors fitted in the rear wheels which have an effective diameter of 850 mm. The volumetric and torque efficiency of the motors are both 0.95. The maximum pressure drop across the motors is 250 bar. Determine:

(a) The required motor displacement.
(b) The fluid flow from the pump at maximum speed.

A.2.26 A rotating feed table with a mass of 50 kg and a radius of gyration of 0.4 m is driven by a hydraulic motor. The table has to be accelerated from rest to a maximum speed of 120 rev/min in 0.5 second. The maximum pressure drop across the motor is 200 bar. Taking the volumetric and torque efficiencies of the motor as 0.96 and 0.95 respectively, determine:

(a) A suitable displacement for the motor.
(b) The actual pump delivery required to drive the motor at maximum speed if the pump efficiencies are the same as those of the motor.

A.2.27 A hydraulically-powered haulage is used to wind a train of wagons up a drift mine. The weight of the train is 5 tonnes, the slope of the track is 1 vertical in 10 measured along the slope length. The haulage rope is layered on the drum but to simplify the calculation assume that the effective diameter remains constant at 1 m. The track has a length from the haulage to the pit bottom of 1500 m. The haulage rope has a weight of 4 kg/m run. The haulage drum has a weight of 1.5 tonnes when the haulage rope is fully unwound and a radius of gyration of 0.5 m.

The train is to be capable of being accelerated from rest at the bottom of the slope to a top speed of 5 km/h in 10 seconds. The maximum pressure drop across the haulage motor is 300 bar and the torque and volumetric efficiencies of the motor are both 95%.

(a) Calculate the motor displacement per revolution.
(b) Should the haulage have to be stopped halfway to the surface, what would be the maximum possible acceleration on restarting? Take into account the change in drum inertia owing to the haulage rope being wound on to the drum.
(c) If the pump supplying the haulage motor has a volumetric efficiency of 92% and an overall efficiency of 87%, calculate the required displacement per revolution when it is driven at 2200 rev/min.

(d) What is the input power needed by the pump if the total pressure loss between the pump and motor is 30 bar?

A.2.28 A closed loop hydraulic system consists of a variable-displacement pump driving a fixed displacement motor as shown diagrammatically in figure A.24.

(a) Show that

$$\frac{\Omega_m(s)}{Y(s)} = \frac{K_p}{d_m}\left(\frac{1}{1+\tau s}\right)$$

where s is the Laplace transform function, K_p is the pump flow constant for a given speed Ω_p, d_m is the motor displacement, and τ is the time constant.
(Neglect the fluid compressibility.) The combined leakage coefficient for the pump and motor is λ.

(b) Determine from first principles the response of the system to a unit step input after a time equal to 3τ. Calculate the value of τ. In a particular system the values are
Load inertia, $I = 100$ Nm s^2
Leakage coefficient, $\lambda = 12 \times 10^{-3}$ l/min/bar
Motor displacement, $d_m = 25$ ml per radian
Pump flow constant, $K_p = 5 \times 10^{-3}$ m^2 s^{-1}

(c) Draw a polar plot for the frequency response of the system to an input sin ωt for values of ω between 1 and 10 radians per second.

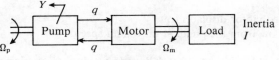

Figure A.24

A.2.29 A dumper truck is to be driven by a reversible flow-controlled hydrostatic transmission. The details of the vehicle and required drive are

Gross weight of vehicle	= 7 tonnes
Number of driven wheels	= 2 (rear wheels)
Rolling resistance	= 100 kg per tonne weight
Rolling radius of wheels	= 0.4 m
Maximum vehicle speed on flat	= 15 km/h
Time to accelerate to maximum speed on flat	= 5 seconds.

The vehicle is to be capable of climbing a maximum gradient of 1 in 4; maximum speed and acceleration on gradient are unimportant.

Variable-displacement pumps are available with maximum theoretical deliveries starting at 20 l/min and increasing in steps of 10 l/min when driven at engine speed.

Fixed-capacity motors are available with capacities starting at 0.1 l/rev in increasing steps of 0.05 l/rev.

The maximum operating pressure of the system is 300 bar. Both pumps and motors have volumetric efficiencies of 0.95, the torque efficiency of the motor is 0.94 and the overall efficiency of the pump 0.9.

(a) Neglecting any pressure drops in pipework select suitable hydraulic pump and motors and determine the input power to the pump.
(b) If the vehicle weight distribution is 70% on the rear wheels, 30% on the front wheels and the coefficient of traction between the drive wheels and the ground is 0.95 determine the maximum gradient up which the vehicle could climb.

A.3 DESIGN PROBLEMS

This section involves circuitry and design calculations to determine sizes of components.

A.3.1 Two hydraulically-powered machines are driven by a single power pack. One machine requires a constant supply of 15 l/min, the other has a demand which varies from 4 to 18 l/min. The machines operate at pressures between 80 and 100 bar and may be used separately or at the same time.
Draw a suitable pumping circuit and calculate the theoretical motor input power required.

A.3.2 A hydraulic power pack is to be used in a test house which undertakes fatigue and reliability investigations. The fluid demand can vary from 5–100 l/min and be at pressures up to 300 bar. The fluid must be pulse free and filtered to $\beta_3 = 75$.
Draw a suitable pumping circuit bearing in mind the importance of system reliability. (If a test is stopped part way through any results will be meaningless.)

A.3.3 A hydraulic cylinder of 1 m stroke is to exert a maximum forward thrust of 0.5 tonne and a maximum retract thrust of 1.5 tonnes. These forces have to be adjustable. The approximate cylinder speed is to be 0.5 m/min extend and 1.5 m/min retract. Draw a suitable circuit showing the power pack, control valves and cylinder. (Do not size any components but indicate the ratio of cylinder bore diameter to cylinder rod diameter which will produce the required relationship between the extend and retract speeds.)

A.3.4 A double-acting hydraulic cylinder has to have the same speed extending and retracting. The speed must be easily adjusted and independent of changes in oil viscosity and the load which always opposes movement. The cylinder must be capable of being positively locked in any position. Any 'kicks' when restarting would be detrimental.
Draw a suitable circuit.

A.3.5 A machine table has to reciprocate continuously for long periods. The operator can select full speed, two-thirds speed and one-third speed. Flow rate into the cylinder for maximum speed is 30 l/min. The speed is to be the same in both directions and within 15% of the selected value. Variations in working pressure can be from 20 to 140 bar.
Design a system for minimum heat generation.

A.3.6 A hydraulic cylinder with a stroke of 1 m is required to exert a forward thrust of 10 tonnes, the retract thrust will normally be 0.1 tonne and to prevent damage to the unit must never exceed 1 tonne. The cylinder is front flange mounted and the piston rod rigidly guided.
If the system pressure is limited to 140 bar calculate a suitable standard size of metric cylinder. If the cycle time is to be 20 seconds, estimate a suitable size of pump. (Neglect all losses.)
Draw a circuit for this system showing the cylinder control valves and power pack.

A.3.7 A double slugger roll crusher is to be hydraulically driven. A layout of the unit is shown in Figure A.25. Two contra-rotating rolls are each driven by a fixed-displacement slow-speed hydraulic motor, the maximum drive speeds being 55 and 65 rev/min for the fixed and sprung rolls respectively. If a piece of non-crushable material enters the rolls and causes them to jam the roll drive must automatically stall off. The sprung roll position is adjusted

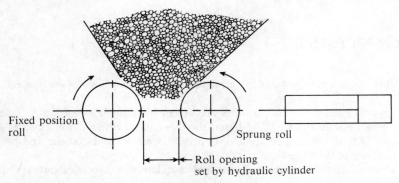

Figure A.25

by a hydraulic cylinder which can be locked to give a minimum distance between rolls but must act as a spring to absorb shock loads. The maximum pressure at the cylinder is to be 200 bar.

Draw a suitable hydraulic circuit. (Note the rolls will have to be reversed if a piece of material becomes jammed between them. Do not draw the electrical control circuit.) The pump is to operate against a maximum pressure of 300 bar. Show the suggested settings for any pressure control valves used.

A.3.8 A machine for hollow grinding blades consists of a sliding table having a 500-mm movement on which the blades are mechanically clamped. The table is hydraulically driven past a fixed position grinding wheel, the blades being positioned so that the forward pass rough grinds the blade and the return pass finish grinds.

Design a hydraulic circuit so that the table moves at the same speed in both directions, the speed to be adjustable between 2 and 4 m/min. The cycle is to be electrically controlled, with a push-button to start. A limit switch signals the cylinder to reverse at the end of the extend stroke and a second limit switch at the end of the retract stroke is used to stop the cylinder and unload the pump. The maximum system pressure is to be 50 bar, and the cylinder is to exert a maximum force of 2 kN in both directions. Assume that the dynamic thrust of the cylinder is 0.9 times the static thrust. Select a standard metric cylinder and determine the relief valve setting for the maximum cylinder thrust of 2 kN. Calculate theoretical pump delivery and pump input power. (Do not draw the electrical circuit.)

A.3.9 A cutting machine as shown diagrammatically in Figure A.26 consists of a rotating cutting head and a sliding horizontal table. The table is driven by a hydraulic cylinder with a stroke of 2 m. The cylinder is to have a rapid approach speed of approximately 3 m/min automatically changing to a slow cutting speed as the pressure increases when the workpiece contacts the cutting head. The slow cutting speed is to be manually adjustable between 10 and 150 mm/min. A retract speed of approximately 3 m/min is required.

The cylinder is trunnion mounted at the rear end and clevis mounted to the table which

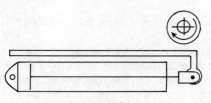

Figure A.26

is rigidly guided. The cylinder has to exert a thrust of 500 kg during rapid approach and retract and 2500 kg on the cutting stroke. The maximum system pressure is to be 70 bar.

Assume actual thrust is 0.9 times theoretical thrust and select a standard metric cylinder. When calculating the buckling length assume the closed length of the cylinder is equal to the stroke.

Determine the maximum flow required from the pump.

A.3.10 A hydraulic boom is powered by two cylinders (Figure A.27). Each cylinder has a stroke of 0.7 m and exerts a maximum thrust of 6 tonnes on extend and 4 tonnes on retract. The maximum pump delivery pressure is 200 bar and the pressure drop through the pipes and valves may be taken as 10 bar (including back pressure). Cylinders are to be operated by manual valves. Design the circuit in such a way that only one cylinder may be operated at one time. The extend time for either cylinder is to be 10 seconds. Assume the extended length of the cylinder between pivots is twice the stroke for buckling load calculations. Determine a suitable metric cylinder size and pump delivery. Draw a hydraulic circuit.

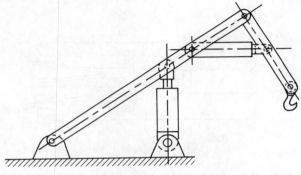

Figure A.27

A.3.11 In a roll mark press the mark, usually a string of characters is formed in relief on the periphery of a roller which is mounted above the machine table as shown in Figure A.28. Components are mounted on the table using quick-release clamps. The table is raised and a

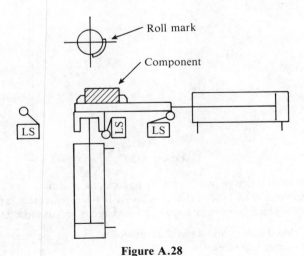

Figure A.28

horizontal cylinder slides the table beneath the roll and the mark is rolled onto the component. On the extend stroke the vertical cylinder has to exert a maximum thrust of 8 tonnes and the horizontal cylinder 2 tonnes. Both cylinders have a stroke of 250 mm. The required marking rate is 3 components per minute. If it takes 5 seconds to remove a marked component and replace it with a new one, determine the pump and cylinder sizes and draw a suitable circuit.

It is important that pressure is maintained on the vertical cylinder during the extend stroke of the horizontal cylinder and it is suggested that a flow of 4 l/min will be satisfactory for this purpose. Maximum system pressure is not to exceed 140 bar.

A.3.12 A hoist or jigger consists of two 3-groove sheave blocks as shown in Figure A.29. A hydraulic cylinder is attached to the inboard block, the outboard block being fixed. Calculate a suitable standard metric cylinder size if the load to be raised by the hoist is 5 tonnes through a distance of 5 m. (The maximum allowable circuit pressure is 160 bar; neglect all losses.) If the 5 m lift is to take 40 seconds, what pump delivery is required?

Draw a suitable circuit so that the hoist can be stopped and locked in any position.

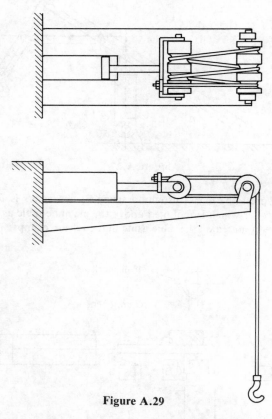

Figure A.29

A.3.13 A large center lathe is driven by a reversible hydrostatic transmission. The chuck speed is to be variable between 40 and 400 rev/min. A 5 kW, 1440 rev/min electric motor is to drive the hydraulic pump which can operate against a maximum pressure of 200 bar.

(a) Design a closed-loop hydrostatic transmission so that as the chuck speed decreases the available torque increases.

(b) Determine suitable pump and motor displacements, the volumetric and torque efficiencies for pump and motor can be taken as 0.93. Assume the total pressure drop in the pipework between the pump and motor is 15 bar.

(c) Calculate the maximum torque available at the motor when running at the minimum speed.

A.3.14 A fairground ride is to have a hydraulic drive. The ride comprises a 15-m diameter disc on which seats are fastened. The disc is horizontal when at rest. Two cylinders are used to tilt the disc to an angle of 45° to the horizontal, the disc being rotated at 4 rev/min. This rotational drive is via a hydraulic motor connected to a rubber tyred wheel 0.6 m diameter which is kept in contact with a 15-m diameter drive track on the disc. The ride must only be tilted out of the horizontal when the disc is rotating. The maximum torque required at the hydraulic motor is 2000 Nm, and the maximum pressure available at the motor is 138 bar. The overall efficiency of the motor is 85% and the volumetric efficiency is 92%.

Design a hydraulic system determining a suitable hydraulic motor displacement. Suggest fail-safe features which can be incorporated in the case of pump failure.

A.3.15 A hydraulically-operated tilting mechanism is to use a rear trunnion mounted cylinder with a rod mounted clevis. The cylinder is to have a stroke of 2.3 m and exert a minimum force of 140 kN in each direction. It is to have a speed of 4.5 m/min on extend and retract and must be capable of being locked in any position against a reversible load.

(a) Draw a suitable circuit and select a standard metric cylinder, assuming a maximum system pressure of 200 bar. What will be the actual operating pressure of the system?

(b) Calculate required pump delivery and input power. (Neglect all losses.)

A.3.16

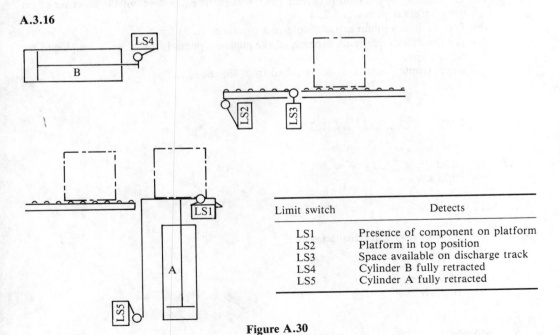

Limit switch	Detects
LS1	Presence of component on platform
LS2	Platform in top position
LS3	Space available on discharge track
LS4	Cylinder B fully retracted
LS5	Cylinder A fully retracted

Figure A.30

A gravity track feeds components to a conveyor transfer point (Figure A.30) which has to lift a load of 3.5 tonnes a height of 3 m to an upper track. When a component is on the lower track it runs under gravity on to the platform on cylinder A. If there is space available on the top track, cylinder A lifts the component level with the top track and cylinder B extends with a maximum thrust of 1000 kg, moving the component onto the upper gravity track. As cylinder A lifts, a mechanical stop prevents further components being fed along the bottom track.

The cylinders are both front flange mounted, the piston rod of cylinder A being rigidly guided while cylinder B piston rod is unguided. The maximum system pressure is to be 140 bar. If the upper track is clear, the time for one complete cycle is to be 60 seconds.

Neglecting all losses in the system determine suitable standard metric cylinder sizes and the pump delivery rate required. Design a suitable circuit using solenoid operated directional control valves.

Limit switches LS1 to LS5 are used to signal component and cylinder positions.

A.3.17 A track-laying vehicle has a hydraulic motor driving each track. It is to be capable of climbing a gradient of 1 vertical in 10 measured along the slope at a top speed of 25 km/h. The vehicle which has a weight of 8 tonnes and a rolling resistance of 85 kg/tonne is to accelerate from rest to 25 km/h in 10 seconds when travelling horizontally. The hydraulic power pack is driven by a diesel engine having a top speed of 2500 rev/min. The maximum system pressure at the power pack is 300 bar and there is a maximum pressure drop in the pipework between the power pack and the motors of 20 bar. The overall and volumetric efficiencies of both the pump and motor units are 80% and 95%.

Draw a circuit for a closed-loop constant torque hydrostatic transmission. The speed of the vehicle is to be adjustable and it is steered by independently altering the speed of each track.

Determine:

(a) The capacity of the motors driving the tracks if the equivalent outside diameter of the track sprocket wheels is 1.2 m.
(b) The maximum pump capacity to give a top speed of 25 km/h.
(c) The theoretical minimum capacity of the make-up pump if it is directly coupled to the diesel engine.
(d) The maximum output power required from the diesel engine.

A.4 SOLUTIONS TO SECTION A.2

A.2.1   (a)   2.22 l/min
            (b)   0.73 kW

A.2.2   12.4 cm^3/rev

A.2.3   (a)   88% of maximum displacement
            (b)   16.3 kW

A.2.4   (a)   23.6 l/min
            (c)   16.2 kW

A.2.5   (a)   30.6 l/min
            (b)   12.24 kW
            (c)   1.68 kW

A.2.6   22 ml/rev

Section A.4 Solutions

A.2.7   (a)   33.95 l/min
           (b)   150.2 bar

A.2.8   (a)   3.32 m/min, 5.09 m/min
           (b)   20.9 s

A.2.9   (a)   15 l/min
           (b)   2.5 l
           (c)   7.45 l

A.2.10 (a)   77 l/min, 7.7 l/min
           (b)   69.3 l/min, 7.7 l/min
           (c)   42.4 ml/rev, 4.7 ml/rev
           (d)   6.4 kW, 5.3 kW
           (e)   10.2 m/min

A.2.11 9.46 s

A.2.12 140 mm bore × 90 mm diameter rod
           Extend pressure required = 177 bar
           Retract pressure required = 181 bar

A.2.13 (a)   80 mm bore cylinder (actual size required is 67 mm bore)
           (b)   15.1 l/min

A.2.14 (a)   3.45 m/min
           (b)   50.8 bar
           (c)   305 bar
           (d)   6.77 l/min

A.2.15 Minimum piston rod diameter 64 mm, standard 70 mm
           Minimum piston diameter 87 mm, standard 100 mm
           Suitable cylinder 100 mm bore × 70 mm rod

A.2.16 (a)   160 bar
           (b)   61.35 l/min
           (c)   Cushion pressure to retard load = 33.9 bar
                 Cushion pressure to balance forward pressure = 332.2 bar.
                 Total cushion pressure = 366.4 bar

A.2.17 (a)   14 to 118 cc/rev, 41.5 l/m
           (b)   11.96 kW

A.2.18 Piston rod diameter for buckling strength = 34.2 mm (36 mm)
           Cylinder bore for extend thrust = 38.5 mm
           Use a 50 mm bore × 36 mm rod cylinder
           Extend pressure = 89 bar
           Retract pressure = 81 bar

A.2.19 (a)   Theoretical sizes 79 mm bore × 35 mm rod (standard cylinder 80 mm × 45 mm)
           (b)   Rapid approach flow = 15.9 l/min; pressing flow 12.56 l/min

A.2.20 (a) 160 bar
(b) 135.4 bar
(c) 27.3 l/min
(d) 21 ml/rev
(e) 9.5 kW

A.2.21 (a) 28.2 ml/rev
(b) 132 bar
(c) 9.76 kW

A.2.22 (a) 7.41 cm^3/rad (46.5 cm^3/rev)
(b) 31 l/min

A.2.23 (a) 3.45 cm^3/rad
(b) 16.86 l/m
(c) 6.6 kW

A.2.24 (a) 995 Nm
(b) 10.1 kW
(c) 213 m/min

A.2.25 (a) 220 ml/rev
(b) 57.8 l/min

A.2.26 (a) 106 ml/rad (66.5 ml/rev)
(b) 8.3 l/min

A.2.27 (a) 1.648 l/rev
(b) 0.572 m/s^2
(c) 27.3 ml/rev
(d) 34.9 kW

A.2.28 (a) See Section 8.3.1 of Chapter 8.
(b) $\Omega_m = 190$ rad/s for unit step input
$\tau = 0.325$
(c)

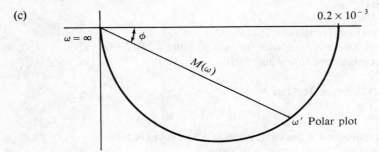

$M(\omega) = \{[1^2 + (0.32)^2]^{1/2}\}^{-1} \times 0.2 \times 10^{-3}$
$\phi = \tan^{-1} 0.32$

A.2.29 (a) Motor displacement = 1.07 l/rev (1.1)
Maximum pump delivery = 230 l/min
Input power to pump = 65.6 kW
(b) Gradient rise 1 in 1.8 along slope

FURTHER READING

History of hydraulics
Pugh, B. (1980) *The Hydraulic Age*, Mechanical Engineering Publications, London.

Fluid mechanics
John, J. E. and Haberman, W. L. (1980) *Introduction to Fluid Mechanics*, Prentice Hall, Englewood Cliffs, NJ.

Roberson, J. A. and Crowe, C. T. (1975) *Engineering Fluid Mechanics*, Houghton Mifflin, New York.

Reference books
Warring, R. H., ed. (1983) *The Hydraulic Handbook*, 8th edn, Trade & Technical Press, Morden, Surrey.

Hydraulic Technical Data, Trade & Technical Press, Morden, Surrey.

Pumps
Lambeck, R. P. (1983) *Hydraulic Pumps and Motors*, Marcel Dekker, New York.

Fluids
Tourret, R. and Wright, E. P. (1979) *Performance Testing of Hydraulic Fluids*, Heyden/Institute of Petroleum, London.

Control and servo
Guillon, M. (1969) *Hydraulic Servo Systems*, Butterworth, London.

Harrison, H. L. and Bollinger, J. G. (1969) *Automatic Controls*, International Textbook Co., London.

McCloy, D. and Martin, H. R. (1980) *Control of Fluid Power*, Ellis Horwood, New York.

Pippinger, J. D. (1984) *Hydraulic Valves and Control*, Marcel Dekker, New York.

Reed, E. W. and Larman, I. S. (1985) *Fluid Power with Microprocessor Control: An introduction*, Prentice Hall, Hemel Hempstead.

Stringer, J. D. (1976) *Hydraulic System Analysis*, Macmillan, London.

INDEX

Page numbers in bold type indicate the main subject reference.

absolute rating, **203**, 207, 211, 250, 296
absorption, 193
accumulator, 36, 48–54, 234, **254**, 272–6, 293, 302
    applications 259–67, 272
    back-up bottles, 258, 275
    bag, 258
    charging, 265–7
    circuit, 36, 40, 53, 64, 100, 276, 277, 305
    dead load, 254–5
    free surface, 256
    gas loaded, 255–9
    isolating valve, 53
    piston, 257, 265
    precharge, 49, 259, 273, 293, 303
    pressure testing, 266
    safety, 265–7, 305
    safety circuit, 100, 267
    sizing, 260, 273
    spring loaded, 255
actuators, **127**
    design information 236
additives, 180, **184**, 192
adiabatic expansion and compression of a gas, 50, 257, 259, 260, 275, 294
adsorption, 194
aeration, 95, 190, 251
aeration control, 191, 251
air blast cooler, 252
air bleed valve, 42, 297
air breather, 199, **211**, 250–1, 296, 301
air bubbles, 183, 190
air motor, 42
air saturation, 191
analog, 328
anti-burst valve, 231
anti-cavitation check valve, 126, 163, 169
anti-foaming additives, 192
anti-lunge device, 74, 76
anti-oxidant, 180
anti-syphon holes, 191, 251
anti-vibration mounts, 245, 301
anti-wear agents, 185

armature, 99, 105, 106, 351, 361
axial piston pump, 22, 25
axial piston motor, 160
ANSI (American National Standards Institute), 110

bacteriological growth, 194
baffles, 252
balanced vane pump, 21
ball motor, 160
basic formulae (summary), **237**
bell housing, 246
bellows, 146
bent axis motor, 160, 162
bent axis pump, 23, 25, 31
beta rating, **203**, 204, 211
bite ring, 218, 220
bladder, 192, 258–9
bleed-off filter, 208
bleed-off flow control, 77
block diagrams, 310, 314, 329, 341, 347, 359, 360
bonded washer, 223
boost pump, 169, 179, 319
brake valve, **66**, 168, 169, 322
Bramah, Joseph, 268
bridge network, 86, 210
British Standard (BS), 139, 181, 214, 220, 288, 297
British Standard Pipe Thread (BSP), 220, 222
bubble point test, 203
buckling length, 147–8, 289
buckling load, 148, 150, 288
bulk modulus, 18, **183**, 262–3, 342–4
bypass filter, 208
bypass flow control, 83, 176
bypass valve, **204**, 206, 213, 269, 303

cam rotor motor, 159
cartridge valve, 99, **112**, 233 (*see also* Valve, cartridge)
cavitation, 163, 189, **191**, 251, 299
centralized hydraulic systems, **253**, 288

centrifugal pump, 15
CETOP, 110, 195, 221, 299
check valve (*see* Valve, check)
    soft seat, 93
chlorinated hydrocarbons, 181
choke pack, 108, 292, 316
circuit breaker, 68
clamps, tube and hose, 230
clean up loop, 207, **210**, 211, 311, 314
clean hydraulic oil, 195, 197
cleanliness standards, 195
cleanliness target levels, 212
clearances in components, 198
closed loop control, 329, 360
closed loop transmission, 35, 168–171, 178
commissioning, 301, 302, 365
component malfunction, 309
compressibility, 183, 342
compression fitting, 218, 219
condition monitoring, 297
constant flow control, 29, 175
constant power control, 28, 175, 179, 286
contamination analysis, 298, 299
contaminant coding, 196, 197
contamination control, **189**
control systems, **328**
conversion coupling, 219
conveyor feed system, 287
copying device, 332, 336
corrosion inhibitor, 180, 185
counterbalancing, 64, 264
counterbalance valve, **64**–6, 76, 95, 297, 309, 315
coupling
    shaft, 246, 303
    tube, 217–25
cracking pressure, 93
crescent seal, 19
cross-line relief, **61**, 167, 292, 322
custom built power packs, 245
cylinder, **127**
    acceleration and deceleration, 140–5
    displacement, 127, 128, 132, 138
    double acting, 127, **133**–51, 153
    dynamic thrust, 136, 241
    cushioning, **142**–5, 155
    cushion pressures, 142
    formulae, 128, 134, 136–8, 140, 241
    maximum speed, 145
    mountings, 146
    operating temperatures, 145
    piston rod buckling strength, 148
    piston rod ends, 146
    piston rod protective covers, 146
    regenerative, 137–9, 149
    seal friction, 136
    speed control, 75
    single acting, **133**, 126, 127, 254, 269, 273
    standard metric, **139**, 142, 151, 285, 288
    static thrust, 136, 241
    strength, 147–50
    stop tube, 148
    synchronization, 89
    telescopic, 130–2
    through rod, 139
    wear, 299, 309

damping orifice, 30, 75
damping ratio, 344–5
dead zone, 331, 352
deceleration valve, 72
decompression, 95–7, 184
Delrin, 93
design
    criteria, **235**
    information, 236
    study, **268**
detent, 105
diaphragm (*see* Flexible separator)
diatomaceous earth, 201
diffuser, 192, 251, 292
digital, 328, 350, 365
DIN standards, 110, 111, 113, 215, 218, 219, 223
directional control valves, **93** (*see also* Valve, directional control)
dirt in systems
    effect, 199
    origin, 194, **198**, 206
dirt fust, 209, 240
    size categories, 199
    tolerance, 20
displacement cylinder, 127–32, 138
dissolved air, 190
dither control, 338, 359, 360
double acting cylinder, 127, **133**–51
dual relief valve, 61

efficiencies
    pump and motor, **16**, 17, 156, **166**–7, 173–9, 243–4, 291–2, 346
    system, 43–56, 271–87
elastomer seals, 101, 128, 145, 187, 259
electrohydraulic servo valve, 338–40
electrostatic contamination, 190
electrostatic filtration, 202
energy contamination, 189
Engler, 181
entrained air, 183, 190, 252
equations of motion, 140
Euler's strut theory, 148

Index

failure
  catastrophic, 198, 200
  degradation, 198, 200
fault
  diagnosis, 306
  finding chart, 312, 315
feedback, 329–33, 338, 353, 361
filter
  capacity, 202, 213, 240
  construction, 206
  contamination time curves, 206
  cold start conditions, 205, 213, 296
  electrostatic, 202
  element
    beta rating, 203
    bubble point test, 203
    changing, 211, 303
    collapse pressure, 204
    condition indicator, 205, 302, 314
    depth type, 201, 202
    flow capacity, 213
    media, 201
    rating 203, 207, 211, 212, 250, 296
    spin-on, 211, 294
    surface type, 201, 202
  housing, 200
  location, 47, **207**–12
  magnetic, 202
  pressure drop through, 213
  sizing, 212–3, 240
fire resistant fluid, 180, 186–7, 194, 190
filtration technology, **200**
first order system, 334
flange connection, 221
flared fittings, 217–9, 223
flapper type servo valve, 339, 361
flexible separator, 192, 199, 250–1, 258–9
float switch, 251, 256
flow control
  bleed off, 77
  bypass, 83–5
  meter in **75**–88, 168, 293
  meter out, **76**–88, 168, 284
  valves, **70** (*see also* Valve, flow control)
flow divider, 89–92
flow meter, 304
flow through an orifice, 70–2, 332
flow velocity, 6, 216
fluid flow, 5, 105, 237
fluids (*see* Hydraulic fluid)
flushing, 302
flushing valve, 170
foaming, 183, 184
form tool, 112
frequency response, 336

gas laws, 49, 257, 275

gaseous contamination, 190
gear motor, 156, 157
gear pump, 19, 31, 45
generated rotor (gerotor), 20, 155, 157

heat dissipation, 190, 248
heat energy, 190
high water-base fluids (HWBF), 186, 188, 189
hi-lo, 39, 63, 271
hose break valve, 231
hydraulic conduits, 214 (*see also* Hydraulic tube *and* hydraulic hose)
hydraulic fluids, **180**
hydraulic fluid
  additives, 180, 184, 185, 192
  characteristics, 188
  classification, 180–1
  compatibility, 188
  conditioners, 252
  contamination, **189**, 298, 304, 305, 309
  contamination control, **189**–213, 296
  future development, 187
  handling, 300
  mineral oil, 181, 185, 288
  oil in water emulsions (HWBF), 186, 188, 189, 252
  operating temperature, 249, 252
  phosphate ester, 181, 187, 190
  sampling, 195, 297, 302
  selection, 187
  storage, 300
  water glycol, 181, 186, 187
  water-in-oil (invert) emulsion, 186–8
  water removal, 193
hydraulic hose, **225**–9
  burst pressure, 226
  end fittings, 227, 229
  failure, 228
  installation, 228, 301
  selection, 228
  suction, 229, 301
  supports, 229, 230, 301
  types, 225
  working pressure, 226
hydraulic motor circuits, **163**
hydraulic motors, **155** (*see also* Motors)
hydraulic power, 8, 9, 17
hydraulic press, 24, 39, 63, 65, 91–3, 95–7, 149
  design study, **268**
hydraulic stiffness, 344
hydraulic symbols, 9, 58, 63, 69, 84, 86, 87, 89, 94, 109
hydraulic tube, **214**
  coned fittings, 220
  connections, 217–25
  fittings, 217–25

flanges, 220
flow rating, 216
installation, 224, 301
pickling, 225
staple connector, 222
weld fittings, 220
working pressures, 215
hydrocapsule, 176
hydrostatic braking, 169
hydrostatic transmissions, 62, 103, **163**
    characteristics, 165, 171
    closed loop, **168**–70, 177, 209, 319, 341
    constant power, 165
    open circuit, **163**–8, 175
hysteresis effect, 361

inhibitors, 185
installation, 224, 301
integrated circuit, 112, 123
intensification, 90, 93, 267
intensifier, **267**, 280
interfaces, 111
ISO (International Standards Organisation), 110, 181, 195, 196
international reference numbers, 110, 195
intravane, 21
invert emulsion, 186–8
isentropic expansion and compression, 50, 257
isothermal expansion and compression, 49, 257, 275, 295

jet pipe servo, 1, 339
JIC (Joint Industrial Council), 217, 222, 245

kickdown sequence valve, 68
kinematic viscosity, 181 (*see also* Viscosity)
KR tube coupling, 219

laminar flow, 6, 215, 239
land (spool), 104, 330
lap, 330, 331
Laplace, 333–8, 342–4, 349–50
last chance filter, 209
leakage coefficient, 341–4, 346–50
    compensation, 263
    control, **214**, 234
    internal, 156, 159, 168, 179, 297, 304, 341
level switch, 251, 256
linear actuators, 127
    transducer, 352
liquid contamination, 192
lock valve, 87–8
logic element, **112**–23 (*see also* Valve, cartridge)
logic valve
    pilot control sources, 119
    poppet type, 113–21

remote switching, 118
spool type, 121–3
switching of multiple elements, 119
logical fault finding, 305, 308, 315–27
London Hydraulic Power Co., 253
lubrication, 101, 184–8

magnetic filter, 202
magnetic separator, 202
magnetism, 190
magnitude ratio, 337
maintenance, routine 299, 303
make-up pump, 163, 168, 320
manifold, 123, 232–4, 301
manual overrides, 107, 108
manual servo, 25–7, 179, 320, 322
mean filtration rating, 203
mechanical efficiency (*see* Torque efficiency)
mechanical feedback, 331
mechanical servo valve, 264, 331, 336
meter in, **75**–88, 168, 293
meter out, **76**–88, 168, 284
microbiological contamination, 194
mineral oil, 181, 185, 288
mobile hydraulic valves, **123**
    arrangements, 124–6
modularized package, 113, 123
monoblock valve, 124
motion control valve, 67, 170
motor, **155**
    axial piston, 160
    ball, 160
    cam rotor, 159
    characteristics, 162, 174
    circuits, **163**, 317, 319
    circuit design examples, **172**, 176
    efficiencies, 166–7, 173–9, 244, 291, 346
    formulae, 163–6, 244, 346
    freewheel, 163, 168
    gear, 156
    gerotor, 157
    leakage, 156, 159, 179, 341
    orbit, 157
    radial piston, 161–3
    sizing, 290
    torque, 164, 178
    vane, 158
    variable displacement, 158–9, 161–6
motors in parallel, 172
motors in series, 171
multi-pass test, 204
multi-pump circuits, **38**–41, 46–8, 65, 271, 284, 292, 316

NFPA (National Fluid Power Association), 110, 299
NPT (National Pipe Thread), 222, 223

natural frequency, 344
needle valve, 71
neon indicators, 107
NG reference number, 111
nominal rating, 203, 206
non-return valve (*see* Valve, check)
notched spool, 104, 116, 352
null point, 331, 361
Nyquist diagram, 337, 338

'O' ring compression fittings, 219
oil cooler, 190, 192, 240, 252
oil heater, 252
oil immersed solenoid, 106
oil-in-water emulsion, 186, 188, 189, 252
oil-in-water micro-emulsion, 189
open circuit (loop) transmission, 35, 163, 167
open loop control, 329
operating temperature, 145, **249**, 298, 306
orbit motor, 157
orifice flow, 70–2, 332, 358
overall efficiency, **17**, **166**–7, 174–6, 179, 243–4, 292, 346
overcenter valve, **65**–7, 168
overlap, 331
oxidation, 189
oxidation inhibitor, 185

parallel connection, 124, 172
particle
 analysis, 197
 counting, 195, 298
 distribution, 196–7
 sizes, 195, 198, 199
particulate contamination, **194**–200, 206, 297, 304 (*see also* Dirt)
Pascal's laws, 4, 268
patch test, 298
phase angle, 337
phosphate ester, 181, 187, 190
pickling, 225, 301
pilot control sources, 109, 119, 271
pilot operated check valve, **94**, 95, 99, 103, 112, 263, 264, 293, 309
pilot pressure unloading, 62
pilot to close check valve, 97
pipe fitting, **217**–25
pipe flow, 238
pipe supports, 229, 230, 301
piston rod buckling, 147–50
piston rod ends, 146
pneumatics, 1, 42, 253
polytropic expansion and compression of a gas, 257
poppet valves, 59, 98–100, 112–121
port identification, 11
port relief valve, 61

pour point, 183, 184
power pack, **245**, 292, 301, 311
Poziflare coupling, 219
prefill valve, **95**, 96, 281, 285
pressure compensation
 pumps, **27**, 54, 175, 272, 276, 284–6, 292
 valves, **73**, 74, 84, 121, 122, 304, 358, 359
pressure control cartridge valves, 120, 121
pressure control valves, **57** (*see also* Valve, pressure control)
pressure drop, 103, 240
 head, 5
 intensification, 90, 93
 line filter, 169, 208, 211, 296
 losses, 240
 of a liquid, 1
 operated directional control valve, 107
 override, 59, 62
 reducing valve, 69, 70, 123
 surge (damping), 262
 switch, 14, 36, 52, 149, 205, 276–9, 320
 zone, 19
priority flow control, 85
properties of fluids, 1
proportional amplifier, 360
proportional flow control, 358–60
 pressure control, 354
 pressure reducing valve, 354
 pressure relief valve, 354
 two stage valves, 356
 versus servo valves, **340**
proportional valves, 121, 123, **350**
proportional valve
 applications, 293, 362–6
 characteristics, 351, 360
 force control, 350
 force, position control, 351
 notched spool, 357
 spool position, 353
pulsation damper, 225, 261
pump characteristics, 45, 175
 delivery, 32
 drives, **41**
 drive speed, 32
 efficiencies, **16**, 17, 176, 243–4, 292, 346
 formulae, **15**–7, 242–3, 342, 346
 location, 245
 noise, 33, 322
 operating pressure, 30–2
 selection, 30, 178, 291
 servo system, **340**
pump types, **15**
 axial piston, 22, 31, 320
 centrifugal, 15
 gear (external), 19, 31, 45, 46
 gear (gerotor), 20
 gear (internal), 19, 31

non-positive displacement, 15
plunger, 24, 31
positive displacement, 15–35
radial piston, 24, 25
reciprocating, 22
rotary, 18
tandem, 46
vane, 20, 21, 31
variable delivery, 21, **25**–31, 41, 175, 286, 292, 302, 345, 363
pumping circuits, 35, 363
push rod, 105

quick release coupling, 230, 231
Q-pump, 33

radial piston pump, 24, 25
radial piston motor, 161–3
ramp 'down', 359
ramp input, 334–6
ramp 'up', 359
Redwood, 181
regeneration, 103
regenerative circuit, 137–9, 149
relative density, 180
relief valve, 27–9, 35, 39, 42, **58**–64, 75–85, 91, 120–3, 168, 176, 270, 277, 284, 302, 309, 314, 320
  ball, 59
  cracking pressure, 59
  cross line, 61, 167, 292, 322
  differential poppet, 59, 60
  direct acting, 59, 358
  filter, 208
  guided piston, 59
  pilot operated, **59**–61, 120, 122, 321
  port relief, 61
  remote operation, 61
  selection, 62
  solenoid controlled, 61, 121
  unloader, 39, 62
remote switching of cartridge element, 118
reservoir
  cooling effect, 248–50
  design, 250–2
  filling, 296–301
  function, 247, 251
  heater, 252
  shape, 248
  size, 33, 247, 249, 295
  types, 247
restrictor cartridge, 116
restrictor check, 97
return line filter, 176, 208, 211, 296, 314
reusable fittings, 227
reverse flow filter, 209
Reynolds number, 239

rotary actuator (*see* motor)
rust inhibitor, 180, 185

safety regulations, 266
sampling valve, 297
sandwich plate (block), 61, 97, 233
Saybolt, 181
scavenging, 169, 170
seal friction, 136
seal protection, 233
sealed reservoir, 192
semi-rotary actuator, 127, **152**, 236
  chain and sprocket, 154
  control, 155
  helical screw, 155
  lever arm, 154
  rack and pinion, 153, 154
  vane, 152, 154
sequence valve, **67**–9, 282, 297
series connection, 125, 170
servo control, 25, 322, **329**, 360–1, 365
servo system stability, 340
servo valve, 302, 339, 361
servo valve filter, 207, 209
sharp edged orifice, 73, 358
shear stability, 182
shut-off valve, 245, 299, 301, 315, 320
shuttle valve, 98, 169, 293
side cylinders, 282, 284
single acting cylinder, 127, 133
sinusoidal input, 336
SAE (Society of Automotive Engineers), 217, 221, 222, 227
soft seat valve, 93
soft-switching, 106, 113, 116
solenoid, 12, 106 (*see also* Valve, directional control)
  controlled relief valve, 61
  manual override, 107
  operation, 105, 351
  operated cartridge valve, 117
  pilot valve, 47, 53, 60, 99, 108
specific gravity, 180
spool
  notched, 104, 116, 352, 357
  shapes, 104
spool valve
  cartridge, 113, 121–3
  center condition, 102–4, 168
  pressure drop, 103
  stroke limiter, 109
  transition states, 101, 102
standby power supply, 253, 262
standard metric cylinders, 139, 140
standards organizations, 110, 180, 217
staple connector, 222
star-delta starter, 41

Index

start-up valve, 42
stop ring, 131
stop tubes, 148
strainer, 192, 207, 251, 315
streamline flow, 6, 215
stroke limiter, 74, 109
suction line filter, 207, 208, 211, 296
suction zone, 19
swaged fitting, 227
swash plate, 22, 25–31, 321
symbols, 9–14 (*see* Hydraulic symbols)
synchronizing valve, 89–90
synthetic fluid, 180, 187
system damping, 340
    efficiency, 271–87
    response, 333–8
    start-up, 302, 305
    stiffness, 191

tandem connection, 126
telescopic cylinder, 130–2
test equipment, 304
test points, 231, 233, 297, 299, 304, 314
thermal expansion, 184, 262
threaded connectors, 217–20, 222–4
three port flow control valve, 83, 85–6
through rod cylinder, 139
torque efficiency, **16**, 17, 166–7, 243–4
transition state (valve spool), 100, 102
transmissions (*see* Hydrostatic transmissions)
tree branching, 314
Triple-lok® tube fitting, 217
trouble shooting, **304**, 310, 314
turbulent flow, 6, 239
two stage directional control valves, **108**, 112, 270, 292, 316, 356
two stage pressure control valves, **59**, 70, 120, 122, 123, 321, 257
two stage sequence valve, 68, 122

underlap, 331, 361
units, 1–3, 237–8
unit step input, 331
unloader valve, 39, 62, 100

valve,
    cartridge, 99, **112**, 233
        flow control, 122
        logic element, **112**–23
        normally closed, 115
        operation, 114
        orifice, 177
        pilot control sources, 119
        poppet type, **113**–21
        pressure control, 121
        pressure reducing, 123
        remote switching, 118
        restrictor poppet, 116
        spool type, 113, **121**–3
        spool valve equivalent, 120
        switching, 118–20
    check (non-return), 10, 36, 39, 40, 65, 67, 71, **93**, 99, 109, 112, 113, 123, 163, 176, 258, 259, 277, 299, 318, 322
        bypass, 204, 206, 213
        pilot operated, **94**, 103, 112, 263, 264, 293, 309
        pilot to close, 97
        prefill, **95**–7, 281, 285
        restrictor, 97
        sandwich, 61, 97, 233
        shuttle, 98, 169, 293
    direct acting, 101, 309
    directional control, **93**
        identification of actuator position, 111
        operators, 105
        pilot operated, 107–9
        poppet solenoid, 99, 112
        sliding spool, **100**, 112
        solenoid operated, 36, 47, 101, 105, 106, 322
        spool center conditions, 102–4, 168
        spool stroke limiter, 109
        spool transition states, 101, 102
        three position, 88, 102
        two stage, **108**, 112, 270, 292, 316, 356
    flow control **70** (*see also* Flow control), 209, 275, 309
        bypass (three port), **83**–6
        deceleration, 72
        divider, 89, 90
        needle, 71
        pressure compensated, **73**–6, 83, 122, 293
        priority, 85, 86
        viscosity compensated, 73
    mobile, **123**
        monoblock, 124
        parallel connection, 124
        series connection, 125
        tandem connection, 126
valve actuation, 101, 105, 111
valve constant, 71, 332
valve interfaces, 111
valve lap, 330, 361
valve pressure control, 12, **57**, 120, 121, 354
    counterbalance, **64**–6, 76, 95, 297, 309, 315
    overcenter, **65**–7, 168
    pressure reducing, **69**, 70, 123
    relief, **58**–64 (*see also* Relief valve)
    sequence, **67**–9, 282, 297
valve, servo systems, **330**
valve stack, 61, 97, 232, 233
vane motor, 158, 159
vane pump, 20, 21, 31

vapor pressure, 186
vector diagram, 337
vehicle suspensions, 259, 264–5
velocity fuse, 231
velocity in pipes, 6, 7, 239
venting, 41, 61–4, 115
vent port, 60–4, 120
viscosity, 6, 73, 102, 152, 159, **181**, 182, 185–9, 288
viscosity index, **181**, 182, 184, 188
viscosity pressure characteristics, 182
viscosity temperature characteristics, 182
viscosity (temperature) compensation, 73, 358
volumetric efficiency, **16**, 156, 166, 173–4, 177, 179, 243, 290–1
volumetric strain, 18, 183
volumetric stress, 18, 183

water-absorbing polymers, 193–4
water based fluids, 180, 186, 194
water-in-oil (contamination), 193, 300
water-in-oil (invert) emulsion, 186, 187, 188
water glycol, 181, 186, 187
water hammer, 262
water removal, 193–4
wear, 298, 309
wet pin solenoid, 105, 106
wire mesh separator, 192, 252
work done, 4, 8, 164
welded connections, 220
welded nipple, 221
wiper ring, 130, 131, 133

zero lap, 331, 332, 361